C0-AZS-551

Lecture Notes on Coastal and Estuarine Studies

25

Clarice M. Yentsch
Frances C. Mague
Paul K. Horan (Eds.)

Immunochemical Approaches to Coastal, Estuarine and Oceanographic Questions

Springer-Verlag
New York Berlin Heidelberg London Paris Tokyo

Editors

Clarice M. Yentsch
Frances C. Mague
Bigelow Laboratory for Ocean Sciences
McKown Point
W. Boothbay Harbor, Maine 04575, USA

Paul K. Horan
Smith, Kline and French Laboratory
Dept. of Cell Biology
709 Swedeland Road, Swedeland, PA 19479, USA

ISBN 3-540-96894-6 Springer-Verlag Berlin Heidelberg New York
ISBN 0-387-96894-6 Springer-Verlag New York Berlin Heidelberg

Printed in Germany

Printing and binding: Druckhaus Beltz, Hemsbach/Bergstr.
2837/3140-543210 – Printed on acid-free paper

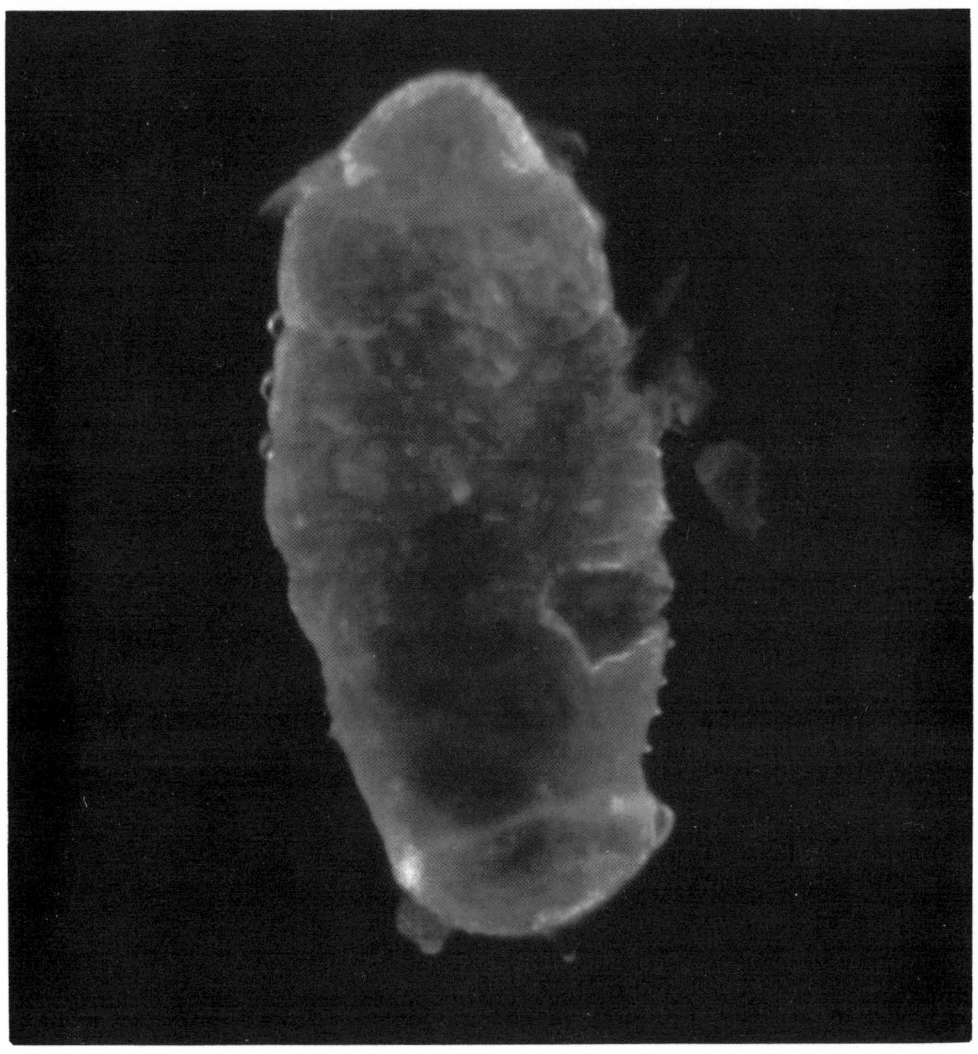

Epifluorescence microscope photograph of a *Capitella* sp. I. larvae stained indirectly with goat anti-rabbit fluorescein isothiocyanate (FITC) stain. The primary incubation was carried out using a 1:100 dilution of polyclonal rabbit antiserum against the larvae of the polychaete worm *Capitella* sp. I. The equipment used was a Reichert Microstar IV epifluorescence microscope with an RITC filter cluster, a 110 W mercury light source and an Olympus OM-4 camera. The photograph was taken by Dwight D. Trueblood, Environmental Sciences Program, UMASS/Boston, Boston, Mass. The *Capitella* sp. I larvae were kindly provided by Dr. Judy Grassle (at the Marine Biological Laboratory, Woods Hole, MA). The polyclonal antiserum was provided by Eugene D. Gallagher, Environmental Sciences program, UMASS/Boston, Boston, MA.

TABLE OF CONTENTS

LIST OF AUTHORS AND ATTENDEES

[a]R.A. AMOS — Bio Metric Systems, Inc
9932 West 74th Street
Eden Prairie, MN 55344

[b]D.M. ANDERSON — Woods Hole Oceanographic Institution
Woods Hole, MA 02543

D.G. BADEN — University of Miami
Rosenstiel School of Marine and Atmospheric Science
4600 Rickenbacker Causeway
Miami, Florida 33149

W.M. BALCH — Institute of Marine Resources A-018
Scripps Institute of oceanography
University of California, San Diego
La Jolla, CA 92093

[a]J.-M. BASSOT — Laboratoire de bioluminescence
CNRS
Gif-sur-Yvette, France

[a]J.C. BEHRENS — Bio Metric Systems, Inc
9932 West 74th Street
Eden Prairie, MN 55344

[b]J. CAMPBELL — Bigelow Laboratory for Ocean Sciences
McKown Point
West Boothbay Harbor, ME 04575

L. CAMPBELL — Department of Oceanography
University of Hawaii, Manoa
Honolulu, HI 96822

[a]W. CAMPBELL — Department of Biological Sciences
Michigan Technological University
Houghton, MI 49931

[b]D.A. CARON — Woods Hole Oceanographic Institution
Woods Hole, MA 02543

P. CASWELL-RENO — Department of Microbiology and Center for Marine Studies
University of Maine
Orono, Maine 04469

[a]R.P. CHAMBERS — Bio Metric Systems, Inc
9932 West 74th Street
Eden Prairie, MN 55344

[b]Y. CHAN — Department of Biology
City College of CUNY
New York, NY 10031

[a]authors who did not attend
[b]attendees who are not authors

[a]C.H. CHEN — Becton Dickinson Immunocytometry Systems
2375 Garcia Ave
Mountain View, CA 94043

[a]L. CHIU — Becton Dickinson Immunocytometry Systems
2375 Garcia Ave
Mountain View, CA 94043

[a]M. CONRAD — Becton Dickinson Immunocytometry Systems
2375 Garcia Ave
Mountain View, CA 94043

T.L. CUCCI — Bigelow Laboratory for Ocean Sciences
McKown Point
W. Boothbay Harbor, ME 04575

[b]S. CYNAR — Scripps Institute of Oceanography
University of California, San Diego
La Jolla, CA 92093

J.F. DALEY — Division of Tumor Immunology
Dana-Farber Cancer Institute
44 Binney Street
Boston, MA 02115

[a]P.H. DUQUETTE — Bio Metric Systems, Inc
9932 West 74th Street
Eden Prairie, MN 55344

R.J. FELLER — Department of Biology
Marine Science Program
Belle W. Baruch Institute for
Marine Biology and Coastal Research
University of South Carolina
Columbia, SC 29208

R.B. FERGUSON — Belle W. Baruch Insitutute for
Marine Biology and Coastal Research
University of South Carolina
Columbia, SC 29208

[b]A.S. FROELICH — Division of Biology and Living Resources
Rosensteil School of Marine and
Atmospheric Science
4600 Rickenbacker Causeway
Miami, FL 33149-1098

[a]E.A. FUNKHOUSER — Department of Biochemistry
Texas A & M University
College Station, TX 77843

A.E. GAGLIARDI — Department of Microbiology
University of Maine
Orono, Maine 04469

E.D. GALLAGHER — Environmental Sciences Program
University of Massachusetts, Boston
Boston, MA 02125

[a]M.G. GLEASON — Department of Zoology
University of California
Berkley, CA 94720

M.F. GOLDFARB | Anatek-EP, 999 Forest Avenue, Portland, ME 04103

[a] J.P. GRASSLE | Marine Biological Laboratory, Woods Hole, MA 02543

P.E. GUIRE | Bio Metric Systems, Inc, 9932 West 74th Street, Eden Prairie, MN 55344

[a] J.W. HASTINGS | Department of Cellular and Developmental Biology, Harvard University, 16 Divinity Avenue, Cambridge, MA 02138

[b] E. HAUGEN | Bigelow Laboratory for Ocean Sciences, McKown Point, W. Boothbay Harbor, ME 04575

[a] U. HEYMAN | Institute of Limnology, University of Uppsala, Uppsala, Sweden

[a] B. HEYMAN | Department of Immunology, Uppsala University Biomedical Center, Uppsala, Sweden

[b] B.HODGKINS | Dean of Applied Sceince, University of Maine - USM, Portland, ME 04101

[a] Y. HOKAMA | Department of Pathology, University of Hawaii, Manoa, Honolulu, HI 96822

[a] S.A.A. HONDA | Department of Pathology, University of Hawaii, Manoa, Honolulu, HI 96822

[a] P.K. HORAN | Smith, Kline and French Laboratory, Department of Cell Biology, 709 Swedeland Road, Swedeland, PA 19479

[a] C.H. JOHNSON | Department of Cellular and Developmental Biology, Harvard University, 16 Divinity Avenue, Cambridge, MA 02138

[b] Z. JONAK | Smith Kline and French Laboratory, Research and Development, 709 Swedeland Road, Swedeland, PA 19479

G.J. JONES | Massachusetts Institute of Technology, 48-213, Cambridge, MA 02139

[a]M.W. JOSEPHSON	Bio Metric Systems, Inc 9932 West 74th Street Eden Prairie, MN 55344
[a]P.A. JUMARS	School of Oceanography WB-10 University of Washington Seattle, WA 98195
[a]M.N. KOBAYASHI	Department of Pathology University of Hawaii, Manoa Honolulu, HI 96822
[a]A. LAGZIEL	Steinetz Marine Biological Laboratory Elat, Israel
[a]V. LIPIPUN	Department of Microbiology and Center for Marine Studies University of Maine Orono, Maine 04469
J.J. LEE	Department of Biology City College of CUNY New York, NY 10031
[b]W. LITAKER	Department of Microbiology and Immunology University of North Carolina at Chapel Hill 609 FLOB Building 231H Chapel Hill, NC 27514
[b]J. LOGAN	Bigelow Laboratory for Ocean Sciences McKown Point W. Boothbay Harbor, ME 04575
[b]E.J. LOVETT	Applied Immunology Program - USM Foundation for Blood Research Box 426 Scarborough, ME 04074
[a]T.J. MENDE	University of Miami School of Medicine PO Box 016129 Miami, Florida 33101 Uppsala, Sweden
[b]F.C. MAGUE	Bigelow Laboratory for Ocean Sciences McKown Point W. Boothbay Harbor, ME 04575
[b]N. MARCUS	Woods Hole Oceanographic Institution Woods Hole, MA 02543
[b]D.L. MIOSKY	Tufts University School of Vet. Medicine and Marine Biological Laboratory Woods Hole, MA 02543
[a]J.T. MIYAHARA	Department of Pathology University of Hawaii, Manoa Honolulu, HI 96822

C.E. MOODY
Department of Microbiology
University of Maine
Orono, Maine 04469

[a] F.M.M. MOREL
Massachusetts Institute of Technology
48-213
Cambridge, MA 02139

[b] I. MORRIS
University of Maryland
Center for Environmental and Estuarine Studies
PO Box 775
Cambridge, MD 21613

[a] J.E. MORRIS
Battelle, Pacific Northwest Laboratory
Biology and Chemistry Department
Richland, WA 99352

[a] L.K. NAKAGAWA
Department of Pathology
University of Hawaii, Manoa
Honolulu, HI 96822

B.L. NICHOLSON
Department of Microbiology and Center for Marine Studies
University of Maine
Orono, Maine 04469

[a] M.-T. NICOLAS
Department of Cellular and Developmental Biology
Harvard University
16 Divinity Avenue
Cambridge, MA 02138
and
Laboratoire de bioluminescence
CNRS
Gif-sur-Yvette, France

M.V. ORELLANA
School of Oceanography and Department of Botany
University of Washington
Seattle, WA 98195

[b] D. PAGE
Department of Chemistry/Biochemistry
Bowdoin College
Brunswick, ME 04011

M.J. PERRY
School of Oceanography and Department of Botany
University of Washington
Seattle, WA 98195

[b] B. PHILLIPS
Turner Association
2247 Old Middlefield Way
Mountain View, CA 94043

[b] D.A. PHINNEY
Bigelow Laboratory for Ocean Sciences
McKown Point
W. Boothbay Harbor, ME 04575

S. POMPONI — Sea Pharm. Inc.
Harbor Branch Oceanographic Institution
5600 Old Dixie Highway
Fort Pierce, FL 33450

D.J. RECKTENWALD — Becton Dickinson Immunocytometry Systems
2375 Garcia Ave
Mountain View, CA 94043

B. REGUERA — Instituto Oceanografico
Orillamar 47
Vigo, Spain

P.W. RENO — Department of Microbiology and
Migratory Fish Research Institute
University of Maine
Orono, Maine 04469

[b]D.W. RICK — EPICS Div., Coulter Corp.
181 Boulder Circle
Glastonbury, CT 06033

D. ROBINS — Institute for Marine Environmental
Research
Prospect Place, The Hoe
Plymouth
Devon, England PL1 3DH

[b]A. ROBLEZ-GONZALEZ — Urbanizacion Verde-Sol, SS, S°-A
El Burgo de las Rozas
28230 Madrid, Spain

[b]R.A. ROBOHM — Northeast Fisheries Center
Milford Laboratory, F/NEC23
212 Rogers Ave
Milford, CT 06460

G. ROESIJADI — University of Maryland
Center of Environmental and
Estuarine Studies
Chesapeake Biological Laboratory
Solomons, MD 20688

[b]R. SELVIN — Bigelow Laboratory for Ocean Sciences
McKown Point
W. Boothbay Harbor, ME 04575

L. SHAPIRO — Bigelow Laboratory for Ocean Sciences
McKown Point
W. Boothbay Harbor, ME 04575

[a]L.K. SHIRAI — Department of Pathology
University of Hawaii, Manoa
Honolulu, HI 96822

J.J. STEGEMAN — Biology Department
Woods Hole Oceanographic Institution
Woods Hole, MA 02543

[b]D. STOECKER — Biology Department
Woods Hole Oceanographic Institution
Woods Hole, MA 02543

[a]A.M. SZMANT — University of Miami
Rosenstiel School of Marine and
Atmospheric Science
4600 Rickenbacker Causeway
Miami, Florida 33149

G.L. TAGHON — Oregon State University
Marine Science Center
Newport, Oregon 97365

[b]S.J. TARAPCHAK — GLERL, NOAA
2300 Wasapenaw Ave
Ann Arbor, MI

G.H. THEILACKER — Southwest Fisheries Center
La Jolla Laboratory
National Marine Fisheries Service, NOAA
PO Box 271
La Jolla, CA 92038

D.D. TRUEBLOOD — Environmental Sciences Program
University of Massachusetts, Boston
Boston, MA 02125

B.B. WARD — Institute of Marine Resources
Scripps Institution of Oceanography
University of California, San Diego
La Jolla, CA 92093

[a]B.A. WATSON — School of Oceanography and
Department of Botany
University of Washington
Seattle, WA 98195

A.M. WOOD — Department of Biology
University of Chicago
915 East 57th St.

C.M. YENTSCH — Bigelow Laboratory for Ocean Sciences
McKown Point
W. Boothbay Harbor, ME 04575
Chicago, IL 60637

G. ZAGURSKY — Department of Biology
Belle W. Baruch Institute for
Marine Biology and Coastal Research
University of South Carolina
Columbia, SC 29208

[a]authors who did not attend
[b]attendees who are not authors

PREFACE

This volume represents a collection of pioneering papers introducing immunochemical techniques to the aquatic sciences. It is the product of a workshop entitled "Immunochemical Approaches to Coastal, Estuarine and Oceanographic Questions" held at the University of Southern Maine, Portland, Oct. 5-7, 1986. Funding from many sources made this workshop possible, and is gratefully acknowledged.

The reader will note great variability in both the quality and scope of the papers which we have divided into four sections:

I. Background material on the immune system (borrowed from a recent NIH publication)

II. Background material related to aquatic sciences

III. Techniques

IV. Applications

Readers familiar with terminology are directed to start with the original contributions related to aquatic sciences, page 21.

Some of the contributions are abstracts only. Others are preliminary, for which we are grateful to the authors for their willingness to share recent sometimes tentative findings with a broader community. Some offer novel and tempting insights, but as they exist, lack adequate standards and controls. Others are thorough and represent major scientific contributions. The promise rests in generating and working with reagents of great specificity enabling orders of magnitude improvements in the sensitivity of our measurements. Authors also address the problems and limitations. We thank the numerous authors and reviewers.

In support of every successful endeavor is someone of unusual savvy. For us, this has been David A. Phinney. His leadership, wisdom and wit have been important to the introduction of flow cytometry/sorting and immunochemistry into aquatic research.

Clarice M. Yentsch
Frances C. Mague
Paul K. Horan

SPONSORS

Anateck, Inc.
Portland, Maine 04106

Bigelow Laboratory for Ocean Sciences
West Boothbay Harbor, Maine 04575

Bowdoin College Chemistry/Biochemistry Department
Brunswick, Maine 04011

Delta Chemicals, Inc.
Searsport, Maine 04974

FMC Corporation
Rockland, Maine 04841

J.J. Mac Isaac Flow Cytometry/Sorting Facility of Bigelow Lab.
West Boothbay Harbor, Maine 04575

Sea Pharm Harbor Branch Foundation
Fort Pierce, Florida 33450

University of Southern Maine
Applied Immunology Program
Portland, Maine 04106

University of Maine at Orono
Sea Grant Program
Orono, Maine 04473

Background of the Immune System

UNDERSTANDING THE IMMUNE SYSTEM

Prepared by the Office of Research
Reporting and Public Response
National Institute of Allergy and Infectious Diseases
National Institutes of Health
Bethesda, Maryland 20205
Bulletin Number 85-529

INTRODUCTION

The immune system is a complex network of specialized organs and cells that has evolved to defend the body against attacks by "foreign" invaders. When functioning properly, it fights off infections by agents such as bacteria and viruses. When it malfunctions, however, it can unleash an enormous variety of diseases, from allergy to arthritis to cancer.

The immune system evolved because we live in a sea of microbes. Like man, these organisms are programmed to perpetuate themselves. The human body provides an ideal habitat for many of them and they try to break in, because the presence of these organisms is often harmful, the immune system will attempt to bar their entry or, if they succeed in getting in, to seek them out and destroy them.

The immune system, which equals in complexity the intricacies of the brain and nervous system, displays several remarkable characteristics. It is able to distinguish between "self" and "nonself." It is able to remember previous experiences and react accordingly: once you have had the mumps, your immune system will prevent you from getting it again. The immune system displays both enormous diversity and extraordinary specificity: not only is it able to recognize many millions of "nonself" molecules, it can produce molecules to match up with and counteract each one of them. And it has at its command a sophisticated array of weapons.

The success of this system in defending the body depends on an incredibly elaborate and dynamic regulatory-communications network. Millions and millions of cells, organized into sets and subsets, pass information back and forth like clouds of bees swarming around a

ANTIGENS

An antigen is any substance -- any of the millions of "nonself" molecules--that can trigger an immune response because the body recognizes it as foreign. A molecule announces itself as "self" or "nonself" by means of intricate and characteristic shapes, called *epitopes*, that protrude from its surface. Almost all antigens carry several different kinds of epitopes--some may carry several hundred--but some epitopes will be more effective than others at stimulating an immune response.

An antigen can be a virus, a bacterium, a fungus, or a parasite, or even a portion or product of one of these organisms. Tissues or cells from another individual, unless it is an identical twin, can also act as antigens, because a transplanted organ is seen as "foreign," the body's natural response is to reject it. The body will even reject nourishing proteins unless they are first broken down, by the digestive system, into their primary building blocks.

In abnormal situations, the immune system can wrongly identify "self" as "nonself." The result can be a so-called autoimmune disease, such as rheumatoid arthritis or systemic lupus erythematosus.

In certain persons, an apparently harmless substance, such as ragweed pollen or cat hair, can provoke the immune system to set off the inappropriate and harmful response known as allergy; in these cases the antigens are known as *allergens*.

THE ORGANS OF THE IMMUNE SYSTEM

The organs of the immune system are stationed throughout the body. They are generally referred to as "lymphoid" organs, because they are concerned with the growth, development and deployment of lymphocytes, the white cells that are the key operatives of the immune system. Lymphoid organs include the bone marrow, the thymus, the lymph nodes, and the spleen, as well as the tonsils, the appendix, and clumps of lymphoid tissue in the small intestine known as Peyer's patches.

Cells that are destined to become lymphocytes are produced in the *bone marrow*, the soft tissue in the hollow shafts of the long bones. Some of these cells, known as stem cells, migrate to the *thymus*, a multilobed organ that lies high behind the breastbone. There they multiply and mature into cells capable of producing an immune response--that is, they become immunocompetent.

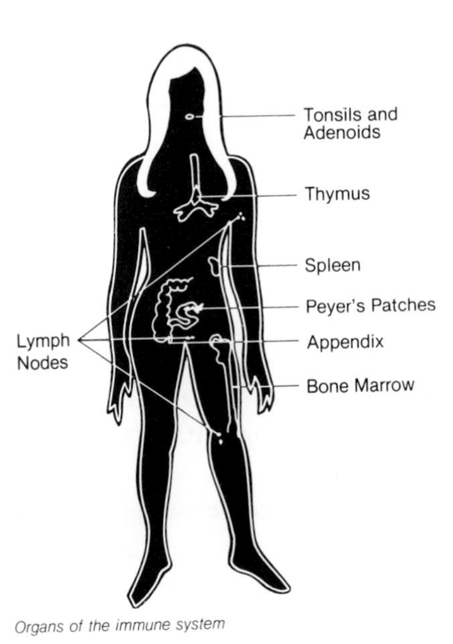

Organs of the immune system

Figure 1.

Stem cells that develop in the thymus are called T cells. Other lymphocytes, which appear to mature either in the bone marrow itself or in lymphoid organs other than the thymus, are called B cells.

Lymph nodes are small bean-shaped structures that are laced throughout the body. Strings of lymph nodes can be found in the neck, armpits, abdomen, and groin.

Each lymph node contains a variety of specialized compartments. Some compartments house B cells, others contain T cells. Yet others are filled with another type of immune cell, the macrophage. Lymph nodes also contain webbed areas that enmesh antigens. Thus, the lymph node brings together the several actors needed to produce an immune response.

Lymph nodes are linked by a network of *lymphatic vessels* similar to the blood vessels. Like a system of small creeks and streams that empty into larger and larger rivers, the vessels of the lymphatic network merge into increasingly larger tributaries. At the base of the neck, the large lymphatic ducts empty into the bloodstream.

The lymphatic vessels carry *lymph*, a clear fluid that bathes the body's tissues. Lymph contains many cells, most of them lymphocytes.

Lymph and the cells and particles it carries, including antigens that have entered the body, drain out of the body's tissues, seeping through the thin walls of the smallest lymphatic vessels. As the lymph passes through lymph nodes, antigens are filtered out and more lymphocytes are picked up. The lymphocytes, along with other assorted cells of the immune system, are carried to the bloodstream, which delivers them to tissues throughout the body. The lymphocytes patrol everywhere for foreign antigens, then gradually drift back into the lymphatic system, to begin the cycle all over again.

During their travels, circulating lymphocytes may spend several hours in the *spleen,* an organ in the abdomen that contains a high concentration of lymphocytes. Persons whose spleens have been damaged by trauma or by diseases such as sickle cell anemia are very susceptible to infection.

THE CELLS OF THE IMMUNE SYSTEM

The immune system stockpiles a tremendous arsenal of immunocompetent cells. By storing just a few cells specific for each potential invader, it has room for the entire array. When an antigen appears, these few specifically matched cells are stimulated to multiply into a full-scale army. Later, to prevent this army from overexpanding wildly, like a cancer, powerful suppressor mechanisms come into play.

Lymphocytes are the white cells that bear the major responsibility for carrying out the activities of the immune system; the immune system contains about one trillion of them. The two major classes of lymphocytes are the T cells (which are processed in the thymus) and the B cells (which grow to maturity independent of the thymus). Both B cells and T cells include a number of different subsets, all with different functions.

B cells secrete soluble substances called *antibodies.* A given antibody exactly matches a specific invading antigen, much like a key matches a lock. Sometimes, when an antibody interlocks with its matching antigen, it can inactivate the antigen. The body is capable of making antibody to millions of antigens.

Every B cell is programmed to make only one specific antibody. When the B cell encounters its triggering antigen, it gives rise to many cells called *plasma cells.* Every plasma cell is essentially a factory for producing antibody. Each of the plasma cells descended from a given B cell -- all members of the same family, or "clone" -- manufactures millions of identical antibody molecules, and secretes them into the bloodstream.

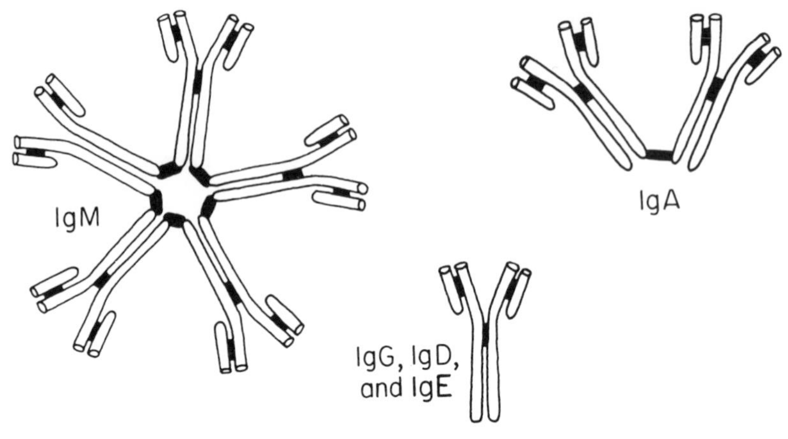

Immunoglobulin structure

Figure 2.

Scientists have identified five classes of antibodies, which are also known as *immunoglobulins* - IgG, IgM, IgA, IgE, and IgD. Each plays a different role in the immune defense strategy. IgG, the major immunoglobulin in the blood, is also able to enter tissue spaces; it works efficiently to coat microorganisms, speeding their uptake by other cells in the immune system. IgM, which usually combines in star-shaped clusters, tends to remain in the bloodstream, where it is very effective in killing bacteria. IgA concentrates in body fluids - like tears and saliva, as well as the secretions of the respiratory and gastrointestinal tracts - guarding the entrances to the body. IgE attaches itself to the surface of specialized immune cells such as mast cells and basophils and, when it encounters its matching antigen, stimulates the specialized cell to pour out its contents. IgD is almost exclusively inserted into cell membranes, where it somehow regulates the activation of the cell.

Antibodies cannot penetrate living cells, so they are ineffective against microorganisms inside cells. However, they circulate freely in the body fluids. Some antibodies disable the toxins produced by bacteria. Others, by coating the surfaces of bacteria, make them palatable to scavenger cells, cells able to engulf and destroy foreign material. Yet others block viruses from entering into cells.

T cells do not secrete antibodies, but their help is essential for antibody production. Some T cells become "helper" cells that turn B cells or other T cells on, while others become "suppressor" cells that turn these cells off.

T cells also secrete *lymphokines,* diverse and potent chemicals that can call into play many other cells and substances, including the elements of the inflammatory response. One class of lymphokines is interferon. (B cells, too, are capable of secreting lymphokines.)

All lymphocytes carry, on their surfaces, receptors that recognize one specific antigen. In the case of the B cell, this receptor is a sample of the antibody that particular B cell is prepared to manufacture. Some T cells use their antibody-like surface receptor to interlock with (and destroy) cells that carry the corresponding antigen, such as microbes and transplanted tissue cells, as well as body cells that have become infected by viruses.

Natural killer cells are granular lymphocytes. As their name suggests, they attack and destroy other cells. They are called "natural" killers because they go into action without prior stimulation by a specific antigen.

Most normal cells are resistant to natural killer cell activity. Most tumor cells, as well as normal cells infected with a virus, however, are susceptible. Thus, the natural killer cell may play a key role in "immune surveillance" against cancer, hunting down any cell that develops abnormal changes.

Macrophages and *monocytes* are large cells that act as scavengers, or phagocytes: they can engulf and digest marauding microorganisms and other antigenic particles. Monocytes circulate in the blood, while macrophages, in a variety of guises, are seeded through body tissues.

Macrophages also play a crucial role in initiating the immune response by "presenting" antigens to T cells in a special way that allows the T cell to recognize them.

In addition, macrophages and monocytes secrete an amazing array of powerful chemical substances. Called *monokines*, these substances are comparable to the lymphokines secreted by lumphocytes. Like lymphokines, monokines help to direct and regulate the immune response.

Granulocytes, like macrophages and monocytes, are phagocytes and thus capable of enveloping and destroying invaders. They contain granules filled with potent chemicals that enable them to digest microorganisms. These chemicals also contribute to inflammatory reactions and are responsible for the symptoms of allergy. Neutrophils, eosinophils, basophils, and mast cells are examples of granulocytes.

COMPLEMENT

The complement system is made up of a series of approximately 20 proteins that circulate in the blood in an inactive form. When the first of the complement substances is triggered--usually by an antibody locked to an antigen in an antigen-antibody complex--it sets in motion a ripple effect. As each component is activated in turn, it acts upon the next in a precise sequence of carefully regulated steps.

This phenomenon, known as the "*complement cascade,*" is responsible for causing mast cells to release the chemicals that produce the redness, warmth, and swelling of the inflammatory response. It can also bring a dramatically rapid death to bacteria and other cells by puncturing their cell membranes.

MOUNTING AN IMMUNE RESPONSE

Infections remain the most common cause of human disease. Produced by bacteria, viruses, parasites, and fungi, infections may range from relatively mild respiratory illnesses, such as the common cold, to debilitating conditions like hepatitis, to life-threatening diseases such as meningitis.

To fend off the threatening horde, the body has devised a series of defenses so intricate they make war games look like child's play. Microbes attempting to enter the body must first find a chink in the body's external protection - the skin and the mucous membranes coated with antibodies. Then they must elude a series of nonspecific defenses - that is, those equipped to attack a variety of antigens, without regard for their specific name-badge epitopes. Many of these microbes will be intercepted by patrolling scavenger cells.

Microbes that escape these traps face the body's specific responses. These may be cellular responses - directed by cells, primarily T cells and their secretions - or humoral responses - produced by antibodies that have been secreted by B cells into the body's fluids, or "humors."

Although immunologists have traditionally made a distinction between cellular and humoral immunity, it is becoming increasingly clear that the two arms of the immune response are closely intertwined. Almost all antigens evoke both a humoral and a cellular response.

The *Cell-mediated immune response* begins with the macrophage. The macrophage picks up the antigen and, perhaps after digesting it, displays the antigen's characteristic markers--its epitopes--on its own surface. Next to the antigen or "nonself" marker the macrophage displays a "self" marker. It takes both of these structures, together, to catch the T cell's attention.

Once triggered, T cells multiply and go into action. Some subsets of T cells become cytotoxic, or cell-killing, cells and set out in search of their specific antigen-targets. Others synthesize and secrete lymphokines. Some of these lymphokines draw other immune cells -- fresh macrophages, granulocytes, and other lymphocytes -- to the site of the infection, while other lymphokines direct the cells' activities once they arrive on the scene.

The B cell is responsible for *humoral immunity*. Some B cells, like the T cells in cellular immunity, need a double trigger -- an antigen plus "helper" factors produced by subsets of T cells. Other B cells can act without T cell help. When stimulated, the B cells swell into plasma cells and begin producing antigen-specific antibodies.

The antibodies lock antigens into antigen-antibody complexes, which trigger the complement cascade or are removed from the circulation by clearing mechanisms in the liver and the spleen.

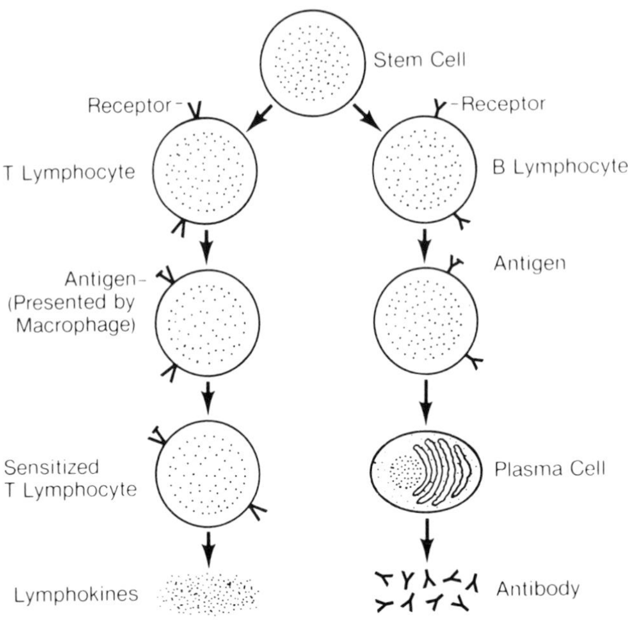

The immune response: cell-mediated (left) and humoral (right)

Figure 3.

Gradually the antigen is overcome and, under the influence of "suppressor" factors produced by yet other subsets of T cells, antibody production wanes.

IMMUNITY--NATURAL AND ACQUIRED

As long ago as the fifth century B.C., Greek physicians noted that people who had recovered from the plague would never get it again -- they had acquired immunity. This is because, whenever T cells and B cells are activated, some subsets become "memory" cells. Then, the next time that an individual encounters that same antigen, the immune system is primed to destroy it.

The degree of immunity depends on the kind of antigen, its amount, and how it enters the body.

Infants are born with relatively weak immune responses. They have, however, natural immunity, protected during the first months of life by means of antibodies they receive from their mothers. The antibody IgG, which travels across the placenta, makes them immune to the same microbes to which their mothers are immune. Children who are nursed also receive IgA antibodies from breast milk.

In addition to acquiring immunity naturally, it is possible to become immunized through the use of vaccines. *Vaccines* contain microorganisms, or parts of microorganisms, that have been altered so that they will produce an immune response but not full-blown disease. Some vaccines are made from microbes, that have been killed. Others use microbes that have been changed slightly (attenuated) so they can no longer produce infection. They may, for instance, be unable to multiply. some vaccines are made from a live virus that has been weakened by growing it for may cycles in animals or tissue cultures.

Recent research has focused on developing vaccines that use only part of the infectious agent. Such subunit vaccines -- which are now available for meningitis and pneumonia -- produce the desired immunity without stirring up separate immune reactions to the many antigens carried, for instance, on a single bacterium.

DISORDERS OF THE IMMUNE SYSTEM

Allergy

The most common types of allergic reactions -- hay fever, some kinds of asthma, and hives -- are produced when the immune system responds to a false alarm. In a susceptible person, a normally harmless substance -- grass pollen or house dust, for example -- is perceived as a threat and is attacked.

Such allergic reactions are related to the antibody known as immunoglobulin E. Like other antibodies, each IgE antibody is specific: one reacts against oak pollen, another against ragweed antigens. The role of IgE in the natural order of things is not known, although some scientists suspect that it developed as a defense against infection by parasitic worms.

When allergic persons are first exposed to an antigen or allergen, they make large amounts of the corresponding antibody. These IgE molecules attach to the surfaces of mast cells (in tissue) or basophils (in the circulation).

When this IgE antibody, sitting on the mast cell or basophil, encounters its specific allergen, the IgE antibody signals the mast cell or basophil to release the powerful chemicals inside it. It is these chemical mediators that cause the symptoms of allergy, including wheezing, sneezing, runny eyes, and itching.

Immune Deficiency Diseases

Lack of one or more components of the immune system results in an immune deficiency disease. These can be inherited, acquired through illness, or produced as an inadvertent side effect of certain drug treatments.

Some children are born with defects in their immune systems. Those with flaws in the B cell component are unable to produce antibodies (immunoglobulins). These conditions, known as agammaglobulinemias or hypogammaglobulinemias, leave the children vulnerable to infectious organisms; such disorders can be combatted with intramuscular injections of immunoglobulins.

Other children, whose thymus is either missing or small and abnormal, lack T cells. The resultant disorders have been treated with thymic or fetal liver transplants.

Very rarely, infants are born lacking all the major immune defenses, this is known as severe combined immunodeficiency disease (SCID). Some children with SCID have lived for years in germ-free rooms and "bubbles." A few SCID patients have been successfully treated with transplants of bone marrow; the immature cells in the bone marrow transplant grow into functioning B and T cells.

People with advanced cancer may have immune deficiencies as a result of the disease process. Transient immune deficiencies can develop in the wake of common viral infections, including influenza, mononucleosis, and measles. Immune responsiveness can also be depressed by blood transfusions, malnutrition, and stress.

In the early 1980's doctors became aware of a growing number of cases of an often fatal disorder named the acquired immune deficiency syndrome (AIDS). Nearly 95 percent of AIDS cases have been found among sexually active homosexual and bisexual men with multiple sex partners, Haitian entrants into the United States, present or past abusers of intravenous drugs, persons with hemophilia, and the children of persons in the categories.

Characterized by very low levels of helper T cells, but normal levels of suppressor T cells, AIDS is now thought to be caused by a transmissible agent, a retrovirus. AIDS can lead to an otherwise rare tumor known as Kaposi's sarcoma and a variety of opportunistic infections including an unusual form of pneumonia caused by *Pneumocystis carinii*. Opportunistic infections are those produced by organisms that do not affect people whose immune systems are working normally. No adequate treatment for AIDS has yet been found.

A defect in even one small component of the complex immune system can have far-reaching effects. For example, in children with cystic fibrosis, mucus-producing glands secrete abnormally thick sticky mucus that clogs air passages. In addition, these children make an enzyme that prevents their macrophages from attacking a particular bacterium, *Pseudomonas aeruginosa*. This bacterium is a frequent cause of pneumonia, permanent lung damage, and even death in these patients.

Immunoproliferative Disorders

Cells of the immune system, like those of other body systems, can proliferate uncontrollably; the result is cancer. Leukemias are caused by the proliferation of white blood cells, or leukocytes. The uncontrolled growth of antibody producing (plasma) cells can lead to multiple myeloma, which affects the bone marrow and the bones. Cancers of the lymphoid organs are known as lymphomas and include Hodgkin's disease. These disorders can be treated -- some of them very successfully -- by drugs and/or irradiation.

Disorders of Immune Regulation

To work properly, the cells of the immune system need to maintain a delicate balance. If suppressor T cells are sluggish, B cells may overproduce antibody; the same is true if helper T cells are overactive.

Thanks to technological advances such as a machine known as the fluorescence-activated cell sorter (FACS) and, more recently, to monoclonal antibody techniques (discussed below), scientists are able to determine the ratio of helper cells to suppressor cells.

In patient with systemic lupus erythematosus, for instance, the B cells themselves are hyperactive while suppressor cells are underactive, although it is not clear which defect comes first. Patients with rheumatoid arthritis, who have a defective suppressor T cell system, continue to make antibodies to a common virus, whereas this response normally shuts down after about a dozen days. Other diseases associated with deficiency in suppressor T cells include multiple sclerosis, severe allergic eczema, and some anemias.

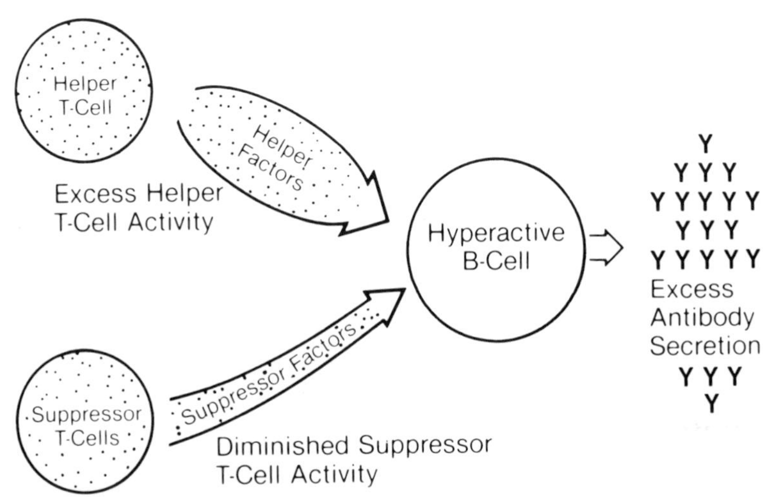

Figure 4. Disease occurs when there is a flaw in immune system regulation. Systemic lupus erythematosus, multiple sclerosis, and rheumatoid arthritis develop when suppressor T-cells are underactive.

Too few helper T cells, in contrast, are seen in some forms of cancer, as well as in AIDS.

Immune Complex Diseases

Immune complexes are clusters of interlocking antigens and antibodies. Under normal conditions immune complexes are removed from the bloodstream by clearing mechanisms in the spleen and liver. In some circumstances, however, immune complexes continue to circulate. Eventually they become trapped in the tissues of the kidneys and other organs. (Just where they end up probably depends on the nature of the antigen in the complex.) There they set off reactions that lead to inflammation tissue damage.

Damage due to immune complexes occurs in many diseases. These include malaria, cancer, gastrointestinal disorders, and bacterial and viral infections. Harmful immune complexes are also formed in autoimmune diseases.

Autoimmune Diseases

Sometimes the immune system's recognition apparatus goes awry, and the body begins to manufacture antibodies directed against the body's own components -- cells, cell components, or specific organs. These are known as *autoantibodies*, and the diseases they produce are called autoimmune diseases. (Not all autoantibodies are harmful; some types appear to be integral to the immune system's regulatory scheme.)

Autoimmune diseases include systemic lupus erythematosus, rheumatoid arthritis, thyroiditis (an autoimmune inflammation of the thyroid gland), mayasthenia gravis, and certain kinds of anemia. Juvenile onset diabetes is probably an autoimmune disease, as is multiple sclerosis.

Some autoantibodies produce harm directly. For instance, autoantibodies to red blood cells can cause anemia. Autoantibodies to platelets and clotting factors in the blood can lead to bruising and even internal bleeding. Autoantibodies to lymphocytes can impair the immune response.

Another type of autoantibody is directed against the components of the cell's nucleus -- DNA, RNA, or proteins. These autoantibodies are called antinuclear antibodies, or ANA's. They can cause serious damage when they link up with "self" antigen to form circulating immune complexes, which become lodged in body tissues and set off inflammatory reactions.

No one knows just what causes an autoimmune disease, but several factors are likely to be involved. These may include viruses, an inherited predisposition, and environmental factors, such as exposure to sunlight, certain chemicals, and some drugs. Sex hormones may be important, too, since most autoimmune diseases are far more common in women than in men.

A wide variety of therapies are being used to combat autoimmune diseases. These include radiation of the lymph nodes, immunosuppressive drugs developed as anticancer agents, and apheresis, a sort of "blood washing" that removes diseased cells and harmful molecules from the circulation.

IMMUNOLOGY AND TRANSPLANTS

Since organ transplantation was introduced over a quarter of a century ago, it has become a widespread remedy for life-threatening disease. Several thousand kidney transplants are performed each year in the United States alone. In addition, physicians have succeeded in transplanting the heart (about 500 heart transplants had been performed by 1982), liver, pancreas, and lung.

The success of a transplant -- whether it is accepted or rejected -- depends on the stubbornness of the immune system. In order for a transplant to "take," the body of the recipient must be made to suppress its natural tendency to get rid of the "foreign" tissue.

Scientists have tackled this problem in two ways. The first is to make sure that the tissues of the donor and the recipient are as similar as possible. The "self" markers, or HLA antigens, on body cells are determined by specific genes on a specific chromosome, chromosome 6. Blood tests called histocompatibility testing can show if the potential donor and the recipient share two or more of these histocompatibility antigens, and thus are likely to make a good "match." The best matches are identical twins; next best are close relatives, brothers and sisters in particular.

The second approach to taming rejection is to lull the recipient's immune system. This can be achieved through a variety of powerful immunosuppressive drugs. One curbs the growth of lymphocytes, another kills lymphocytes, and yet another stimulates antibodies that inactivate lymphocytes from producing a lymphokine that is necessary for other T cells, including cytotoxic T cells and helper cells, to grow and develop.

In addition, blood transfusions prior to transplantation surgery quiet the immune system in some unknown way. As an extra precaution, some transplant experts also remove the patient's spleen.

Not surprisingly, such an assault on the immune system leaves the transplant recipient susceptible to both infection and cancer. Although such patients need careful medical follow-up, many of them are able to lead active and essentially normal lives.

FRONTIERS IN IMMUNOLOGY

Molecular Genetics

Molecular genetics is the study of gene structure. By plucking interesting genes from their cells of origin and reinserting them into bacterial cells, scientists can obtain and analyze quantities of genes, including the genes involved in immune responses.

Among other advances, this area of research has provided valuable insights into the ways that antibodies work -- insights that will help us understand the inappropriate antibody responses that occur in autoimmune diseases. Some day this technology may be used to correct gene defects. Other potential uses include making prenatal diagnoses, developing vaccines, and correcting defective lymphocytes.

Hybrid Cells

Through a finesse known as hybridoma technology, scientists are now able to obtain, in quantity, substances secreted by cells of the immune system. A "hybridoma" is created by fusing two cells, a secreting cell from the immune system and a long-lived cancer cell, within a single membrane. The resulting hybrid cell can be cloned, producing many identical offspring. Each of these daughter clones will secrete, over a long period of time, the immune cell product -- antibody if a B cell was used, a lymphokine if a T cell was used. Because the secretions trace their origins o a single cell, they are pure, and in the case of antibodies, they recognize a single, specific antigen.

Cell hybridization holds enormous potential in the prevention, diagnosis, and treatment of disease. Antibodies produced by a hybridoma, which are called monoclonal antibodies, have been used to distinguish among different types of lymphomas and leukemias, permitting physicians to tailor therapy accordingly. They are being used to determine the ratio of helper T cells and suppressor T cells in immune disorders. They may also make it possible to develop totally specific immunoglobulins that can be used in vaccines to protect people against diseases like influenza.

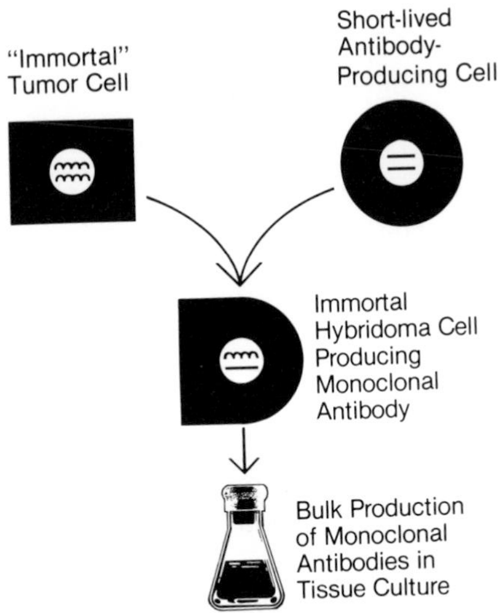

Figure 5. Cell fusion results in hybridoma, which produces monoclonal antibodies.

Immunoregulation

Research into the delicate and complex checks and balances that regulate the immune response is leading not only to a appreciation of the events involved in normal immunity, but also to abnormalities of immune functions. Eventually it may be possible to treat diseases such as systemic lupus erythematosus by selectively suppressing parts of the immune system that are overactive and selectively stimulating those that are underactive.

GLOSSARY

Acquired Immune Deficiency Syndrome (AIDS) - A disease characterized by breakdown of the body's immune system.

Agammaglobulinemia - Total lack of immunoglobulins.

Allergen - Any substance that causes an allergy.

Allergy - The inappropriate and harmful response of the immune system to normally harmless substances.

Antibody - Protein molecules that are produced and secreted by certain types of white cells in response to stimulation by an antigen.

Antigen - Any substance that provokes an immune response when introduced into the body.

Appendix - An organ of the immune system.

Attenuated - Descriptive of a microbe that has been changed slightly so that it no longer causes disease.

Autoantibody - An antibody that reacts against a person's own tissue.

Autoimmune disease - A disease that results when the body's immune system produces harmful autoantibodies.

Bacterium - A microscopic organism composed of a single cell. Many bacteria can cause disease in man.

Basophil - A special white blood cell, called a granulocyte, filled with granules of toxic chemicals, that can digest microorganisms. Like the mast cell, its counterpart in the tissue, basophils are responsible for the symptoms of allergy.

B cells - White blood cells of the immune system derived from bone marrow and involved in the production of antibodies; they are also called B lymphocytes.

Bone marrow - Soft tissue located in the cavities of the bones. Responsible for producing blood cells.

Complement - A complex series of blood proteins involved in the immune response.

Complement cascade - A precise sequence of events, usually triggered by an antigen-antibody complex, in which each component of the complement system is activated in turn, resulting in inflammation and destruction of microbes.

Eosinophil - A special white blood cell, called a granulocyte, that can digest microorganisms. Plays a role in allergic reactions.

Epitope - A characteristic shape or marker on an antigen's surface.

Fungus - Member of a class of relatively primitive vegetable organisms including mushrooms, yeasts, rusts, molds, and smuts.

Granulocytes - A cell of the immune system filled with granules of toxic chemicals that enable them to digest microorganisms. Basophils, neutrophils, eosinophils, and mast cells are examples of granulocytes.

Helper T cells - A subset of T cells that turn-on antibody production.

Histocompatibility tests - Immunologic tests for comparing the antigens on the tissue of a transplant donor with those of the recipient. The more closely matched the better the chance for the transplant to take.

Hybridoma - A hybrid cell, created by fusing a lymphocyte with a cancer cell, that secretes a pure immune substance, either a lymphokine or an antibody specific for only one antigen.

Hypogammaglobulinemia - Lower than normal levels of immunoglobulins.

Immune complex - Large molecules formed when antigen and antibody bind together.

Immune response - The activity of the immune system against foreign substances.

Immunocompetent - Capacity to develop an immune response.

Immunoglobulin - Another name for antibody.

Leukocytes - All the white blood cells.

Lymph - A transparent, slightly yellow fluid containing primarily lymphocytes. Lymph is composed of tissue fluids collected from all parts of the body and returned to the blood via the lymphatic vessels.

Lymph nodes - Small bean-sized organs of the immune system, distributed widely throughout the body. An outpost for B lymphocytes.

Lymphocytes - Small white cells, normally present in the blood and in lymphoid tissue, that bear the major responsibility for carrying out the functions of the immune system.

Lymphokines - Powerful substances, produced and released into the bloodstream by T lymphocytes and capable of stimulating other cells in the immune system.

Macrophage - A scavenger cell found in the tissues, able to destroy invading bacteria or other foreign material.

Mast cells - Special cells found in the tissue. Contain granules of chemicals responsible for the symptoms of allergy.

Microbes - Minute living organisms, including bacteria, protozoa, and fungi.

Microorganism - A microscopic plant or animal.

Molecule - The smallest unit of matter of an element or compound.

Monoclonal antibodies - Antibodies, produced by hybridomas, specific for only one antigen.

Monocyte - A large white blood cell that acts as a scavenger, capable of destroying invading bacteria or other foreign material.

Monokines - Powerful chemical substances that are secreted by monocytes and macrophages and help direct and regulate the immune response.

Natural killer cells - Large granular lymphocytes that attack and destroy other cells such as tumor cells and those infected with viruses or other microbes.

Neutrophil - A special white blood cell, called a granulocyte, that can digest microorganisms.

Opportunistic infection - An infection in immune-suppressed persons, caused by organisms that do not usually affect people with normal immune systems.

Organism - Individual living thing.

Parasite - A plant or animal that lives, grows, and feeds on or within another living organism.

Peyer's patches - A special collection of lymphoid glands in the intestinal tract.

Phagocytes - Cells that ingest microbes or other cells and foreign particles.

Plasma cells - Antibody-producing cells descended from B cells.

Scavenger cells - Any of a diverse group of cells that have the capacity to engulf and destroy foreign material, dead tissues, or cells.

Severe Combined Immunodeficiency Disease (SCID) - A disease in which infants are born lacking all major immune defenses.

Spleen - An organ in the abdominal cavity. It is an important site of antibody production.

Stem cells - Cells from which all blood cells derive.

Subunit vaccine - A vaccine produced from only part of an infectious agent.

Suppressor T cells - Subset of T cells that "turn off" antibody production.

T cells - White blood cells that are processed in the thymus. The produce lymphokines and are responsible, in part, for carrying out the immune response. They are also called T lymphocytes.

Thymus - A central lymphoid organ important in the development of immune capability.

Tonsils - Prominent oval masses of lymphoid tissue on each side of the throat.

Vaccine - A substance that contains the antigen of an organism and that stimulates active immunity and future protection against infection by that organism.

Virus - Submicroscopic microbe causing infectious disease. Can reproduce only in living cells.

Background

WHAT DOES THE IMMUNOLOGICAL APPROACH OFFER AQUATIC RESEARCH? AN OVERVIEW

S.A. Pomponi
Sea Pharm. Inc.
Harbor Branch Oceanographic Institution
5600 N. Old Dixie Highway
Fort Pierce, FL 33450

C.M. Yentsch
Bigelow Laboratory for Ocean Sciences
West Boothbay Harbor, ME 04575
and
Bowdoin College Chemistry Department
Brunswick, ME 04011

P.K. Horan
Smith, Kline & French Laboratories
Department of Immunology
709 Swedeland Road, Swedeland, PA 19406

INTRODUCTION

Chemical reagents have permitted the probing of cells, metabolic rates and processes, all of which make up an understanding of the world's living resources on land, and freshwater as well as estuarine, coastal and oceanic waters. During the past decade, there has been an explosion of immunochemical reagents and techniques, primarily in the biomedical sciences. Within the past few years there has been keen interest in adapting some of this methodology to the study of marine organisms. The greatest incentive is derived from the fact that one can tag and trace something of interest in a complex mixture -- without elaborate separation and isolation procedures. Thus, for the aquatic research community the approach is ideal. Oceans, lakes, and estuaries are often dilute mixtures, but they are always complex mixtures of autotrophs, heterotrophs, larval stages, adult forms, detritus, and fecal pellets.

The interest is becoming a reality. Several innovative projects involving immunochemical approaches to aquatic questions are now underway. These projects represent a wide range of applications and include: 1) identification (localization, quantitation) of specific cell types in various water masses; 2) identification (localization, quantitation) of specific proteins/enzymes within single cells; 3) identification and monitoring development of specific larvae; 4) detection and quantitation of concentrations of marine toxins, e.g. from dinoflagellates; 5) determination of evolutionary linkages by cross-reactivity; and 6) specific tags and tracers for selective feeding experimentation.

In this chapter we will present a section on the evolution of immunity as it is currently understood; a sketch of the brief history of antibody research and its eventual introduction into the aquatic research community; and an immunologist's comments on some current immunology research and forward thinking for aquatic researchers. While the details of the tools are not included in this volume, it is prudent for those whose interest is aroused to keep abreast of these ideas.

AN EVOLUTIONARY PERSPECTIVE

The Evolution of Immunity

The immune response in mammals can be either cell-mediated or humoral. Lymphocytes are involved in both types of immune response. One facet of cell-mediated response begins when an antigen is phagocytized by a macrophage which then displays the antigen's epitope, or marker, on its surface. This non-self marker, together with the macrophage's own self marker, stimulates the T-lymphocyte into activity. Some T-cells kill the antigens directly; others synthesize and secrete lymphokines, which, in turn, stimulate other immunocompetent cells into activity at the site of infection. The humoral response involves a different lymphocyte, the B-cell. B-cells, by their antibody-binding capacity, are stimulated to divide. This so-called clonal expansion produces many more B-cells. They are then further stimulated into activity by an antigen, develop into plasma cells and begin excreting specific antibodies to the antigen. The antibodies form complexes with the antigens, which stimulate the complement system, a series of about 20 proteins that circulate in the blood. Lysed cells are then cleared from the circulatory system through the liver and spleen.

Comparative immunology of phylogenetic groups, ranging from the most primitive to the most advanced, have not yet clarified the evolutionary sequence of vertebrate immunity. They have demonstrated, however, that immunocompetence exists in invertebrates, ranging from sponges to echinoderms.

Defensive cells, or immunocytes, occur in all metazoans. Immunocyte types range from archaeocytes (and derivatives) in sponges and amoebocytes in coelenterates, to coelomocytes and haemocytes in annelids, molluscs, and arthropods, to primitive lymphoid cells in echinoderms and tunicates.

Cellular aspects of alloimmune (same species) reactions have been examined in detail for the most primitive metazoan phylum, the sponges (Buscema and Van de Vyver, 1984), in which a surprising degree of diversity is found. This diversity can occur within a genus, a species, or even an individual. Using allograft experiments, three types of cellular immune responses have been observed: phagocytosis, cytotoxicity, and encapsulation. Archaeocytes, collencytes, and perhaps spherulous cells are involved in cytotoxic responses. Data suggest that the immune response involves cell specialization with archaeocytes responsible for rapid response via cytotoxicity and for phagocytosis, and collencytes responsible for constructing a collagen barrier (Buscema and Van de Vyver, 1984).

This diversity does not occur in more advanced metazoans such as coelenterates (Kolenkine, 1971), nemertines (Langlet and Bierne, 1977), annelids (Valembois *et al.*, 1982), or arthropods (Lackie, 1977, 1979).

In higher invertebrates, immunocytes may develop within or aggregate into tissues. Although there is no evidence for homology with vertebrate lymphoid tissue, these cellular aggregates have been termed "lymphoid" (Roitt *et al.*, 1985). Examples include the haemal glands of earthworms, white bodies and branchial spleens of molluscs, haemopoietic tissues of arthropods, axial organs of echinoderms, and lymph nodules associated with the gills of tunicates (Roitt *et al.*, 1985).

Immunity implies a type of "memory" - - once stimulated by an antigen, the individual's immune system "remembers" so that when exposed to the same antigen again, it can destroy the invading cells. Until recently, it was believed that immune memory evolved in vertebrates (Marchalonis, 1977). If we consider immunocompetence in broader terms, however, specific alloimmune memory has been demonstrated in invertebrates, ranging from sponges to echinoderms (Hildemann, *et al.*, 1979; Twarog, 1974; Twarog *et al.*, 1975).

Immunocompetence is more broadly defined as an adaptive defense, with or without immunoglobulin molecules, and involving three components: (1) a cytotoxic reaction; (2) a more rapid response after a second challenge (memory); and (3) selective reactivity (Hildemann *et al.*, 1980). These three criteria have been demonstrated in many invertebrate phyla (Table 1), primarily through grafting experiment involving the same or different individuals, species, or genera (isografts, allografts, and xenografts).

Xenograft rejection occurs in all invertebrates, including protozoans, but this probably involves biochemical or enzymatic incompatibility rather than immunocompetence. Allogeneic cell-surface polymorphism occurs in protozoans, but it is a mating response rather than an immune reaction. Immunocompetence has, therefore, only been demonstrated to date in metazoans.

The primary difference between invertebrate and vertebrate immune memory may be its duration (Bigger *et al.*, 1982). Accelerated rejections of allografts can occur from 2-4 weeks after initial sensitization in sponges, to as long as 6 months in echinoderms. Long-term memory may be characteristic of higher invertebrates and lower vertebrates (Bigger *et al.*, 1982). The short-term memory in lower invertebrates may merely be a consequence of the memory being effected by immunocytes with shorter life spans or more rapid turnover rates than in vertebrates.

It is difficult to predict immunocompetence within a phylum, genus, species, or individual. Investigations of the immune responses in sponges suggest that the diversity of responses found within this phylum may depend upon the type of allogeneic challenge or the experimental conditions (Hildemann *et al.*, 1980; Bigger *et al.*, 1982; Buscema and Van de Vyver, 1984).

Buscema and Van de Vyver (1984) suggest that the diversity of immune responses in sponges may be a function of their primitive level of organization, and that responses become more specialized with increasing phylogenetic complexity.

Similarities in cell-mediated immunity among vertebrates and invertebrates suggest that they share a common ancestry (Bigger *et al.*, 1982), but there is no experimental evidence to support such a hypothesis. The functional similarities of invertebrate immunocytes to vertebrate T-cells make it tempting to speculate on the origins of T- and B-cells. Comparative studies of the molecular basis of cell recognition are needed to clarify the evolution of lymphocytes, histocompatibility antigens, and immunoglobulins. Research on sponge cell recognition (for review, see Muller, 1982) may form the basis for such comparative investigations.

IMMUNOPOTENTIALITIES OF INVERTEBRATES*

	1	2	3	4	5	6	7
Protozoans			x				
Sponges	x		x	x	x	?	
Coelenterates	x		x	x	x	?	
Flatworms	x		x	?	?	?	
Nemertines	x		x			?	
Annelids	x	x	x	x	x	x	?
Molluscs	x	x	x	x	?	?	?
Arthropods	x	x	x	?	?	?	?
Echinoderms	x	x	x	x	x	x	?
Tunicates	x	x	x	x	?	x	?

1. immunocytes
2. "lymphoid" tissue
3. allogeneic cell-surface polymorphism
4. allograft rejection
5. alloimmune memory
6. T mitogen reactivity
7. inducible antisomes

*from Roitt *et al.*, 1985 and Hildemann *et al.*, 1980.

AN AQUATIC RESEARCH PERSPECTIVE

History of antibody research

Most living organisms in aquatic systems exist as independent, single-cell entities in a fluid medium. In even the most nutrient-impoverished oligotrophic waters of the world, very small cells occur at nearly one million cells per liter. In more eutrophic waters, numbers per unit volume are several orders of magnitude higher. It is no wonder that the examination and characterization of individual cells were preoccupations of early naturalists interested in aquatic ecosystems. For the most part, cells, whether occurring as solitary units or as multicellular larvae or adult organisms, are in the 1-100 μ size range. In its infancy, the era of aquatic research was paralleled by an explosion of chemical assessment methods. Thus there was a thrust to improve biological sampling through chemical measurements in dynamic aquatic systems. These bulk assays were highly informative, yet give no information about: a) the contribution and alteration of the optics of the water mass, b) the contribution by rare events and forms, c) the contribution from the smallest cells (1-3 μm and less), d) the distribution of properties among cells, and e) any index of variability (Yentsch and Pomponi, 1986).

Thus, researchers are now going back to a careful evaluation of the cells present in the aquatic system of interest, with tools far advanced from those used by the early naturalists. But the questions remain much the same. Predictive ecology will depend upon an understanding of the various species components (plus knowledge of any molecular idiosyncrasies), and their variability, not merely a mean value for a population or community. The promise lies in a broad range of techniques designed to measure properties and rate processes of individual cells and individual organisms. As summarized by Yentsch and Pomponi (1986), appropriate application of automated individual particle analysis can become coupled with immunochemistry. Many pressing questions should result in new interpretations and understanding, examples of which are: allometry, ataxonomy, distinctive properties, cell metabolism, cell growth, cell duplication and division, cell recognition, differentiation and development, and cell-to-cell interaction.

Visualization of an antigen-antibody response was achieved using fluorescent dyes as early as 1933 (Visser and Van den Engh, 1982). This accomplishment opened the doors for automated analysis, an aspect considered critical for both routine and experimental work. There were many early problems such as: a) high background fluorescence, therefore a signal to noise problem, b) nonspecific staining, and c) considerable loss of specificity of the labeled antibody with time. These three problems are now routinely overcome. Additionally, major breakthroughs in the design of microscopes, optical filters, quality light sources such as lasers, and automated flow cytometer instrumentation permit new approaches to both cell-surface and histochemical antibody detection.

Cell biology has witnessed a rapid rise in the popularity of antibody techniques over the past decade. Assays include: fluorescent assays measured via epifluorescent microscopy or a flow cytometer; immunogold and immunosilver labeling detectable with transmission electron microscopy (TEM); and bulk assays such as radio-immunoassays (RIA), measured using a scintillation counter; and enzyme-linked immunoassays (ELISA), measured using a spectrophotometer or crude colorimeter.

The success of the antibody technology rests in several aspects:

1. Specific antibodies to virtually any macromolecule can be produced experimentally - even when the chemistry is not well understood.
2. Specific antibodies to small molecules can be produced experimentally if the small molecule is bonded to a macromolecule, thus a hapten conjugate.
3. The technique is very sensitive and depending on the eventual goal, either a) a polyclonal antibody, b) a polyclonal antibody of restricted heterogeneity, c) a homogeneous monoclonal antibody or d) a cocktail of two or more monoclonal antibodies might be selected for use.
4. Fluorescent images allow one to map out the molecular anatomy of the cell to a resolution of about 0.5 µm (Fujiwara and Pollard, 1980), and immunogold TEM images allow one to map out the molecular anatomy of the cell to a resolution of about 0.05 µm (Nicolas *et al.*, 1985; and this volume).

Monoclonal antibodies have received considerable attention. In 1975 there was a major breakthrough. George Köhler, a postdoctoral fellow in Cesar Milstein's laboratory at Cambridge University, fused together two cell types, cancer cells with cells producing specific desirable antibodies. He was surprised to observe that the new "immortal" cell type continued to produce large quantities of the antibody.

These cells, termed hybridomas, can be cloned; single cells can be selected and cultured. The product is called a monoclonal antibody. Thus antibody can be produced on demand. When the demand is high, many cultures can be grown and harvested. When the demand is low, the cell lines can be kept alive, but at rest in low temperatures. See Köhler and Milstein, (1975) for the classic report of this discovery.

The technique is elegant in its simplicity, and its influence over many areas of scientific investigations is profound. An echo of this breakthrough is beginning to be heard in aquatic research. The production of monoclonal antibodies is independent and continuous after the animal initially injected is sacrificed. Monoclonals are typically screened for complete specificity. The screening process is very labor intensive and the initial product amount is generally far less than traditional polyclonal antibodies. Choosing monoclonal antibodies is, therefore, not always an advantage, despite the immortality of the antibody producing cells. The most difficult dilemma for aquatic researchers may well be whether to invest in monoclonal or polyclonal antibodies to best achieve their goals.

Published antibody research in the aquatic sciences to date has been dominated by surface labelling of polyclonal antibodies. Ward and Perry (1980) introduced the concept by discriminating nitrifying bacteria. Campbell *et al* (1983) followed with various serotypes of cyanobacteria. In each case, different antibody reactive types have been identified and mapped in water masses, both vertically and horizontally over broad geographic regions. Diversity dynamics of marine bacteria have been described by Dahle and Laake (1981;1982). Lee (this volume) has identified diatoms as symbionts. Feller and co-workers have prepared evolutionary schemes for invertebrates using multiple antibody cross-reactivity. Additionally, predator-prey relationships can be evaluated based on immunological reaction and identification (Feller and Ferguson, this volume; Theilacker, this volume; Zagursky and Feller, this volume).

One of the most exciting applications exploiting cell surface protein tags is the marking of various larval species in fresh collections from the natural environment. The antigen, in this case, was prepared to proteins from the adult of the species. The identification and segregation of certain members of a look-alike community will permit important knowledge of abundance, life history, cell differentiation and development of a species in nature (Gallagher *et al.*, this volume; Gallagher *et al.*, abstract, this volume). See frontispiece of this volume.

Pathology and toxicology have benefited by the antibody techniques. There has been a monoclonal antibody assay developed for shellfish MSX disease (Reinisch *et al.*, 1983), for oxidase activity altered by toxic substances (Stegeman *et al.*, 1986;this volume), for eel virus (Lipipun *et al.*, this volume) and salmon virus (Moody *et al.*, this volume). There are immunoassays in development for marine Paralytic Shellfish Poisoning (PSP) toxins (Guire *et al.*, this volume), Ciguatera toxins (Hokama *et al.*, this volume) and Florida Red Tide toxins (Baden *et al.*, 1985).

The three major advances permitted by specific molecule markers are:

A) Rate estimates without incubation - very important novel approaches to primary productivity estimates (RuBPCase-antibody in Orellana *et al.*, this volume) and nutrient uptake and assimilation estimates (nitrate reductase-antibody in Balch *et al.*, this volume).

B) Cell molecular mapping for bioluminescent organelles by anti-luciferase localization (Nicolas *et al.*, 1985; this volume); 700 chlorophyll *a*-protein complex (Vierling and Alberte, 1983); saxitoxin (D. Anderson, personal communication) using the antibody of Guire *et al.*, (this volume).

C) Quantitation and detection of specific toxic molecules of public health interest such as saxitoxin (Guire *et al.*, this volume) and ciguatoxin (Hokama *et al.*, this volume).

D) Identification of specific pigments in microalgae (Alberte and Friedman, 1986; Friedman and Alberte, 1986; in review); Shapiro, this volume) which should open the door to a new chemotaxonomy and establish evolutionary schemes for microalgae.

In a category of its own is the work of Baden *et al.* (this volume) where sodium channels in membranes are studied by using antibodies to *Ptychodiscus brevis* (Florida red tide) and *Gambierdiscus toxicus* (Ciguatera poisoning). The investigations have led to novel approaches. Similarly, Jones *et al.* (this volume) exploit membrane phenomenon and competitive antibody interaction in trace metal binding studies.

In summary, until now immunological approaches have exploited cell surface and molecular polymorphism, primarily responses of prokaryote and eukaryote unicells to antigens that they normally would not encounter. With the exception of Robert Anderson and colleagues (personal communication), there has not been substantial exploitation of the "natural" immune memory. We need now to explore the molecular and genetic basis for non-vertebrate immune responses and the far reaching potential of using immunochemistry in aquatic research.

AN IMMUNOLOGICAL COMMENT

Initial uses of Immunological Techniques

The use of immunological probes in aquatic research has been on the incline since 1980. However, one of the major impediments to the use of these probes in aquatic research is the fact that these antibodies must be produced in non-marine organisms. Thus, marine biologists must either convince researchers familiar with mammalian systems to produce the immunological probes, or the marine biologists themselves, must learn the immunological techniques. It is clear that as a number of very interesting immunological probes are produced and the data are published, demonstrating their superiority as marker reagents, then the desire to produce immunochemical probes will increase.

One of the major advantages of the use of immunochemical probes is the ability to detect specific antigens on and in marine organisms. The antibody molecules themselves are not easily detected unless some reporter molecule is covalently bound to the antibody molecule. Generally, fluorescent molecules or radioactive atoms are bound to the antibody molecule, making antigen detection a routine matter. These labeled antibodies can be detected using batch techniques (Friefelder, 1982) or using single cell methodology such as that described for flow cytometry (Muirhead *et al.*, 1985).

The value of using immunochemical probes for the detection of high frequency antigens is the specificity of the methodology. Low frequency antigens can be detected because immunochemical probes can be made very sensitive using amplifying schemes for the reporter molecules. One such scheme utilizes fluorescein molecules bound inside a liposome with the antibody molecule bound to the outside (Truneh *et al.*, in press). In this way, one achieves as many as one thousand fluorescein molecules bound to every antibody molecule.

Immunochemical techniques can be used to detect antigens on the surface of the cell membrane as well as antigens which are intracellular. The intracellular antigens are usually made accessible by permeabilizing membranes. The permeabilization can take place using a variety of methods, but in general mild fixation using paraformaldehyde or alcohol is the method of choice. With this technology it is possible to use multiple color fluorescence to correlate the intensity of two or more antigens within cells (Lanier *et al.*, 1986). In this way, it is possible to study the life cycle distribution of a given biochemical molecule and its relationship to cell function.

In another application the antibody molecule used to detect these antigens may have the capability of binding complement. The binding of complement at 37°C results in killing the cell. Thus, it is possible to purify heterogeneous cell populations by adding antibodies to an undesired cell population within the population studied and applying complement. The antibody plus complement mixture punches holes in the membrane of the cells which are unwanted, killing them.

In many cases of antibody production hetero-antisera will be used because of the ease of generating such reagents. To produce these reagents it is necessary to inject purified antigen subcutaneously in the presence of an adjuvant (Colwell *et al.*, 1986). This methodology usually requires multiple injections at weekly intervals so that the animal is hyper-immunized. The advantage of a hetero-antiserum is that the reagent produced has a very high sensitivity but low specificity. That is, the animal makes antibody to any chemical it has seen during its life-time or antigen within the immunizing reagent, as well as to the antigen of interest. Furthermore, the specificity of the hetero-antisera recognize many different sites of the antigen molecule makes them very sensitive for diagnostic testing. Thus, if we are looking for intact antigen molecules or fragments of a molecule, hetero-antisera would find it easier to detect the molecule of interest even in the face of partial degradation. Typically, antibodies within a hetero- antisera would find it easier to detect the molecule of interest even in the face of partial degradation. Typically, antibodies within a hetero-antiserum have a much higher affinity than monoclonal antibodies making them useful to purify specific biochemicals. Covalently binding the purified hetero-antiserum to columns makes it possible to pass cell supernatants across the column which then results in the binding of the biochemical interest to the antibody bound to the column (Andrews, 1984). After washing off all the non-specifically bound reagents, the specifically bound protein is released by the addition of a chaotrophic buffer. It should also be pointed out that this procedure is used in most biopharmaceutical facilities to assist in the purification of biomolecules.

Monoclonal antibodies will have a very special place in marine science. This is due to the fact that it is possible to produce a monoclonal antibody which recognizes only a single epitope of an antigen. Most antigens when injected into the mouse will have thousands of different binding sites for antibodies. However, the monoclonal production process allows us to select out a cell that can be grown in culture which produces an antibody which recognizes only one epitope on the antigen used as immunogen (Kennett *et al.*, 1980). It should be pointed out that these monoclonal antibodies while having high specificity have low sensitivity. The reason there is a very low sensitivity is that for every antigen there is at the most only one antibody molecule bound to that antigen. However, when using hetero-antisera there may be ten or twenty antibodies bound to the antigen depending on the size of the antigen molecule. Using a single monoclonal antibody dies not permit the binding of more than one monoclonal per antigen. If one were to use a cocktail of five or six monoclonal antibodies, then the sensitivity would be increased because there would be four or five monoclonal antibodies bound to each antigen. Another fact of life is that monoclonal antibodies tend to have a much lower affinity than antibodies in hetero-antisera. This may have something to do with the mechanism of producing monoclonal antibodies, but it is not clear why this fact remains so. Monoclonal antibodies can also be used as agents to assist in the isolation of specific genes. Monoclonal antibodies to the gene product are bound covalently to a sephadex column where they bind the gene product to an affinity column, trapping the mRNA onto the affinity column with the gene product. The mRNA is then rescued and reverse transcribed to DNA permitting the isolation of the specific gene. This methodology will be extremely valuable in marine sciences when molecular genetic approaches are added to the biochemical methodology currently available in marine sciences.

Of note, is the fact that monoclonal antibodies can also be used as probes of cell function. Receptors on the surface of the cells are the "sensors" which the cell uses to detect the outside environment. Monoclonal antibodies which bind to these receptors can be used to activate or inactivate cells (Springer, 1985). The value of these monoclonal antibodies in mammalian sciences has been incalculable. It is assumed that the same value will be achieved in the marine sciences.

There are a number of potential problems with the use of mammalian immunoglobulins in marine systems. Not the least of which is the problem of the F_c portion of the immunoglobulin molecule. The F_c portion has a tendency to non-specifically stick to the surfaces of cells. Thus, non-specific reactions may be so strong using marines cells and mammalian immunoglobulins that the specific reactions will be masked. Another potential problem is the fact that marine systems exist in much higher salt levels which have the tendency to reduce the binding constant of the mammalian monoclonal antibody. An assay which is developed at physiological levels for mammalian cells may not work in physiological cell conditions for marine organisms. This problem can only be solved by the purification in marine immunoglobulin molecules which operate normally at higher osmolar conditions.

Longer Term Development of Immunological Methods

As can be seen in the previous section, the first applications of immunological techniques in marine sciences will attempt to make use of the methodologies developed in mammalian sciences. However, it won't be long before marine biologists have the experience with these techniques and begin to study invertebrate immunology in its own right. Many studies will involve the development of monoclonal antibodies from the humoral aspects of invertebrate species. These antibodies may be more valuable because they can operate in the physiologic conditions of the cells being studies. Additionally, just understanding how humoral responses develop will be of extreme importance to controlling pathologic organisms in a number of marine species. The humoral responses will probably be investigated before cellular responses because of the ease with which the humoral studies can be carried out relative to the cellular responses. Cellular studies require much more patience and the ability to culture immune cells from invertebrate species. Once the culturing techniques have been mastered, then it is likely the long term cell culture systems using invertebrate cells will permit dissection of the cellular responses in these species.

In mammalian systems, one of the first methods used to study cellular immune problems was to deplete the animal of a specific cell type. Antibody plus complement directed at a specific cell type made it possible to deplete the animal of T helper cells, for example. *In vitro* one can remove B-cells by passing the lymphocytes through a nylon wool column, thereby, providing an enriched T cell population. These same types of studies will be extended to invertebrate biology providing enormous increase in the amount of information on the role of each cell type in the immune system. Additionally, it is expected that a series of strain variants which have either increased or decreased responsiveness to specific immune challenges will be generated. These cloned animals will provide further dissection of immune competence.

As more is learned about the mechanisms of the immune response in invertebrates, there will come a time when it is recognized that the control of these cells is carried out by a series of molecules known as biological response modifiers. Molecules like the interleukins 1,2, and 3 of the mammalian systems will find correlates within the invertebrate biology.

The ability to enhance or decrease immune responsiveness will be the hallmark of immunomodulator characterization. These methodologies will then be applied to study pathologic states and determine if there is the possibility to affect the course of a disease. The end result is that these studies will allow investigators to purify immunobiological molecules and then develop a series of agonist or antagonists which may have partial function in mammalian systems. Thus, study of the marine immune system may provide important clues to the genetic stability of the biological response modifiers.

The Longest Term Application of Immunologic Methods

After many years of study as outlined in the previous two sections, the research efforts in immunobiology of marine organisms will come full circle and attempts will be made to use the biological response modifiers developed for invertebrate systems to see if they have any affect on mammalian immune responses. From the general outline just described, it is clear that the marine sciences are entering an era of exciting research by the use of very specific and well characterized immunological probes. These methodologies will make available hundreds of specific probes which will enlighten humans' understanding of the marine world. We acknowledge NSF grant OCE86-03830.

REFERENCES

Alberte, R.S. and A.L. Friedman. 1986. A diatom-specific signature: use of immunological techniques. Abstract. EOS. 67: 1056.

Andrews, D.W. 1984. Use of monoclonal antibody immunoaffinity column to purify subsets of human HLA-DR antigens. IN: Methods in Enzymology. J.J. Langone, H. Van Vunakis and G. DiSabato. (eds.). 108: 600.

Baden, D.G., T.J. Mende and L.E. Brand. 1985. Cross-reactivity in immunoassays directed against toxins isolated from *Ptychodiscus brevis*. IN: Toxic Dinoflagellates. D. Anderson, A. White and D.G. Baden (eds.). Elsevier, NY. pp. 363-369.

Bigger, C.H., P.L. Jokiel, W.H. Hildemann and I.S. Johnston. 1982. Characterization of alloimmune memory in a sponge. J. Immunol. 129: 1570-1572.

Buscema, M. and G. Van de Vyver. 1984. Cellular aspects of alloimmune reactions in sponges of the genus *Axinella. II. Axinella verrucosa* and *Axinella damicornis*. J. Expt. Zool. 229: 19-32.

Campbell, L., E.J. Carpenter and V.J. Iacono. 1983. Identification and enumeration of marine *Chroococcoid* cyanobacteria by immunofluorescence. Appl. Environ. Microbiol. 46: 553-559.

Colwell, D.E., S.M. Michalek and J.R. McGee. 1986. Method for generating high frequency of hybridomas producing monoclonal antibodies. IN: Methods in Enzymology. J.J. Langone and H. VanVunakis (eds.). Academic Press, NY. 121: 42.

Dahle, A.B. and M. Laake. 1981. Diversity dynamics of marine bacteria: immunofluorescence stain on membrane filters. J. Appl. Microbiol. 43: 169-179.

Dahle, A.B. and M. Laake. 1982. Diversity dynamics of marine bacteria studies by immunofluorescent staining on membrane filters. Appl. Environ. Microbiol. 43: 169-176.

Friedman, A.L. and R.S. Alberte. 1986. Biogenesis and light regulation of the major light harvesting chlorophyll-protein of diatoms. Plant Physiol. 80: 43-51.

Friedman, A.L. and R.S. Alberte. (submitted) Phylogenetic distribution of the major diatom light-harvesting pigment-protein determined by immunological methods.

Friefelder, D. 1982. Immunological Methods. W.H. Freeman & Co., NY. Chapter 1. p. 323.

Fujiwara, K. and T.D. Pollard. 1980. J. Cell. Biol. 87(2): 222A.

Hildemann, W.H., C.H. Bigger and I.S. Johnson. 1979. Histocompatibility reactions and allogeneic polymorphism among invertebrates. Transplantation Proceedings 11: 1136-1142.

Hildemann, W.H., C.H. Bigger, P.L. Jokiel and I.S. Johnston. 1980. Characteristics of immune memory in invertebrates. IN: Phylogeny of immunological memory. J. Manning (ed.). Elsevier/North-Holland Biomedical Press, Amsterdam, pp. 9-14.

Kennett, R.H., T.J. McKearn and K.B. Bechton. Editors. 1980. Monoclonal Antibodies. Hybridomas: a new dimension in biological analysis.

Köhler, G. and C. Milstein. 1975. Continuous cultures of fused cells secreting antibody of predefined specificity. Nature. 256: 495-497.

Kolenkine, X. 1971. Les histocompatibilitiés intraspécifique et interspécifique chez les hydres d'eau douce. Arch. Zool. Exp. Géner. 112: 63-70.

Lackie, A.M. 1977. Cellular recognition of "non-self" in insects. IN: Developmental Immunobiology. J.B. Soloman and J.D. Horton (eds.). Elsevier/North-Holland Biomedical Press, Amsterdam, pp. 75-81.

Lackie, A.M. 1979. Cellular recognition of foreignness in two insect species, the American cockroach and the desert locust. Immunobiology 36: 909-914.

Langlet, C. and J. Bierne. 1977. The immune response to xenografts in nemertines of the genus *Lineus* IN: Developmental Immunobiology. J.B. Solomon and J.D. Horton, eds., Elsevier/North-Holland Biomedial Press, Amsterdam, pp. 17-26.

Lanier, L.L., E.G. Engleman and P. Gatenby. 1983. Correlation of functional properties of human lymphoid cell subsets and surface marker phenotypes using multiparameter analysis and flow cytometry. Imnol. Rev. 74: 143.

Marchalonis, J.J. 1977. Immunity in Evolution. Arnold, London, 238 pp.

Muller, W.E.G. 1982. Int. Rev. Cytol. 77: 129-181.

Muirhead, K.A., P.K. Horan and G. Poste. 1985. Flow cytometry: present and future. Bio/technology. 3: 337-356.

Nicolas, M-T., C.H. Johnson, J-M. Bassot and J.W. Hastings. 1985. Immunogold labeling of organelles in the bioluminescent dinoflagellate *Gonyaulax polyedra* with antiluciferase antibody. Cell Biol. Internl. Repts. 9: 797-802.

Reinisch, C.L., A.M. Charles and J. Froutner. 1983. Dev. Comp. Immunol. 7: 33-39.

Roitt, I., J. Brostoff and D. Male. 1985. Immunology. C.V. Mosby, St. Louis.

Springer, T.A. Editor. 1985. Hybridoma technology in the biosciences and medicine. Plenum Press. NY.

Stegeman, J.J., R.J. Kloepper-Sams and J.W. Farrington. 1986. Monooxygenase induction and chlorobiphenyls in the deep sea fish *Coryphaenoides armatus*. Science. 231: 1287-1289.

Truneh, A., P. Machy and P.K. Horan. (in press). Antibody-bearing liposomes as multicolor immunofluoro markers for flow cytometry and imaging. J. Imnol. Methods.

Twarog, B.M. 1974. "Immunity" to paralytic shellfish toxin in bivalve molluscs. Proc. Second Intnl. Coral Reef Symposium. Great Barrier Reef Committee, Brisbane, Australia. pp. 505-512.

Twarog, B.M., T. Hidaka and H. Yamaguchi. 1972. Resistance to tetrodotoxin and saxitoxin in nerves of bivalve molluscs. Toxicon. 10: 273-278.

Twarog, B.M. and H. Yamaguchi. 1975. Resistance to paralytic shellfish toxins in bivalve molluscs. Proc. of the First Intnl. Conf. on Toxic Dinoflagellate Blooms. Mass. Sci. Tech. Fdn. Wakefield, MA. pp. 381-393.

Valembois, P. 1973. Quelques aspects phylogénéteiques de la réaction d'incompatibilité aux greffes chez les Métazoaires. Ann. Biol. 12: 1-26.

Vierling, E. and R.S. Alberte. 1983. P_{700} chlorophyll a-protein, Purification, Characterization, and Antibody preparation. Plant Physiol. 72: 625-633.

Visser, J.W.M. and G.R. Van den Engh. 1982. IN: Immunofluorescence Technology Selected Theoretical and Clinical Aspects, Wick. (ed.). Elsevier, Amsterdam. pp. 95-128.

Ward, B.B. and M.J. Perry. 1980. Immunofluorescent assay for the marine ammonium-oxidizing bacterium *Nitrosococcus oceanus*. Appl. Environ. Microbiol. 39: 913-918.

Yentsch, C.M. and S.A. Pomponi. 1986. Automated individual cell analysis in aquatic research. Internal. Rev. Cytol. 105: 183-243.

MOLECULAR BIOLOGY, SINGLE CELL ANALYSIS AND QUANTITATIVE GENETICS: NEW EVOLUTIONARY GENETIC APPROACHES IN PHYTOPLANKTON ECOLOGY

A. Michelle Wood

Department of Biology
University of Chicago
915 East 57th St.
Chicago, IL 60637

and

Bigelow Laboratory for Ocean Sciences
McKown Point
West Boothbay Harbor, ME 04575

INTRODUCTION

It is widely recognized that phytoplankton in the marine environment experience variation in a range of environmental variables over many temporal and spatial scales (Harris, 1980, 1986; Lewis and Platt, 1982; Ducklow, 1984). Adaptation, has, therefore received considerable attention as a research topic during the last several decades. The conceptual basis for much of this work has emphasized a nongenetic approach to the adaptive process; most examinations of the tolerance of individual phytoplankton species to environmental variation follow the approaches pioneered by Barker (1935), Braarud (1951), and Provasoli and Pintner (1953), in which the response of single clonal representatives of a species to changes in the experimental variable(s) of interest are studied under controlled conditions. From these studies, optimum conditions for growth are inferred and used to interpret patterns of distribution and abundance observed in nature (e.g., Braarud, 1961; Guillard, 1968; Eppley *et al.*, 1969; Titman, 1976; Tilman, 1977; Kilham *et al.*, 1977; Brand and Guillard, 1981; Brand *et al.*, 1983, 1986).

There have, however, been a variety of problems with this approach (cf. discussion in Brand, 1984) and a growing body of evidence which shows the existence of considerable genetic variation within phytoplankton species. Since much of this variation is expressed as phenotypic variation in ecologically important characters, it seems likely that microevolutionary processes (e.g., changes in gene frequency within populations, cf. Fisher, 1958; Dobzhansky, 1970; Lewontin, 1974; Wright, 1978) resulting from natural selection, are an important component of the adaptive strategy of many phytoplankton species. In other words, rather than the relative abundance or persistance of a species in a community being determined by the degree to which prevailing environmental conditions are optimal for the species'mean phenotype, the persistance and success of a species may be more closely related to the genetic diversity within the species, the degree to which this diversity is available for selection at the phenotypic level, and the rate of evolutionary response to selection.

In experimental terms, this implies that studies of physiological ecology must continue to shift their emphasis from measurement of mean phenotypes to measurement of phenotypic variance, particularly genetically determined phenotypic variance, and studies of natural phytoplankton populations should begin to include an examination of the genetic structure of the population.

Unfortunately, until recently, estimates of the genetic component of variation in phenotypic characters within phytoplankton species were almost impossible to obtain (See discussion of difficulties in obtaining genetic information for marine species in Harding and Tebble, 1963 and Gooch, 1975). The advent of molecular methods for identifying genotypes, our increasing ability to culture marine phytoplankton and to measure a variety of traits on an individual basis, and theoretical developments in the field of evolutionary biology suggest that we are on the threshold of a new era in phytoplankton ecology - one in which these new techniques enable us to investigate directly the interplay between environment and genotype as mediated by selection.

In this paper, I will review some of the evidence which suggests that microevolutionary processes are an important component of the adaptive strategy of phytoplankton species, describe several ways in which molecular techniques can improve our ability to evaluate the genetic structure of a population, introduce the general theory of quantitative genetics as it is being applied to evolutionary biology, and suggest ways in which immunochemical methods greatly expand the range of characters to which this theory can be applied.

EVIDENCE FOR MICROEVOLUTION IN RESPONSE TO SELECTION

A microevolutionary response to environmental variation requires heritable phenotypic variation which affects fitness. The evidence for this with regard to phytoplankton can be divided into two categories: 1) demonstrations that species are genetically polymorphic for characters directly related to fitness (or important components of fitness), and 2) demonstrations that the genetic polymorphism which exists within species appears to have resulted from natural selection. In the latter case, it is proper to view morphotypes as ecotypes, but the data must show 1) that populations experiencing apparently different selection pressure show genetic differences and 2) that genotypes from the same selection regime share phenotypic characteristics consistent with the apparent direction of selection.

It should be noted in this regard that genetic variation between populations, when shown by enzyme electrophoresis or other molecular methods (Lewontin, 1985), suggests that there may be a potential for evolution by natural selection because it reveals the existence of genetic variability. It is not necessarily an indication that an evolutionary response to selection has occurred since molecular evolution appears to be primarily a result of mutation and drift (Kimura, 1983).

Intraspecific genetic variation

Essentially all studies of phytoplankton genetics have relied on techniques involving cultures established from single cell isolates. Among the earliest was Lewin's (1955) demonstration that there were different physiological races of *Navicula pelliculosa*; intraspecific genetic variation has since been demonstrated for growth rate (Hayward, 1968; Terry *et al.*, 1983) and many other characters which are undoubtedly important components of fitness: vitamin requirements (Lewin and Lewin, 1960; Hargraves and Guillard, 1974); nutrient uptake kinetics (Kilham, 1975; Terry *et al.*, 1983); periodicity of cell division (Nelson and Brand, 1979); trace metal tolerance (Brand *et al.*, 1986); temperature and salinity tolerance (Guillard *et al.*, 1974; Hargraves and Guillard, 1974; Brand, 1984); biochemical composition under conditions of light limitation (Terry *et al.*, 1983; Alberte *et al.*, 1984; Gallagher *et al.*, 1984; Gallagher and Alberte, 1985; Wood *et al.*, 1985; Wood and Bidigare, unpublished data); luminescence (Schmidt *et al.*, 1978); toxicity (Alam *et al.*, 1979; Schmidt and Loeblich, 1979; Cembella *et al.*, *1986*); and photosynthesis versus irradiance relationships in diatoms (Gallagher *et al.*, 1984; Gallagher and Alberte, 1985) and cyanobacteria (Alberte *et al.*, 1984; Barlow and Alberte, 1985; Wood, 1985; Glover *et al.*, 1986b, 1987; Glibert *et al.*, 1986).

The basic assumptions in these studies are that each clonal culture is composed of many individuals of identical genotype, that variation within a single clone is, thus, nongenetic, and variation among different clones is primarily genetic. It should be noted that evolution can occur in clones (Lukavsky, 1979; Necas, 1979). This is most problematic with diatoms since diatoms are diploid during vegetative growth; new genotypes can be generated by sexual recombination within the culture (Murphy, 1978). Also, the progressive diminution in size associated with asexual reproduction in diatoms (Drebes, 1977) means that between-clone variation in size-dependent characters may have a significant nongenetic component of variation depending on the range of sizes represented among the different clones. For dinoflagellates and coccolithophores, which are haploid and do not change size during asexual growth, the clonal approach is relatively straightforward, although the possible generation of new genotypes by unequal recombination or other forms of mutation must be kept in mind.

Several workers have evaluated the genetic structure of different populations of phytoplankton species by comparing properties of numerous clones isolated from each population. Soudek and Robinson (1983) analyzed the population structure of *Asterionella formosa* by examining the electrophoretic banding patterns of 101 clones isolated from 32 different lakes or rivers. Electrophoretically detectable differences were not observed among clones isolated from the same populations even if they were obtained at different seasons or in different years; however there was considerable genetic variation between populations from different lakes. This was not a response to two obvious sources of selection since there was no correlation between the electrophoretic type of populations from lakes of similar trophic status or lakes in the same geographic region; the genetic divergence among populations can most plausibly be explained as a result of mutation and random genetic drift.

Various methods, including enzyme electrophoresis, HPLC toxin analysis, and quantitative DNA determinations have been used to show that isolates of cells in the *Protogonyaulax tamarensis/catenella* species complex show genetic similarity within populations and between populations isolated from the same geographic regions, but not between populations isolated from different geographic regions (Cembella and Taylor, 1985; 1986). These dinoflagellates, which can cause paralytic shellfish poisoning, were shown to have considerable genetic polymorphism in toxin composition; the relative proportions of different toxins in individual clones remained constant through the growth cycle, indicating that toxin composition is a stable genetically determined character (Cembella *et al.*, 1986).

Enzyme electrophoresis was used to elucidate extremely interesting patterns of genetic relatedness among clones of the marine dinoflagellate genus *Heterocapsa* by Watson and Loeblich (1983). Included in this study were six strains of *H. pygmaea*: Isolate 145 from the Mediterranean and Isolates 7 and 515 from the Gulf of Mexico as well as three subcultures of Isolate 7 which had been maintained in culture for several years prior to the genetics experiments. By cluster analysis, it was shown that Isolate 515, Isolate 7 and its subcultures formed a genetically related group which was very distinct from Isolate 145, suggesting that geographic distance between the two source populations resulted in genetic divergence. More remarkable, however, was their finding that the genetic distance between Isolate 7 and one of its subcultures was almost as great as that observed between isolates from two other species of *Heterocapsa*. This finding indicates a high rate of spontaneous non-lethal mutation in *H. pygmaea* and provides the first quantitative data on mutation as a source of genetic variability in phytoplankton species.

Genetic variability in coccolithophores has been investigated by Brand (1981; 1982) who used an innovative approach based on direct estimates of clonal growth rate which, for asexually reproducing unicellular organisms, can be viewed as a direct measure of total fitness. He showed considerable local genetic variation (e.g., between clones isolated from the same sample) within different populations of several species. Comparison of population mean and variances showed no significant difference between different local populations sampled from the same water mass, but highly significant differences between populations sampled from different water masses were observed. This latter result, which was also obtained for the three dinoflagellate species included in the studies, is suggestive of a selective influence acting on populations entrained within water masses of relatively homogeneous internal condition and is supported by the results of a subsequent study on genetic variability in the reproductive rates of *Prorocentrum micans* (Dinophyceae). Genetic variability in a population of *P. micans* from the Gulf of Maine was compared with that of a population from adjacent waters over George's Bank (Brand, 1985). While ample opportunity exists for gene flow between these two regions, the population from the Gulf of Maine showed significantly more genetic variability in clonal growth rate than the one from George's Bank, a situation which was interpreted as reflecting differences in rates of sexual reproduction between the two regions, but which, according to the data presented, may as easily reflect selection for specific Gulf of Maine genotypes by the unusual environmental conditions on George's Bank.

The only intensive investigation of genetic variation in a single phytoplankton population over time is that carried out by Gallagher (1980), who studied electrophoretic banding patterns in 457 clones of the diatom *Skeletonema costatum* isolated prior to and during the winter-spring and summer-fall bloom periods in Narragansett Bay in 1977 and 1978. The pooled results, based on analysis at five loci, showed that specific electrophoretic types dominated the winter-spring and summer-fall bloom communities. The genetic distance between genotypes dominating the community during these different blooms was greater than that normally observed between different species of terrestrial plants. It should be noted that *S. costatum* is a dominant species in Narragansett Bay and can comprise more than 90% of the total phytoplankton during bloom periods (Gallagher, 1980 and references therein). This study demonstrated that maintenance of a dominant position in the phytoplankton community by *S. costatum* was not due to extreme phenotypic plasticity in a few genotypes, but rather a phenomenon associated with major changes in the genetic structure of the population. An examination of physiological traits among and within electrophoretic types showed significant differences between genotypes which were abundant during different seasons (Gallagher, 1982) and strongly supported Gallagher's hypothesis (1980) that natural selection is the driving force determining the cyclical nature of genotypic frequencies in *S. costatum* in Narragansett Bay.

Evidence for adaptive genetic variation between populations (ecotypy)

One of the oldest empirical methods used to evaluate the role of natural selection in the evolution of wild populations is based on the hypothesis that, if natural selection is operating, geographic or temporal variation in trait expression should follow geographic or temporal variation in selective environmental factors. The null hypothesis is that phenotypic characters vary independently of environmental factors. Additionally, comparisons of homologous traits between closely related species can be used to test the hypothesis that natural selection has influenced phenotypic evolution. If so, traits in closely related species should be similar if the species do not compete with each other, and should show divergence if they do compete with each other. If selection is not operating, homologous traits will vary independently among closely related species. Endler (1986, pp. 53-62) discusses the caveats inherent in these two approaches, both of which have been used extensively to evaluate ecotypic differentiation across the neritic-oceanic boundary in the *Thalassiosira pseudonana/oceanica* species complex; a group of closely related species found in neritic (*T. pseudonana* and *T. weisflogii*) and oceanic (*T. oceanica*) waters.

Work on the *T. pseudonana/oceanica* species complex, recently reviewed in other contexts (Bonin *et al*., 1986; Shapiro and Guillard, 1987), dates from Guillard and Ryther's (1962) demonstration that oceanic isolates were more stenothermal and stenohaline than neritic isolates. Their observation that genetic polymorphisms were apparently adaptive in the habitats from which the genotypes were isolated was the first evidence for ecotypy in a phytoplankton species and has been supported by subsequent work on the *T. pseudonana/oceanica* species complex. Other phenotypic characters which vary in an apparently adaptive way between isolates obtained from neritic and oceanic waters include vitamin specificity (Guillard, 1968), nitrate half-saturation constants (Carpenter and Guillard, 1971), sensitivity to chlorinated hydrocarbons (Fisher *et al*., 1973), temperature-dependent growth rate (Goldman and Carpenter, 1974; Brand *et al*., 1981), kinetics of silicic acid uptake and dissolution rates (Guillard *et al*., 1973; Nelson *et al*., 1976), trace metal sensitivity and nutrition (Fisher, 1977; Murphy and Belastock, 1980; Murphy *et al*., 1982, 1984), and relative importance of C-3 and C-4 carboxylating pathways in photosynthesis (Morris, 1980). Electrophoretic banding patterns of neritic and oceanic isolates also show considerable genetic variation within and among populations and are suggestive of a genetic cline for this species complex across the neritic-oceanic boundary in the western North Atlantic (Murphy and Guillard, 1976; Brand *et al.*, 1981).

In a study of nitrate uptake kinetics in *Biddulphia aurita*, a neritic diatom, Underhill (1977) found that isolates from more oligotrophic waters had lower half saturation constants and lower rates of maximum uptake than isolates from more eutrophic waters. These results show an apparent response to selection similar to that observed across the neritic/oceanic boundary described for the *T. pseudonana/oceanica* species complex. Latitudinal selection may also occur in phytoplankton; temperate isolates of several diatom species show lower temperature optima for growth than cospecific isolates from tropical waters (Figures 12.1 and 12.2 in Guillard and Kilham 1977).

The most compelling evidence for ecotypic differentiation in phytoplankton is the intraspecific variation observed in tolerance of chemical pollutants. It is reasonable to assume that the appearance of high concentrations of these materials in the environment is a relatively recent development in the evolutionary history of most phytoplankton species. Thus, the correlation (observed in several species) between tolerance of trace metals or industrial waste and the history of pollution in the environment from which different genotypes were isolated (Jensen *et al.*, 1974; Fisher and Frood, 1980; Murphy and Belastock, 1980; Murphy *et al.*, 1982) suggests that evolutionary processes will proceed rapidly in phytoplankton populations experiencing strong directional selection.

MOLECULAR METHODS FOR EXAMINING POPULATION GENETIC STRUCTURE

The most unequivocal demonstration of an evolutionary response to selection in natural populations ideally involves direct estimation of gene frequencies before and after selection by genotypic classification of individuals within a population. Until recently, such data were nearly impossible to obtain for microbial populations; immunochemical and other molecular markers appear to offer a means of rapidly identifying genotypes of microorganisms within a population.

Molecular markers for individual genotypes - serotype.

Immunofluorescence offers a rapid means of serotyping individuals in microbial populations; because surface antigens are generally stable genetically determined characters, individuals assigned to different serotypes can be viewed as genetically different. While this approach has been used to investigate the population structure of nitrifying bacteria (Ward and Carlucci, 1985 and references cited therein) and freshwater cyanobacteria (Fliermans and Schmidt, 1977), its only application to marine phytoplankton has been the work of Campbell (this volume). Her finding (Campbell *et al.*, 1983; Glover *et al.*, 1986a) that different serotypes of marine *Synechococcus* spp. predominate in coastal and oceanic waters and that the oceanic serotype predominates in oceanic waters entrained on the continental shelf indicates that coastal and oceanic populations of this taxa are genetically distinct.

While different serotypes may be viewed as genetically distinct, the existence of genetic variability within a serotype creates problems with the use of serological markers for population genetic studies. The extensive work which has been done on the relationship between genetic diversity and serotype in *Escherichia coli* and other pathogenic microbes (cf. reviews by Hartl and Dykhuizen, 1984; Selander, 1985, 1987; and Achtman and Pluschke, 1986) provides insight into the magnitude of the problem and suggests several solutions. Serotyping of *E. coli* is based on immunospecificity for three antigenic determinants: O, H, and K. Initial studies involving 88 isolates showed that there was extremely high genetic variability within the eight O serogroups; each contained an average of 8.25 electrophoretic types (Ochman *et al.*, 1984). Subsequent work with 261 isolates confirmed this finding but also showed that genetic diversity within serogroup was significantly reduced if isolates within a serogroup also shared a common K antigen and nearly eliminated for serogroups in which the isolates had the same O, H, and K antigenic determinant (Caugant *et al.*, 1985). These results imply a close genetic relationship between the marine *Synechococcus* clones which shared two or more antigenic sites and suggest that if antibodies are prepared to several antigenic determinants within a species of phytoplankton, subpopulations of genetically related individuals can be identified from natural samples with some degree of confidence. Electrophoretic studies with reference cultures should, of course, be used to confirm the genetic relatedness of serogroups; such data can also provide insight into the degree of recombination among genotypes (Selander and Levin, 1980; Selander and Whittam, 1983; Whittam *et al.*, 1983).

Molecular markers for individual genotypes--biochemical polymorphism.

The studies of *E. coli* discussed above also showed that, among isolates with the same O:K serotype, genetically homogenous subgroups could be identified on the basis of the electrophoretic mobility of outer membrane proteins (OMPs). Figure 1 shows the genetic distance between several groups of isolates with the same O:K serotype; note that the genetically divergent isolates with and O1:K1 serotype have different OMP patterns, but that isolates of the O1:K1 and O2:K1 serotype which have the same OMP pattern are genetically similar. These results suggest that serotyping based on surface antigenic specificity and determination of the state of genetically determined protein polymorphism, if used together, can provide relatively precise genetic classification of individuals in a heterogenous population.

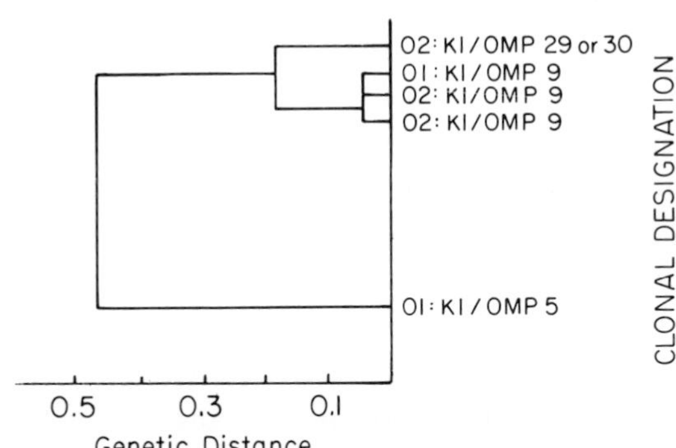

Figure 1. Genetic distance between electrophoretic types of *Escherichia coli* as determined by protein electrophoresis of 108 isolates from diverse sources (after Achtman and Pluschke, 1986 with permission from the Annual Reviews of Microbiology, Vol. 40, 1986, Annual Reviews, Inc.). Each electrophoretic type is composed of isolates with similar O:K serotype and outer membrane protein (OMP) pattern as indicated by the clonal designation.

This approach can be applied directly to the marine *Synechococcus* since multiple surface antigens clearly exist and genetically determined pigment polymorphism can be detected for individual cells by epifluorescence microscopy and flow cytometry

(Wood *et al.*, 1985; Olsen *et al.*, 1986; Iturriaga *et al.*, 1986). In cases where protein polymorphism does not produce readily detectable phenotypic differences, typing may be achieved by immunofluorescence. In *Gonyaulax tamarensis*, for example, there is genetically determined polymorphism in toxin composition (Cembella *et al.*, 1986); for population genetic studies of this species group, antibodies can be prepared to the different toxins and used to evaluate the relative performance of different genotypes in natural or artificial populations. It would be particularly interesting to use fluorescent antibodies prepared against such ecologically interesting biochemical polymorphisms to determine the relative frequency of different genotypes before and after a grazing event or other major bout of selection.

Molecular markers for individual genotypes--genomic characterization.

Direct analysis of genomic variation has been used to estimate the genetic relationship between human populations (Wallace *et al.*, 1985; Bonné-Tamir *et al.*, 1986), bacterial communities (Lane *et al.*, 1985; Stahl *et al.*, 1984), and populations of terrestrial (Kreitman, 1983; Kreitman and Aguade, 1986; Avise, 1986) and aquatic (Avise *et al.*, 1984; Bermingham and Avise, 1986; Bagshaw and Warner, 1986) animals, but these methods have not been applied to problems in phytoplankton population genetics. The rapidly developing methods for use of ribosomal RNA (rRNA) sequencing in microbial ecology and evolution (cf. review by Olsen *et al.*, 1986) may ultimately be of use in evaluating the genetic structure of phytoplankton populations before and after selection, but the present sensitivity of the methods do not allow discrimination between limited taxonomic groups (e.g., species or genera). This may be partially due to the high degree of conservation in rRNAs across phylogenetic groups. The general approach used by Stahl *et al.* (1985) in their study of a hot spring microbial community might be applicable to population genetic studies with phytoplankton if mitochondrial DNA, which evolves more rapidly than rRNA (Avise, 1986), were sequenced instead of rRNA. It is also likely that restriction fragment length polymorphism in chromosomal DNA can be used to evaluate the relative frequency of different genotypes in a population if care is taken to calibrate the sample with probes for conserved genes of known copy number. These approaches may require separation of individuals in the study species from sources of contaminant DNA or RNA, but recent advances in flow cytometry (cf. Trask *et al.*, 1982; Yentsch and Yentsch, 1984; Wood *et al.*, 1985; Yentsch and Pomponi, 1986) indicate that this could be accomplished relatively easily for many species.

In situ hybridization offers the potential for direct genetic classification of individuals based on the presence or absence of specific nucleic acid sequences. Radioactively labelled nucleic acid probes can be detected by microautoradiography or labelled with fluorescent compounds and detected by fluorescence microscopy. Probes labelled with 2-acetyl aminofluorine (AAF) have been used for *in situ* hybridization and identified using antibodies to the AAF (Trask *et al.*, 1986) as have antibodies to avidin-biotin complexes on biotin substituted nucleic acid probes (Brigati, 1983). These techniques, which are in the early stages of development, hold considerable promise for population genetic studies since they can be combined with automated image analysis to provide rapid identification and enumeration of specific genotypes in natural and experimental populations.

APPROACHES TO THE EVOLUTION OF QUANTITATIVE TRAITS

Many characters of interest to the evolutionary ecologist are not coded for by a single gene, but are the result of the combined action of many genes, each with relatively small effect. For these polygenic or quantitative characters, phenotypic variation is continuous between genotypes and the response of a population to selection must be inferred from changes in the phenotypic means and variances of characters in the population (Arnold, 1983; Endler, 1986).

The difference between selection and an evolutionary response to selection must be emphasized at this point (cf. Fisher, 1958; Haldane, 1954; Arnold and Wade, 1984a; and Endler, 1986). Selection, acting on the phenotypes present in a population, results in changes in the phenotypic distribution (mean, variance, and covariance) of traits within a generation. This may, or may not, result in a change in gene frequency in subsequent generations, depending on the pattern of inheritance of the selected traits. In other words, if there is phenotypic variation among individuals and, if there are fitness differences associated with different phenotypes, directional selection can produce major changes in the average phenotype population within a generation. These two conditions are sufficient for selection, but evolutionary change will occur only if there is also a consistent relationship between parent and offspring for each selectable trait (i.e., a consistent genetic difference between selected and unselected phenotypes).

A theoretical framework for the description and measurement of the response to selection on quantitative characters was first derived in the early part of this century (Fisher, 1918; Wright, 1921; Haldane, 1924) and has been developed extensively for animal and plant breeding (Simmonds, 1979; Falconer, 1981), but its application to natural selection is relatively recent (cf. Manly, 1985). Natural selection acting on single quantitative characters in a population can be evaluated according to models developed by Lande (1976). As demonstrated in his analysis of equine tooth evolution (Lande, 1976) and in Reyment's (1982a) analysis of the evolution of marine ostracods, these models can be applied to strictly phenotypic data from natural populations without cohort analysis if reasonable estimates of heritability (h^2 or H^2, see below) and generation time are available. Longitudinal data on phenotypic means and variances can also be analyzed nonparametrically (Lande, 1977; Manly, 1985) to test the hypothesis that phenotypic variation in a population is nonrandom (i.e., presumably varying in response to selection).

Analysis of univariate response to selection can be misleading even if the strength of selection and heritability are known, because genetic correlation between characters can result in apparently anomalous changes in the character presumed to be the focus of selection. Genetic correlation, which has been detected between morphological traits in marine diatoms (Wood *et al.*, 1987), can be caused by either pleiotropy or linkage disequilibrium (cf. Falconer, 1981; Suzuki *et al.*, 1985), and means that selection on one character will lead to a correlated evolutionary response in other characters. The fitness of the selected phenotype will depend on the multivariate character complex. The actual pattern of phenotypic evolution in the population will, therefore, not depend on the relative fitness of different phenotypes associated with genetic variability in a particular character upon which selection appears to be acting, but also on the variance and covariance of fitness associated with correlated characters. Lande's models have been extended to cases of multivariate evolution (Lande, 1979, 1980; Via and Lande, 1985) and computational techniques have been derived for the measurement of multivariate selection in natural populations (Lande and Arnold, 1983; Arnold and Wade, 1984a,b). These methods have been most thoroughly applied to sexually reproducing populations in which individual cohorts can be identified (Arnold and Wade, 1984b; Arnold, 1986; Kalisz, 1986), but they have also been applied to marine microfossils (Reyment, 1982b,c; 1983).

Estimating genetic variability in quantitative characters.

Polygenic or quantitative characters include essentially all traits which show continuous variation and are measured or counted. Typically studied characters are growth rate, size and shape, but physiological traits like stress tolerance, enzyme activity and concentration of biochemical components can also be appropriately studied as quantitative characters (Laurie-Ahlberg *et al.*, 1980; Laurie-Ahlberg, 1985; Salzman, 1986). The expression of these traits is often strongly influenced by nongenetic factors and, for phytoplankton, which can be reproduced clonally, simple analysis of variance techniques can be used to determine the relative significance of environmental and genetic sources of variation in the character. Such data provide the basis for estimating the potential for evolution of the character in response to selection.

The basic theory for estimation of the genetic component of variance in a quantitative character is straightforward. For any character X, the phenotype of the i'th individual (X_i) can be expressed:

$$X_i = \mu + G_i + E_i$$

where μ is the population mean phenotype, G_i is the deviation from the mean due to genetic causes and E_i is the deviation from the mean due to environmental causes. For the entire population, the total phenotypic variance (V_P) has a genetic (V_G) and an environmental component (V_E):

$$V_P = V_G + V_E + V_{GE}$$

where:

$$V_G = V_A + V_D + V_I$$

and V_A is the additive effect of different genes, V_P is the effect of dominance between different alleles at the same locus, V_I results from the interaction of different genes, and V_{GE} is the nonadditive genotype-environment interaction. More complete treatment of the theory and its practical application can be found elsewhere (Falconer, 1981; Becker, 1984; Manley, 1985), but, for a single character, the proportion of the total phenotypic variance due to genetic factors is generally termed "heritability" and is defined, in the "broad sense" as:

$$H^2 = V_G / V_P$$

and in the "narrow sense" as:

$$h^2 = V_A / V_P$$

Broad-sense heritability, which is normally used to calculate the response to selection in asexually reproducing organisms (Simmonds, 1979), can be obtained easily for characters which can be measured on single cells in species which can be grown in clonal culture. If a variety of genotypes (e.g., clonal lines) are grown with replication in a common environment, phenotypic variation between clones of the same genotype provide an estimate of the magnitude of nongenetic (e.g., microenvironmental and developmental) sources of variation in the character. Variation among genotypes provides an estimate of the magnitude of genetic sources of variation in the character. The individual measurements provide the error variance needed to test within-clone variance for statistical significance and replicate cultures provide the error variance needed to test the between clone variance (cf. Scheffé 1959). If the entire experimental design is repeated under more than one set of environmental conditions, the significance of environment-genotype interaction can also be evaluated (cf. Cochran and Cox, 1957).

This approach has been used to estimate the heritability of variation in several taxonomically important morphological characters in an Antarctic diatom (Wood *et al.*, 1987). As shown in Table 1, the analysis of data from this experiment required correction of between-clone variance for non-genetic sources of variation between clones resulting from covariance of the characters with diameter. As mentioned in an earlier section of this review, valve size in diatoms decreases during asexual reproduction so that differences in diameter between clones of diatoms can represent a significant nongenetic source of variation for any size-dependent character. While this correction decreased between-clone variance by nearly 50% for some characters, the heritability calculated for all characters was sufficient to support rapid evolution under conditions of sustained directional selection.

With present technology, a wide variety of characters can be measured on individual phytoplankton cells (Muirhead *et al.*, 1985; Yentsch and Pomponi, 1986) and immunochemical methods promise to further expand the types of characters for which heritability can be estimated. Developing technology for the staining and quantitative analysis of intracellular antigens (Clevenger and Epstein, 1984; Jacobberger *et al.*, 1986) suggest that antibodies to ecologically important molecules (e.g., nitrate reductase, Balch *et al.*, this volume; ribulose bis-phosphate carboxylase, Orellana *et al.*, this volume; chlorophyll a:c binding protein, Freidman and Alberte, 1984,

Table 1. Components of the total phenotypic variance ($V_p = V_{clone} + V_{flask} + V_{ind}$) and heritability ($H^2$) calculation for morphological traits in the antarctic diatom *Thalassiosira tumida* Janisch (Hasle). Duplicate cultures of ten clones were grown in a common environment; number of marginal labiate processes (LP) and diameter were measured for twenty individuals from each culture by light microscopy and a number of central strutted processes (CSP) and average number of satellite pores per strutted process (AVSAT) were measured on twenty individuals from each culture by scanning electron microscopy. For details, cf. Wood *et al.*, 1987.

	V_{clone}	V_{flask}	V_{ind}	r^2	V_G	H^2
No. CSP	3.835	0.479	8.053	0.7496	1.680	0.14±0.06
AVSAT	0.048	0.002	0.109	-0.0889	0.048	0.30±0.11
No. LP	0.356	0.061	1.061	0.6600	0.201	0.14±0.06
Diameter	125.264	0.791	3.326	-	-	-

V_{clone} = variance among clones

V_{flask} = average variance between individual cultures within clones

V_{ind} = average variance between individuals within individual flasks or cultures

r^2 = correlation (among clones) of characters with diameter

V_G = variance among clones corrected for differences in diameter, or genetic variance, $V_G = (1-r^2)\,V^2_{clone}$

H^2 = $V_G/(V_{clone} + V_{flask} + V_{ind})$ (after Falconer, 1981, p. 155); because in every case V_{flask} was more than an order of magnitude smaller than V_{ind}, approximate standard errors were computed from Fisher's formula for the sampling variance of the intraclass correlation (Robertson, 1959).

1986) can be used to evaluate genetic variance in the amount of enzyme or protein per cell, rate of synthesis, and response time of cells to environmental cues which induce the synthesis of specific molecules.

Heritability estimates obtained experimentally provide tremendous insight into the possibility of evolution in a quantitative character as a result of natural selection. Applying the values of H^2 obtained in such studies to specific models of evolution in natural populations must, however, be done with great caution because heritability estimates strictly apply only to the environment in which they were measured (Falconer, 1981; Manly, 1985). In wild populations, the effect of the natural environment on the components of phenotypic variance and the stability of H^2 as a measure over generational time are unknown. Lande (1976; 1979) has argued that under certain conditions heritability can remain constant over many generations in obligately sexual populations, but analysis of species which alternate asexual reproduction with intermittent episodes of sexual reproduction has shown that changes in heritability may be an important feature of the genetic structure of the population (Lynch and Gabriel, 1983).

An additional caution to be considered in applying H^2 to models of evolution in natural populations is that broad-sense heritability, obtained from experiments with clonal asexually reproducing cultures, is only properly used to estimate the response to selection in asexually reproducing populations. While it is likely that many phytoplankton populations experience long periods of continuous asexual division, information on phenotypic covariance between parent and offspring are needed to estimate the narrow-sense heritability, h^2, of quantitative characters in sexually reproducing populations (for outstanding examples of the estimation of h^2 in natural populations, cf. papers by Boag, 1983 and Grant, 1983). Both h^2 and H^2 are needed to fully characterize the response to selection in facultatively sexual species and are essentially impossible to obtain for most phytoplankton species given our present understanding of factors regulating sexuality in these organisms. An adequate theory for evolution of quantitative characters in facultatively sexual species is not yet developed. Information on the synchronization of sexual reproduction in phytoplankton would thus be of particular value for evolutionary genetic studies since there is considerable theory developed for species with heterogonic life cycles (Maynard-Smith, 1978; Lynch, 1984).

SUMMARY AND CONCLUDING REMARKS

The data reviewed in the earlier sections of this paper show considerable evolutionary potential in a variety of phytoplankton taxa and support the notion that changes in the frequency of different genotypes within a phytoplankton population can result from natural selection on genetically determined phenotypic variation. Because of their short generation times, marine phytoplankton are capable of showing a rapid response to selection when measured in absolute time scales. This evolutionary component of the adaptive repertoire of phytoplankton species should be specifically considered with regard to the effect of newly introduced selective agents (e.g., pollutants) as well as to naturally occurring changes in the selection regime which result from entrainment into different water masses, seasonal changes in temperature or illumination, or other factors influencing survival and reproduction.

This can be done by monitoring the response of the genetic structure of phytoplankton communities to selection using direct molecular methods for examining population genetic structure or by applying quantitative genetic theory to phenotypic data depending on the nature of the characters of particular interest. It is also possible, using clonal cultures, to follow the response of a character to selection as has been done with zooplankton (Marcus, 1985) and to estimate the genetic variability in a character.

In conclusion, it should be noted that phytoplankton species may be ideal organisms for research on a number of basic questions in evolutionary biology. They can be grown in controlled experimental conditions, have short generation times, and, because they are unicellular, growth measurements provide a relatively direct measure of fitness. The colonial habit of certain species makes them potentially well suited for studies of group selection (cf. Wade, 1978; Dawkins, 1982; Brandon, 1982; Sober, 1984; Endler, 1986); their patterns of dispersal are such that many aspects of habitat diversity and selection can be examined in either experimental or natural populations (cf. Maynard-Smith and Hoekstra, 1980; Endler, 1977;

1986); and the long fossil record of many extant species means that evolutionary hypotheses generated from population genetic studies can be tested on geological time scales.

ACKNOWLEDGEMENTS

The development of this review greatly benefitted from the input of a number of scientists from a wide variety of disciplines. I particularly thank S. Arnold, J. Bagshaw, W. Balch, L. Campbell, J. Coyne, S. Douglas, H. Glover, S. Golden, R. Guillard, S. Kalisz, R. Lande, L. Mets, K. Muirhead, L. Shapiro, and T. Whittam for useful discussion, and C.M. Yentsch for her continued encouragement. I also thank M. Keller, R. Lande, L. Shapiro, and C.M. Yentsch for their comments on the manuscript. This is Bigelow Laboratory Contribution 86040; its preparation was supported by DOE Contract DE-AC02-81ER60014.

REFERENCES

Achtman, M. and G. Pluschke. 1986. Clonal analysis of descent and virulence among selected *Escherichia coli*. Ann. Rev. Microbiol. 40: 185-210.

Alam, M.I., C.P. Hsu and Y. Shimizu. 1979. Comparisons of toxins in three isolates of *Gonyaulax tamarensis* (Dinophyceae). J. Phycol. 15: 106-110.

Alberte, R.S., A.M. Wood, T.A. Kursar and R.R.L. Guillard. 1984. Novel phycoerythrins in marine *Synechococcus* spp. Plant Physiol. 75: 732-39.

Arnold, S.J. 1983. Morphology, performance, and fitness. Amer. Zool. 23: 347-361.

Arnold, S.J. 1986. Laboratory and field approaches to the study of adaptation. IN: Predator-prey relationships. M.E. Feder and G.V. Lauder (eds.). Chicago/London. University of Chicago Press. pp. 157-177.

Arnold, S.J. and M.J. Wade. 1984a. On the measurement of natural and sexual selection: theory. Evolution. 38: 709-719.

Arnold, S.J. and M.J. Wade. 1984b. On the measurement of natural and sexual selection: applications. Evolution. 38: 720-734.

Avise, J.C. 1986. Mitochondrial DNA and the evolutionary genetics of higher animals. Phil. Trans. R. Soc. Lond. B. 312: 325-342.

Avise, J.C., E. Bermingham, L.G. Kessler and N.C. Saunders. 1984. Characterization of mitochondrial DNA variability in a hybrid swarm between subspecies of bluegill sunfish (*Lepomis macrochirus*). Evolution. 38: 931-934.

Bagshaw, J.C. and A.H. Warner. 1986. Restriction fragment length polymorphism in isolated populations of *Artemia*. Am. Soc. Biol. Chem. Fed. Proc. 45: 1883.

Barker, H.A. 1935. The culture and physiology of marine dinoflagellates. Arch. Mikrobiol. 6: 157-81.

Barlow, R.G. and R.S. Alberte. 1985. Photosynthetic characteristics of phycoerythrin-containing marine *Synechococcus* spp. Mar. Biol. 86: 63-74.

Becker, W.A. 1984. Manual of Quantitative Genetics. Pullman. Academic Enterprises.

Bermingham, E. and J.C. Avise. 1986. Molecular zoogeography of freshwater fishes in the southeastern United States. Genetics. 113: 939-965.

Boag, P.T. 1983. The heritability of external morphology in Darwin's ground finches *Geospiza* on Isla Daphne Major Galapagos Ecuador. Evolution. 37: 877-894.

Bonin, D.J., M.R. Droop, S.Y. Maestrini and M-C. Bonin. 1986. Physiological features of six micro-algae to be used as indicators of seawater quality. Cryptogamie, Algologie. 7: 23-83.

Bonné-Tamir, B., M.J. Johnson, A. Natali, D.C. Wallace and L.L. Cavalli-Sforza. 1986. Human mitochondrial DNA types in two Israeli populations--a comparative study at the DNA level. Am. J. Hum. Gen. 38: 341-351.

Braarud, T. 1951. Taxonomical studies of marine dinoflagellates. Nytt. Mag. Nat. Vidensk. 88: 43-48.

Braarud, T. 1961. Cultivation of marine organisms as means of understanding environmental influences on populations. IN: Oceanography. M. Sears (ed.). Publ. No. 67. Am. Assoc. Adv. Sci. Washington D.C. pp. 271-298.

Brand, L.E. 1981. Genetic variability in reproduction rates in marine phytoplankton populations. Evolution. 35: 1117-1127.

Brand, L.E. 1982. Genetic variability and spatial patterns of genetic differentiation in the reproductive rates of the marine coccolithphores *Emiliana huxleyi* and *Gephyrocapsa oceanica*. Limnol. Oceanogr. 27: 236-245.

Brand, L.E. 1984. The salinity tolerance of forty-six marine phytoplankton isolates. Est. Coast. Shelf Sci. 18: 543-556.

Brand, L.E. 1985. Low genetic variability in reproduction rates in populations of *Prorocentrum micans* ehrenb. (Dinophyceae) over Georges Bank. J. Exp. Mar. Biol. Ecol. 88: 55-65.

Brand, L.E. and R.R.L. Guillard. 1981. The effects of continuous light and light intensity on the reproduction rates of twenty-two species of marine phytoplankton. J. Exp. Mar. Biol. Ecol. 50: 119-132.

Brand, L.E., R.R.L. Guillard and L.S. Murphy. 1981. A method for the rapid and precise determination of acclimated phytoplankton reproduction rates. J. Plankt. Res. 3: 193-201.

Brand, L.E., L.S. Murphy, R.R.L. Guillard and H-t. Lee. 1981. Genetic variability and differentiation in the temperature niche component of the diatom *Thalassiosira pseudonana*. Mar. Biol. 62: 103-110.

Brand, L.E., W.G. Sunda and R.R.L. Guillard. 1983. Limitation of marine phytoplankton reproductive rates by zinc, manganese, and iron. Limnol. Oceanogr. 28: 1182-1198.

Brand, L.E., W.G. Sunda and R.R.L. Guillard. 1986. Reduction of marine phytoplankton reproduction rates by copper and cadmium. J. Exp. Mar. Biol. Ecol. 96: 225-250.

Brandon, R.N. 1982. The levels of selection. IN: P.S.A. 1982, Vol. I. P. Asquith and T. Nickles. (eds.). East Lansing. Philosophy of Sci. Assoc. pp. 315-322.

Brigati, D.J., D. Myerson, J.J. Leary, B. Fpalholz, S.Z. Travis, C.K.Y. Fong, G.D. Hfiung and D.C. Ward. 1983. Detection of viral genomes in cultures cells and parafin-imbedded tissue sections using biotin-labelled hybridization probes. Virology. 126: 32-50.

Campbell, L., E.J. Carpenter and V.J. Iacono. 1983. Identification and enumeration of marine chroococcoid cyanobacteria by immunofluorescence. Appl. Env. Microbiol. 46: 533-559.

Carpenter, E.J. and R.R.L. Guillard. 1971. Intraspecific differences in nitrate half-saturation constants for three species of marine phytoplankton. Ecology. 52: 183-185.

Caugant, D.A., B.R. Levin, I. Orskov, F. Orskov, C.V. Eden and R.K. Selander. 1985. Genetic diversity in relation to serotype in *Escherichia coli*. Inf. and Immun. 49: 407-413.

Cembella, A.D. and F.J.R. Taylor. 1985. Biochemical variability within the *Protogonyaulax tamarensis/catenella* species complex. IN: Toxic Dinoflagellates. D.M. Anderson, A.W. White and D.G. Baden. (eds.) North Holland/New York. Elsevier. pp. 55-60.

Cembella, A.D. and F.J.R. Taylor. 1986. Electrophoretic variability within the *Protogonyaulax tamarensis/catenella* species complex: pyridine linked dehydrogenases. Biochem. Syst. and Ecol. 14: 311-323.

Cembella, A.D., J.J. Sullivan, G.L. Boyer, F.J.R. Taylor and R.J. Andersen. 1986. Variation in paralytic shellfish toxin composition within the *Protogonyaulax tamarensis/catenella* species complex: Red Tide dinoflagellates. Bioch. Syst. and Ecol. (in press).

Clevenger, C.V. and A.L. Epstein. 1984. Identification of a nuclear protein component of interchromatin granules using a monoclonal antibody and immunogold electron microscopy. Exp. Cell Res. 151: 194-207.

Brand, L.E., R.R.L. Guillard and L.S. Murphy. 1981. A method for the rapid and precise determination of acclimated phytoplankton reproduction rates. J. Plankt. Res. 3: 193-201.

Brand, L.E., L.S. Murphy, R.R.L. Guillard and H-t. Lee. 1981. Genetic variability and differentiation in the temperature niche component of the diatom *Thalassiosira pseudonana*. Mar. Biol. 62: 103-110.

Brand, L.E., W.G. Sunda and R.R.L. Guillard. 1983. Limitation of marine phytoplankton reproductive rates by zinc, manganese, and iron. Limnol. Oceanogr. 28: 1182-1198.

Brand, L.E., W.G. Sunda and R.R.L. Guillard. 1986. Reduction of marine phytoplankton reproduction rates by copper and cadmium. J. Exp. Mar. Biol. Ecol. 96: 225-250.

Brandon, R.N. 1982. The levels of selection. IN: P.S.A. 1982, Vol. I. P. Asquith and T. Nickles. (eds.). East Lansing. Philosophy of Sci. Assoc. pp. 315-322.

Brigati, D.J., D. Myerson, J.J. Leary, B. Fpalholz, S.Z. Travis, C.K.Y. Fong, G.D. Hfiung and D.C. Ward. 1983. Detection of viral genomes in cultures cells and parafin-imbedded tissue sections using biotin-labelled hybridization probes. Virology. 126: 32-50.

Campbell, L., E.J. Carpenter and V.J. Iacono. 1983. Identification and enumeration of marine chroococcoid cyanobacteria by immunofluorescence. Appl. Env. Microbiol. 46: 533-559.

Carpenter, E.J. and R.R.L. Guillard. 1971. Intraspecific differences in nitrate half-saturation constants for three species of marine phytoplankton. Ecology. 52: 183-185.

Caugant, D.A., B.R. Levin, I. Orskov, F. Orskov, C.V. Eden and R.K. Selander. 1985. Genetic diversity in relation to serotype in *Escherichia coli*. Inf. and Immun. 49: 407-413.

Cembella, A.D. and F.J.R. Taylor. 1985. Biochemical variability within the *Protogonyaulax tamarensis/catenella* species complex. IN: Toxic Dinoflagellates. D.M. Anderson, A.W. White and D.G. Baden. (eds.) North Holland/New York. Elsevier. pp. 55-60.

Cembella, A.D. and F.J.R. Taylor. 1986. Electrophoretic variability within the *Protogonyaulax tamarensis/catenella* species complex: pyridine linked dehydrogenases. Biochem. Syst. and Ecol. 14: 311-323.

Cembella, A.D., J.J. Sullivan, G.L. Boyer, F.J.R. Taylor and R.J. Andersen. 1986. Variation in paralytic shellfish toxin composition within the *Protogonyaulax tamarensis/catenella* species complex: Red Tide dinoflagellates. Bioch. Syst. and Ecol. (in press).

Clevenger, C.V. and A.L. Epstein. 1984. Identification of a nuclear protein component of interchromatin granules using a monoclonal antibody and immunogold electron microscopy. Exp. Cell Res. 151: 194-207.

Cochran, W.G. and G.M. Cox. 1957. Experimental Designs, 2nd ed. New York/London. Wiley.

Dawkins, R. 1982. The Extended Phenotype: The Gene as a Unit of Selection. San Francisco. Freeman.

Dobzhansky, Th. 1970. Genetics of the Evolutionary Process. New York. Columbia Univ. Press.

Drebes, G. 1977. Sexuality. IN: The Biology of Diatoms. D. Werner. (ed.). Berkeley/Los Angeles. Univ. Calif. Press. pp. 250-283.

Ducklow, H.W. 1984. Geographical ecology of marine bacteria: physical and biological variability at the mesoscale. IN: Current Prespectives in Microbial Ecology. M.J. Klug and C.A. Reddy. (eds.). Washington D.C. Am. Soc. Microbiol. pp. 22-31.

Endler, J. 1977. Geographic Variation, Speciation, and Clines. Princeton. Princeton Univ. Press.

Endler, J. 1986. Natural Selection in the Wild. Princeton. Princeton Univ. Press.

Eppley, R. W., J.N. Rogers and J.J. McCarthy. 1969. Half-saturation constants for uptake of nitrogen and ammonium by marine phytoplankton. Limnol. Oceanogr. 14: 912-20.

Falconer, D.S. 1981. Introduction to Quantitative Genetics, 2nd ed. London. Longman.

Fisher, N.S. 1977. On the differential sensitivity of estuarine and open-ocean diatoms to exotic chemical stress. Am. Nat. 111: 871-895.

Fisher, N.S. and D. Frood. 1980. Heavy metals and marine diatoms: influence of dissolved organic compounds on toxicity and selection for metal tolerance among four species. Mar. Biol. 59: 85-93.

Fisher, N.S., L.B. Graham, E.J. Carpenter, and C.F. Wurster. 1973. Geographic differences in phytoplankton sensitivity to PCBs. Nature. 241: 548-549.

Fisher, R.A. 1918. The correlation between relatives on the supposition of Mendelian inheritance. Trans. Roy. Soc. Edinb. 52: 399-433.

Fisher, R.A. 1958. The genetical theory of natural selection. 2nd. ed. New York. Dover.

Fliermans, C.B. and E.L. Schmidt. 1977. Immunofluorescence for autecological study of a unicellular bluegreen alga. J. Phycol. 13: 364-368.

Friedman, A.L. and R.S. Alberte. 1984. A diatom light-harvesting pigment-protein complex. Pl. Physiol. 76: 483-489.

Friedman, A.L. and R.S. Alberte. 1986. Biogenesis and light regulation of the major light harvesting chlorophyll-protein of diatoms. Pl. Physiol. 80: 43-51.

Gallagher, J.C. 1980. Population genetics of *Skeletonema costatum* (Bacillariophyceae) in Narragansett Bay. J. Phycol. 16: 464-474.

Gallagher, J.C. 1982. Physiological variation and electrophoretic banding patterns of genetically different seasonal populations of *Skeletonema costatum* (Bacillariophyceae).

Gallagher, J.C. and R.S. Alberte. 1985. Photosynthetic and cellular photoadaptive characteristics of three ecotypes of the marine diatom, *Skeletonema costatum* (Grev.) Cleve. J. Exp. Mar. Biol. Ecol. 94: 233-250.

Gallagher, J.C., A.M. Wood and R.S. Alberte. 1984. Ecotypic differentiation in a marine diatom. I. Influence of light intensity on the photosynthetic apparatus. Mar. Biol. 82: 121-134.

Glibert, P.M., T.M. Kana, R.J. Olsen, D.L. Kirchman and R.S. Alberte. 1986. Clonal comparisons of growth and photosynthetic responses to nitrogen availability in marine *Synechococcus* spp. J. Exp. Mar. Biol. Ecol. 101: 199-208.

Glover, H.E., L. Campbell and B. Prezelin. 1986a. Contribution of Synechococcus spp. to size-fractioned primary productivity in three water masses in the Northwest Atlantic. Mar. Biol. 91: 193-203.

Glover, H.E., M.D. Keller and R.R.L. Guillard. 1986b. Light quality and oceanic ultraphytoplankters. Nature. 319: 142-143.

Glover, H.E., M.D. Keller and R.W. Spinrad. 1987. The effects of light quality and intensity on photosynthesis and growth of marine eukaryotic and prokaryotic phytoplankton clones. J. Exp. Mar. Biol. Ecol. (in press).

Goldman, J.C. and E.J. Carpenter. 1974. A kinetic approach to the effect of temperature on algal growth. Limnol. Oceanogr. 19: 756-766.

Gooch, J.L. 1975. Mechanisms of evolution and population genetics. IN: Marine Ecology, Vol. II, Pt. 1. O. Kinne. (ed.). Chichester/New York/Brisbane/Toronto. Wiley. pp. 349-409.

Grant, P.R. 1983. Inheritance of size and shape in a population of Darwin's finches, *Geospiza conirostris.* Proc. Roy. Soc. Lond. B. 220: 219-236.

Guillard, R.R.L. 1968. B12 Specificity of marine centric diatoms. J. Phycol. 4: 59-64.

Guillard, R.R.L., E.J. Carpenter and B.E.F. Reimann. 1974. *Skeletonema menzelii* sp. nov., a new diatom from the Western Atlantic Ocean. Phycologia. 13: 131-138.

Guillard, R.R.L. and P. Kilham. 1977. The ecology of marine planktonic diatoms. IN: The Biology of Diatoms. D. Werner. (ed.). Berkeley/Los Angeles. Univ. Calif. Press. pp. 372-469.

Guillard, R.R.L., P. Kilham, and T. A. Jackson. 1973. Kinetics of silicon-limited growth in the marine diatom *Thalassiosira pseudonana* Hasle and Heimdal (=*Cyclotella nana* Hustedt). J. Phycol. 9:233-237.

Guillard, R.R.L. and J. Ryther. 1962. Studies of marine planktonic diatoms I. *Cyclotella nana* Hustedt and *Detonula confervacea* (Cleve) Gran. Can. J. Microbiol. 8: 229-239.

Haldane, J.B.S. 1924. A mathematical theory of natural and artificial selection. Trans. Camb. Phil. Soc. 23: 19-40.

Haldane, J.B.S. 1954. The measurement of natural selection. Proc. IX Intl. Cong. Genet. 1: 480-487.

Harding, J.P. and Tebbles, N. 1963. Speciation in the sea. London. The Systematics Association.

Hargraves, P. and R.R.L. Guillard. 1974. Structural and physiological observations on some small marine diatoms. Phycologia. 13: 163-172.

Harris, G.P. 1980. Spatial and temporal scales in phytoplankton ecology. Mechanisms, methods, models, and management. Can. J. Fish. Aq. Sci. 37: 877-900.

Harris, G. P. 1986. Phytoplankton Ecology. London/New York. Chapman and Hall.

Hartl, D.L. and D.E. Dykhuizen. 1984. The population genetics of *Escherichia coli*. Ann. Rev. Genet. 18: 31-68.

Hayward, J. 1968. Studies on the growth of *Phaeodactylum tricornutum*. IV. Comparison of different isolates. J. Mar. Biol. Assoc. U.K. 48: 657-666.

Iturriaga, R., J.H. Morrow and D.A. Kiefer. 1986. Application of microphotometry to the study of optical properties of marine particulates. EOS. 67:973.

Jacobberger, J.W., D. Fogleman and J.M. Lehman. 1986. Analysis of intracellular antigens by flow cytometry. Cytometry. 7: 356-364.

Jensen, A., B. Rystad and S. Melsom. 1974. Heavy metal tolerance of marine phytoplankton. I. The tolerance of three algal species to zinc in coastal seawater. J. Exp. Mar. Biol. Ecol. 15: 145-157.

Kalisz, S. 1986. Variable selection on the timing of germination in *Collinsia verna* (Scrophulariaceae). Evolution. 40: 479-491.

Kilham, S.S. 1975. Kinetics of silicon-limited growth in the freshwater diatom *Asterionella formosa*. J. Phycol. 11: 396-399.

Kilham, S.S., C.L. Kott and D. Tilman. 1977. Phosphate and silicate kinetics for the Lake Michigan diatom *Diatoma elongatum*. J. Great Lakes Res. 3:93-99.

Kimura, M. 1983. The Neutral Theory of Molecular Evolution. New York. Cambridge Univ. Press.

Kreitman, M. 1983. Nucleotide polymorphism at the alcohol dehydrogenase locus of *Drosophila melanogaster*. Nature. 304: 412-417.

Kreitman, M. and M. Aguade. 1986. Genetic uniformity in two populations of *Drosophila melanogaster* as revealed by the filter hybridization of four nucleotide-recognizing restriction enzyme digests. Proc. Nat. Acad. Sci. U.S.A. 83: 3562-3566.

Lande, R. 1976. Natural selection and random genetic drift in phenotypic evolution. Evolution. 30: 314-334.

Lande, R. 1977. Statistical tests for natural selection on quantitative traits. Evolution. 31: 442-444.

Lande, R. 1979. Quantitative genetic analysis of multivariate evolution, applied to brain-body size allometry. Evolution. 33: 402-416.

Lande, R. 1980. The genetic covariance between characters maintained by pleiotropic mutations. Genetics. 94: 203-215.

Lande, R. and S.J. Arnold. 1983. The measurement of selection on correlated characters. Evolution. 37: 1210-1226.

Lane, D.J., D.A. Stahl, G.J. Olsen, D.J. Heller and N.R. Pace. 1985. Phylogenetic analysis of the genera *Thiobacillus* and *Thermomicrospira* by 5S rRNA Sequences. J. Bact. 163: 75-81.

Laurie-Ahlberg, C.C. 1985. Genetic variation affecting the expression of enzyme-coding genes in *Drosophila*: an evolutionary perspective. IN: Isozymes. Current Topics In Biol. and Med. Res. 12: 33-88.

Laurie-Ahlberg, C.C., G. Maroni, G.C. Bewley, J.C. Lucchesi and B.S. Weir. 1980. Quantitative genetic variation of enzyme activities in natural populations of *Drosophila melanogaster*.

Lewin, J.C. 1955. Physiological races of the diatom, *Navicula pelliculosa*. Biol Bul. 109: 343.

Lewin, J.C. and R.A. Lewin. 1960. Autotrophy and heterotrophy in marine littoral diatoms. Can. J. Microbiol. 6: 128-134.

Lewis, M.R. and T. Platt. 1982. Scales of variability in estuarine ecosystems. IN: Estuarine Comparisons. Proc. 6th Bien. Int. Est. Res. Conf. V. Kennedy. (ed.). New York. Academic Press. pp. 3-20.

Lewontin, R.C. 1974. The Genetic Basis of Evolutionary Change. New York. Columbia University Press.

Lewontin, R.C. 1985. Population genetics. Ann. Rev. Genetics. 19: 81-102.

Lukavsky, J. 1979. Algal cultures from the point of view of population genetics. IN: Algal Assays and Monitoring Eutrophication. P. Marvan, S. Pribil and O. Lhotsky. (eds.) Stuttgart. E. Schweizerbart'sche Verlagsbuchhandlung. pp. 137-139.

Lynch, M. 1984. The genetic structure of a cyclical parthenogen. Evolution. 38: 186-203.

Lynch, M. and W. Gabriel. 1983. Phenotypic evolution and parthenogenesis. Am. Nat. 122: 745-764.

Manly, B.F.J. 1985. The Statistics of Natural Selection on Animal Populations. London/New York. Chapman and Hall.

Marcus, N. 1985. Population dynamics of marine copepods: the importance of genetic variation. Bul. Mar. Sci. 37: 684-690.

Maynard-Smith, J. 1978. The Evolution of Sex. Cambridge. Cambridge Univ. Press.

Maynard-Smith, J. and R. Hoekstra. 1980. Polymorphism in a variable environment: how robust are the models? Genet. Res. Cambr. 35: 45-57.

Morris, I. 1980. Paths of carbon assimilation in marine phytoplankton. IN: Primary productivity in the sea. P.G. Falkowski (ed.). New York, Plenum Press. pp. 139-159.

Muirhead, K.A., P.K. Horan and G. Poste. 1985. Flow cytometry: present and future. Bio/technology. 3: 337-356.

Murphy, L. S. 1978. Biochemical taxonomy of marine phytoplankton by electrophoresis of enzymes. II. Loss of heterozygosity in clonal cultures of the centric diatoms *Skeletonema costatum* and *Thalassiosira pseudonana*. J. Phycol. 14: 247-250.

Murphy, L. S. and R.A. Belastock. 1980. The effect of environmental origin on the response of marine diatoms to chemical stress. Limnol. and Oceanogr. 25: 160-165.

Murphy, L.S. and R.R.L. Guillard. 1976. Biochemical taxonomy of marine phytoplankton by electrophoresis of enzymes. I. The centric diatoms *Thalassiosira pseudonana* and *T. fluviatilis*. J. Phycol. 12: 9-13.

Murphy, L.S., R.R.L. Guillard and J.F. Brown. 1984. The effects of iron and manganese on copper sensitivity in diatoms: differences in the responses of closely related neritic and oceanic species. Biol. Oceanogr. 3: 187-201.

Murphy, L.S., R.R.L. Guillard and J. Gavis. 1982. Evolution of resistant phytoplankton strains through exposure to marine pollutants. IN: Ecological Stress and the New York Bight: Science and Management. G.F. Mayer. (ed.). Columbia. Estuarine Research Federation. pp. 401-412.

Necas, J. 1979. Genetic variability and the resulting nonhomogeneity in algal populations. IN: Algal Assays and Monitoring Eutrophication. P. Marvan, S. Pribil and O. Lhotsky. (eds.). Stuttgart. E. Schweivzerbart'sche Verlagsbuchhandlung. pp. 141-152.

Nelson, D.M. and L.E. Brand. 1979. Cell division periodicity in 13 species of marine phytoplankton on a light:dark cycle. J. Phycol. 15: 67-75.

Nelson, D.M., J.J. Goering, S.S. Kilham and R.R.L. Guillard. 1976. Kinetics of silicic acid uptake and rates of silica dissolution in the marine diatom *Thalassiosira pseudonana*. J. Phycol. 12: 246-252.

Ochman, H., R.A. Wilson, T.S. Wittam and R.K. Selander. 1984. Genetic diversity within serotypes of *Escherichia coli*. Proc. 4th Int. Symp. on Neonatal Diarrhea. Saskatchewan. VIDO Publications. University of Saskatchewan. pp. 202-224.

Olsen, G.J., D.J. Lane, S.J. Giovannoni and N.R. Pace. 1986. Microbial ecology and evolution: a ribosomal RNA approach. Ann. Rev. Microbiol. 40: 337-365.

Olson, R.J., S.W. Chisholm and E.R. Zettler. 1986. Dual-beam flow cytometry for distinguishing between phytoplankton pigment types. EOS. 67: 973.

Provasoli, L. and I.J. Pintner. 1953. Ecological implications of in vitro nutritional requirements of algal flagellates. Ann. N.Y. Acad. Sci. 56: 839-851.

Reyment, R.A. 1982a. Quantitative-genetic analysis of evolution in two late Cretaceous species of ostracods. Proc. III N. Am. Paleontol. Conv. J. Paleontol. 56(2nd Suppl.): 21-22.

Reyment, R.A. 1982b. Application of quantitative genetics to evolutionary series of microfossils. IN: Nordic Symp. of Appl. Statistics and Data Processing. Copenhagen. Forlag NEUCC. pp. 307-325.

Reyment, R.A. 1982c. Phenotypic evolution in a Cretaceous foraminifer. Evolution. 36: 1182-1199.

Reyment, R.A. 1983. Phenotypic evolution in microfossils. Evolutionary Biol. 16: 209-254.

Robertson, A. 1959. Experimental design in the evaluation of genetic parameters. Biometrics. 15: 219-226.

Salzman, A. 1986. Analysis of natural selection on a quantitative physiological trait: salt tolerance in Western Ragweed (*Ambrosia pfilostachya*). Ph.D. Dissertation. Department of Biology. University of Chicago.

Scheffė, H. 1959. The Analysis of Variance. New York. Wiley.

Schmidt, R.J. and A.R. Loeblich III. 1979. Distribution of paralytic shellfish poison among Pyrrhophyta. J. Mar. Biol. Assoc. U.K. 59: 479-487.

Schmidt, R.J., V.D. Gooch, A.R. Loeblich III and J.W. Hastings. 1978. Comparative study of luminescent and nonluminescent strains of *Gonyaulax excavata*. J. Phycol. 14: 5-9.

Selander, R.K. 1985. Protein polymorphism and the genetic structure of natural populations of bacteria. IN: Population Genetics and Molecular Evolution. T. Ohta and K. Aoki. (eds.). Tokyo/Berlin. Japan Sci. Soc. Press/Springer Verlag. pp. 85-106.

Selander, R.K. and B.R. Levin. 1980. Genetic diversity and structure in *Escherichia coli* populations. Science. 210: 545-547.

Selander, R.K., D.A. Caugant and T.S. Whittam. 1987. Genetic structure and variation in natural populations of *Escherichia coli*. IN: *Escherichia coli* and *Salmonella typhimurium*, Cellular and Molecular Biology. J.L. Ingraham *et al*. (Eds.). ASM Publications, Washington, D.C. (in press)

Selander, R.K. and T.S. Whittam. 1983. Protein polymorphism and the genetic structure of populations. IN: Evolution of Genes and Proteins. M. Nei and R.K. Koehn. (eds.). Sunderland. Sinauer.

Shapiro, L.P. and R.R.L. Guillard. 1987. Physiology and ecology of the marine eukaryotic ultraplankton. IN: T. Platt and W. Li. (eds.). Physiological Ecology of Picoplankton. Can. J. Fish. Aq. Sci., Bul. 214. (in press.).

Simmonds, N.W. 1979. Principles of Crop Improvement. New York. Longman.

Sober, E. 1984. The Nature of Selection: A Philosophical Inquiry. Cambridge, MA. MIT Press.

Soudek, D. and G.G.C. Robinson. 1983. Electrophoretic analysis of the species and population structure of the diatom *Asterionella formosa*. Can. J. Bot. 61: 418-433.

Stahl, D.A., D.J. Lane, G.J. Olsen and N.R. Pace. 1984. Analysis of hydrothermal vent-associated symbionts by ribosomal RNA sequences. Science. 224: 409-411.

Stahl, D.A., D.J. Lane, G.J. Olsen and N.R. Pace. 1985. Characterization of a Yellowstone hot spring microbial community by 5S rRNA sequences. Appl. and Env. Microbiol. 49: 1379-1384.

Suzuki, D.T., R.C. Lewontin and J. Miller. 1985. An Introduction to Genetic Analysis. New York. Freeman.

Terry, K.L., J. Hirata and E.A. Laws. 1983. Light-limited growth of two strains of the marine diatom *Phaeodactylum tricornutum* Bohlin: chemical composition, carbon partitioning and the diel periodicity of physiological processes. J. Exp. Mar. Biol. Ecol. 68: 209-227.

Tilman, D. 1977. Resource competition between planktonic algae: an experimental and theoretical approach. Ecology. 58: 338-348.

Titman, D. 1976. Ecological competition between algae: an experimental confirmation of resource-based competition theory. Science. 192: 463-465.

Trask, B.J., G.J. van den Engh and J.H. Elbershuizen. 1982. Analysis of phytoplankton by flow cytometry. Cytometry. 2: 258-264.

Trask, B., G. van den Engh, J. Landegent, N. Jansen in del Wal and M. van der Ploeg. 1985. Detection of DNA sequences in nuclei in suspension by *in situ* hybridization and dual beam flow cytometry. Science. 230: 1401-1403.

Underhill, P.A. 1977. Nitrate uptake kinetics and clonal variability in the neritic diatom *Biddulphia aurita*. J. Phycol. 13: 170-176.

Via, S. and R. Lande. 1985. Genotype-environment interaction and the evolution of phenotypic plasticity. Evolution. 39:505-522.

Wade, M.J. 1978. A critical review of the models of group selection. Quart. Rev. Biol. 53: 101-113.

Wallace, D.C., K. Garrison and W.C. Knowler. 1985. Dramatic founder effects in Amerindian mitochondrial DNAs. Am. J. Phys. Anthro. 68: 149-155.

Ward, B.B. and A.F. Carlucci. 1985. Marine ammonia- and nitrite-oxidizing bacteria: serological diversity determined by immunofluorescence in culture and in the environment. Appl. and Env. Microbiol. 50: 194-201.

Watson, D.A. and A.R. Loeblich III. 1983. The application of electrophoresis to the systematics of the marine dinoflagellate genus *Heterocapsa*. Biochem. Syst. and Ecol. 11: 67-71.

Wood, A.M. 1985. Adaptation of photosynthetic apparatus of marine ultraphytoplankton to natural light fields. Nature. 316: 253-255.

Wood, A.M., P.K. Horan, K. Muirhead, D.A. Phinney, C.M. Yentsch and J.B. Waterbury. 1985. Discrimination between types of pigments in marine *Synechococcus* spp. by scanning spectroscopy, epifluorescence microscopy, and flow cytometry. Limnol. Oceanogr. 30: 1303-1315.

Wood, A.M., R. Lande and G.A. Fryxell. 1987. Quantitative genetic analysis of morphological variation in an antarctic diatom grown at two light intensities. J. Phycol. (in press).

Whittam, T.S., H. Ochman and R.K. Selander. 1983. Multilocus genetic structure in natural populations of *Escherichia coli*. Proc. Nat. Acad. Sci. 80: 1751-1755.

Wright, S. 1921. Systems of mating. Genetics. 6: 111-178.

Wright, S. 1978. Evolution and the Genetics of Populations. Variability Within and Among Natural Populations. Chicago. University of Chicago Press. Chicago. Vol. 4.

Yentsch. C.M. and S.A. Pomponi. 1986. Automated individual cell analysis in aquatic research. Int. Rev. Cytol. 105: 183-243.

Yentsch, C.S. and C.M. Yentsch. 1984. Emergence of optical instrumentation for measuring biological properties. Oceanogr. Mar. Biol. Ann. Rev. 22: 55-98.

Techniques

THE PRODUCTION OF MONOSPECIFIC ANTISERA TO SOFT-BOTTOM BENTHIC TAXA

Eugene D. Gallagher
Environmental Sciences Program
University of Massachusetts, Boston
Boston, MA 02125

Peter A. Jumars
School of Oceanography WB-10
University of Washington
Seattle, WA 98195

Gary L. Taghon
Oregon State University
Marine Science Center
Newport, Oregon 97365

INTRODUCTION

In 1979 we began field tests of the hypothesis that predation on larvae and juveniles controls benthic community structure and dynamics. We picked an easily accessible intertidal community of low diversity, the Skagit flats in northern Puget Sound, for our experiments (Gallagher *et al.*, 1983). We also developed serological methods to document and quantify predation on larvae and juveniles. In this paper we will describe the methods used to produce monospecific antisera to soft-bottom benthic taxa.

We initially adapted Ouchterlony's serological assay for the study of soft-bottom benthos to identify masticated juveniles and larvae in the guts of predators (Feller *et al.*, 1979). Ouchterlony's double-diffusion assay is limited by its sensitivity and specificity and is often not sensitive enough to detect single, small prey organisms in the guts of predators. Most of the serological assays more sensitive than Ouchterlony's cannot be used with cross-reacting, polyspecific antisera. Most of the antisera produced to marine benthic invertebrates are cross-reactive (Feller *et al.* 1979; Feller and Gallagher, 1982; Feller, 1984a & b; Feller *et al.*, 1985; Feller *et al.* this volume). Cross-reactions occur when the antiserum produced to one species reacts with protein extracts from other species. Most of the cross-reactions are due to shared antigenic determinants or epitopes, among phylogenetically related taxa (Feller *et al.*, 1979; Feller and Gallagher, 1982), though some are due to contamination of the homogenates used to elicit antibody production.

In this paper we describe a three-step procedure for obtaining non cross-reactive antisera to small invertebrates. For each step we use immunoelectrophoresis. First, the unique antigens of a taxon are identified on immunoelectrophoretic gels. Second, a unique antigen is purified and used for subsequent immunization. Finally, the antiserum produced is checked for specificity and sensitivity.

METHODS AND RESULTS

General Comments

Serological assays detect the reactions between antibodies, in this case rabbit Immunoglobulin G (IgG), and their homologous antigens. Antibody and antigen molecules precipitate when they meet at their equivalency points in agarose gels. This white precipitate, called precipitin, is often visible in the clear agarose gel but can be highlighted by protein staining. In the Ouchterlony double-diffusion assay, the process of simple radial diffusion from point sources drives the antigen and antibody molecules together. In the more sensitive electrophoretic assays, the movement of charged molecules in an electric field drives the charged antigen and antibody molecules together. Most immunoelectrophoretic assays detect antigens that are negatively charged at pH 8.6, the approximate isoelectric point of IgG. These negatively charged antigens move through a buffered (pH 8.6) agarose gel towards the anode, while the neutrally charged antibody molecules remain stationary. When the antigen and antibody molecules precipitate, they can be viewed in the clear agarose gel layer that contained the antibody molecules. Staining with the general protein stain, Coomassie Blue R, enhances greatly the visibility of the precipitin reactions.

We will present methods for performing seven types of serological assays: (1) Ouchterlony micro-double diffusion, (2) rocket electrophoresis, (3) line electrophoresis, (4) split-line electrophoresis, (5) rocket-line electrophoresis, (6) two-dimensional immunoelectrophoresis, and (7) counter immunoelectrophoresis.

Axelsen and Bock (1972) provide an excellent, readable, general review of immuno-electrophoretic techniques. More detailed reviews will be cited in the descriptions of each assay.

The standard reagents and procedures for the assays are described in Table 1. The immunoelectrophoresis chamber, which was made by the Medical Instruments Div. (University of Washington), can accommodate 18 25-mm X 75-mm microscope slides or 12 43-mm X 43-mm lantern slides.

Table 1. Reagents and protocols used for immunoassay.

Agarose Sticks: 1 % agarose

for 1 tube:		
5.0	ml	distilled H_2O
0.05	g	agarose
0.001	g	Na azide

Add agarose and azide to cold water and heat agarose and water to boiling in a water bath.

Veronal Buffer: pH 8.6, ionic strength = 0.1

0.017 M	Veronal (=Barbital)
0.083 M	Na Veronal (=Na Barbital)

Dissolve veronal in distilled water while heating distilled water to boiling. Add Na Veronal while cooling.

Veronal and Triton X-100 (V & T) Buffer

for 1 l		
950.	ml	Veronal buffer
50.	ml	10 % Triton X-100

Use high-quality Triton X-100 (e.g., Calbiochem™).

TES - Saline Buffer: pH 7.3

0.25 M TES stock solution:	
0.25 M	TES
1.0 M	NaOH - Add slowly to adjust pH to 7.3 at room temperature.

4 l TES-Saline buffer:		
80.0	ml	0.25 M TES stock solution
35.04	g	NaCl
3920.0	ml	distilled H_2O
0.4	g	Na azide*

*Delete if TES-saline buffer is to be used to prepare antigen solutions for immunizations

Table 1. continued

Staining Solutions

Destain solution:	450. ml 95% ethanol 450. ml distilled H_2O 100. ml glacial acetic acid
Coomassie Blue	1.0 g Coomassie Blue R. (Sigma) 200. ml Destain solution

Stain slides for 5-10 minutes and destain for 5-30 minutes

Precoated Slides

Dilute agarose sticks 1:2 with distilled H_2O. Pour 1 ml; per 25 X 75mm slide, 1.5 ml per 43 X 43 mm slide, or 2 ml per 50 X 75 mm slide; let set. Dry agarose gel in 100° C oven for 1 h.

Slide Preparation

Dilute agarose sticks 1:2 with Veronal and Triton X-100 buffer (0.5% agarose-V&T) and heat in a water bath. Warm pipettes before pipetting agarose solutions.

Ouchterlony:	Pour 1.2 ml of 0.5% agarose-V&T per 25 X 75 mm precoated slide; let gel.
Rocket, Line and Rocket-line:	Pour 2.5 ml of 0.5% agarose-V&T on each pre-coated 25 X 75 mm slide; let gel.
Split-line:	Pour 5.0 ml of 0.5% agarose-V&T on a clean (not precoated) 50 X 75 mm slide; let gel.
Counter:	Dilute V&T buffer 1:2 before adding to agarose sticks. Pour 2.5 ml of 0.5% agarose-V&T diluted) on each precoated 25 X 75 mm slide; let gel.

Ouchterlony's double-diffusion assay

General reviews of double diffusion assays can be found in Ouchterlony (1968), Crowle (1973) and Oudin (1980). Antigen extracts and antisera (approximately 20 l each) diffuse radially from pinholes at the base of wells cut in plastic templates (Figure 1A) which rest on agarose-covered slides (Table 1). Antibody and antigen molecules precipitate at their zones of equivalence in the buffered agarose gel. After 48 h in a humidified chamber, the templates are removed, and the slides are immersed for 24 h in TES-saline buffer (with Na azide) to remove soluble protein. After a 4-h rinse in distilled water, the slides are dried at 50°C overnight and stained (Table 1). Dark-field illumination provides the best light for counting the precipitin lines, which occur between the center and peripheral wells. Eight antigen-antibody reactions can be analyzed per slide. The maximum number of distinct lines that can be detected per reaction is about 15.

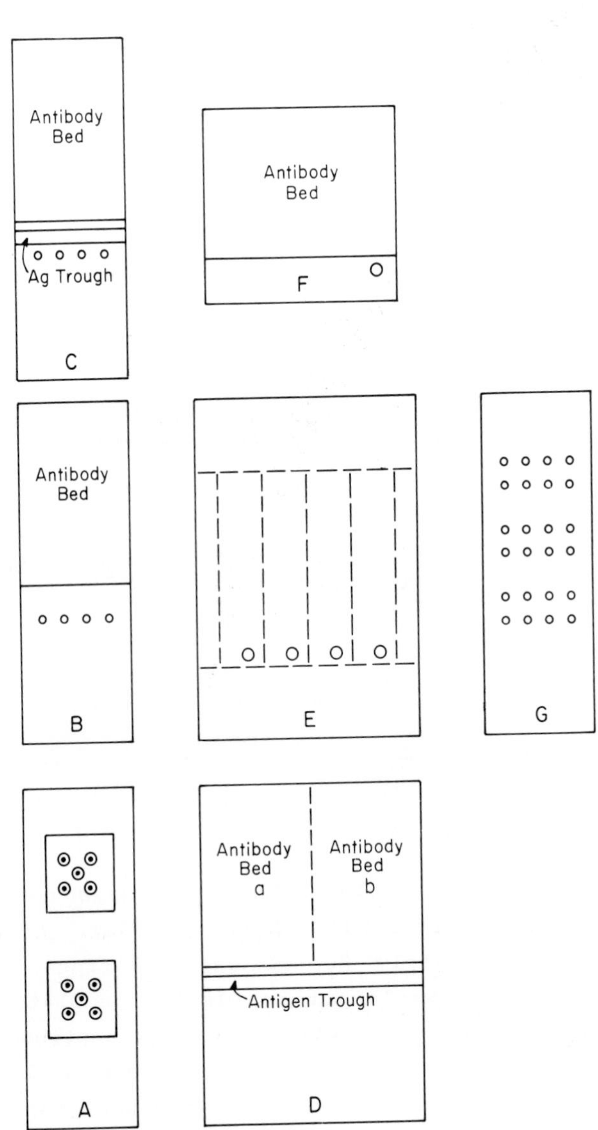

Figure 1. Templates for immunoassay. The sizes of the slides are 25 x 75 mm (A,B,C,and G), 50 X 75 mm (D and E), and 43 X 43 mm (F). Methods for preparing the slides are described in Table IV-1. For all the immunoelectrophoresis slides (B-G), the slides are oriented with the anode at the top of the figure.

A: Ouchterlony's double diffusion assay. Two plastic templates containing 4 wells with pinholes are placed on a precoated slide covered with an agarose + V & T layer.
B: Rocket electrophoresis. The upper 4 cm of the slide is covered with a mixture of antiserum and agarose + V & T. Antigen extract is pipetted into the 4 wells.
C: Line and rocket-line electrophoresis. The upper 4 cm is covered with agarose + V & T mixed with antiserum. The antigen trough is filled with antigen extract mixed with agarose + V & T. The 4 wells, punched only for rocket-line applications, are filled with antigen extract.
D. Split-line electrophoresis. Two separate antiserum mixtures are pipetted onto the upper portion of the slide. A mixture of antigen extract and agarose + V & T is pipetted to fill the antigen trough.
E: First-dimension slide for two-dimensional electrophoresis. The same antigen extract is pipetted into each of the 4 wells cut in the agarose + V & T layer. After electrophoresis, the first-dimension gels are cut as indicated and transferred to a 43 X 43 mm slide.
F: Second dimension of two-dimensional electrophoresis. After the first-dimension gel has been transferred to the slide, a mixture of antiserum and agarose + V & T is pipetted onto the upper 3.3 cm of the slide.
G: Counter electrophoresis. Wells are cut as indicated. The well closer to the anode (top of the figure) of each of the 12 pairs of wells is filled with antiserum: the well closer to the cathode is filled with antigen.

Rocket Electrophoresis

Laurell and McKay (1981) review the theory and development of this technique; our technique is a slight modification of that of Kenny and Foy (1975). The upper 3-cm agarose layer and underlying agarose precoat of a prepared slide (Table 1) is removed with a razor blade and replaced with 1 ml of antibody bed. The antibody bed is a 1:1 mixture of antiserum (up to 20% of total volume) diluted with V & T buffer and 1% agarose. Antigen extracts to be assayed (5 to 15 µl) are placed in 3 or 4 wells (1- or 3-mm diameter) punched in the agarose layer (Figure 1B). We can place up to 18 slides in the electrophoresis chamber at one time. The electrophoresis buffer is V & T. Whatmann^{TM} filter-paper wicks are inserted in the buffer vessels and touch the anodic and cathodic edges of the slides. In rocket electrophoresis we run the slides for 4 hr at 3 V/cm potential; the potential is measured directly on the agarose with a voltmeter (e.g., Tandy™ voltmeter). Because uniform voltage cannot always be achieved, we adjust the voltage so that the minimum voltage drop across any one slide exceeds 3 V/cm for 4 hr. We wash and stain the slides as described in Table 1.

In rocket electrophoresis the antigen-antibody precipitates form "rockets" above the antigen wells. With polyspecific antisera dozens of rockets can be observed above an antigen well, but a monospecific serum will precipitate to form only one rocket. The height of a rocket is directly proportional to the antigen concentration and indirectly proportional to the antibody concentration (Axelsen and Bock, 1972; Laurell and McKay, 1981).

Line Electrophoresis

Line electrophoresis was developed by and reviewed by Krøll (1973a; 1981a; 1981b). The initial preparation of the slides and preparation of antibody beds follows the protocol used for rocket electrophoresis. An antigen trough, cut in the agarose layer (Figure 1C), is filled with an antigen bed consisting of a 1:1 mixture of antigen extract - V & T buffer and 1% agarose. Electrophoresis, washing, and staining follow the procedures used in rocket electrophoresis.

Instead of rockets, the antigen-antibody precipitates in line electrophoresis form horizontal lines. Each line is the result of at least one antigen-antibody precipitate. We have observed several different antigens, separable using two-dimensional electrophoresis, precipitate at the same vertical location in the antibody bed. Therefore, one line does not necessarily mean one antigen.

Either the antigen or antibody concentrations can be adjusted to obtain satisfactory spacing of the lines. An equal dilution of both the antigen and antibody solutions will not change the vertical positions of the lines, but the lines will be much fainter. Precipitin lines will not be detectable if the antigen and antibody solutions are too dilute.

Split-Line Electrophoresis

This technique was described by Krøll (1973a; 1981a; 1981b). Line electrophoresis and split-line electrophoresis differ in that 50 X 75 mm microscope slides are used (Table 1), and two contiguous antibody beds are poured on the same slide (Figure 1D).

Rocket-Line Electrophoresis

This technique, described by Krøll (1981b), is a straightforward combination of line and rocket electrophoresis. Just as the fusion of two lines indicates a reaction of identity in split-line electrophoresis, the fusion of a rocket and a line also indicates a reaction of identity in rocket-line electrophoresis. To perform rocket-line electrophoresis, the slide is prepared as in line electrophoresis, and wells are punched and filled beneath the antigen trough (Figure 1C).

Rocket-line electrophoresis detects reactions of identity between the antigens which form rockets and those which form lines. A line displaced anodically in the form of a rocket above a well indicates that the same antigen was present in both the antigen trough and well.

Two-dimensional immunoelectrophoresis

Our procedures for two-dimensional immunoelectrophoresis are modifications of those of Thirkill and Kenny (1974; 1975). Differences in electrophoretic mobility separate the antigens in the first dimension, and electrophoresis into a stationary antibody bed separates the antigen-antibody complexes in the second dimension. This technique separates antigens better than line electrophoresis. Two antigens precipitating at the same location in line electrophoresis could separate on the basis of relative electrophoretic mobility in two-dimensional immunoelectrophoresis.

Thirkill and Kenny (1974; 1975) performed the first-dimension runs on separate 43-mm X 43-mm slides, but we process 4 first-dimension samples (5 - 15 l each) on 1 50 X 75 mm slide (Table 1, Figure 1E). The four samples on each slide should contain the same antigen mixture to prevent contamination. In many first-dimension runs, we also add 3 l of bovine serum albumin (BSA, 1 mg protein/ml). Negatively charged antigens move in an electric field at characteristic rates relative to BSA; antigens can be identified by their relative electrophoretic mobilities in the first dimension of two-dimensional electrophoresis. Voltages and times for the first-dimension electrophoresis vary for different antigen solutions. Many soft-bottom benthic taxa have antigens with electrophoretic mobilities far exceeding that of BSA, so often the first-dimension electrophoresis times are short.

After the first-dimension electrophoresis, we transfer the 10 X 43 mm strips to precoated 43 X 43 mm lantern slides (sold by custom photography stores). We pour 1.4 ml of antibody bed, often including an antiserum to BSA, onto the clear portion of each slide (Figure 1F). Usual electrophoresis times and voltages in the second dimension are 6 to 8 h at 2 V/cm. We wash, dry, and stain the slides as previously described. If the slides are to be used for the purification of antigen-antibody precipitates, we wash the slides in TES-saline without sodium azide; these slides are neither dried nor stained.

Counter immunoelectrophoresis

The technique is described by Kenny and Foy (1975). In the previously described immunoelectrophoretic procedures, the neutrally charged antibody molecules move cathodically in an electroendosmotic flow of buffer (Kenny and Foy, 1975). Counter immunoelectrophoresis takes advantage of the electroendosmotic flow to "drive" antibody molecules into the antigen molecules which are moving anodically in the electric field. Agarose containing a high concentration of sulfide groups (Sea Plaque™ HEEO agarose) enhances the amount of electroendosmosis. We prepare the slides (Table 1) and punch pairs of wells for the antigen solutions and antisera (Figure 1G). We pipette an antigen solution into the well of each pair located nearest the cathode and pipette an antiserum into the well located nearest the anode. Though hundreds of reactions could conceivably be run at once (12 reactions X 18 slides), electrophoresis should begin before substantial portions of the antigen and antibody solutions begin to diffuse from the wells; only about 20 to 30 reactions can be performed at any one time. Electrophoresis is performed at 6 V/cm for 30 minutes. The antigen-antibody precipitates are usually visible at the end of the run. The precipitates can be more easily seen after the slides have been washed, dried and stained.

Preparation of oligospecific sera to *Mya arenaria*

The goal of these experiments was to produce an antiserum to a single type of antigen in *M. arenaria;* such antisera are termed monospecific. A monospecific antiserum could be used with sensitive immunoelectrophoretic assays to assay the gut contents of suspected predators of *M. arenaria*. The large size of individual *M. arenaria*, the East Coast soft-shelled clam, provided an abundant supply of antigen for serological analysis and made it the logical target for a first attempt at a monospecific antiserum. We were not successful in producing a monospecific serum to *M. arenaria*, but the electrophoretic techniques proved their usefulness for producing monospecific antisera to other organisms.

Since the polyspecific antiserum to *M. arenaria* reacted with many other soft-bottom benthic taxa (Feller *et al*. 1979; Feller & Gallagher, 1982), we had to identify an antigen that was found in only *M. arenaria*. We used counter immunoelectrophoresis to assay extracts from the Skagit community (Feller *et al*., 1979) in order to exclude those taxa which did not react with the *M. arenaria* antiserum.

Two-dimensional immunoelectrophoresis of *M. arenaria* (Figure 2A) was not used to isolate a unique antigen because the technique could not resolve the many antigen-antibody precipitates. We therefore used rocket-line electrophoresis to assay for common antigens and line electrophoresis to obtain enough of the selected immunoprecipitates to immunize a pair of rabbits. We assayed antigen solutions for cross-reactions using rocket-line immunoelectrophoresis and the polyspecific antiserum to *M. arenaria* (Feller *et al*., 1979). A reaction of identity between a line produced by *M. arenaria* and a rocket formed by a cross-reacting antigen solution revealed that the antigen was shared between the cross-reactive taxa and *M. arenaria*. We were able to identify one line that was unique to *M. arenaria* and was also present in an extract of *M. arenaria* spat (kindly provided by L. Watling, University of Maine). This line was designated line 4 (Figure 2B). We then used line immunoelectrophoresis to isolate the antigen which produced line 4.

The principle of using immunoprecipitates to immunize rabbits was independently developed by Crowle (1973) and Vaitukaitis *et al*. (1971). More recent descriptions of this immunization procedure can be found in Krøll (1981b) and Vaitukaitis (1981). Our procedures

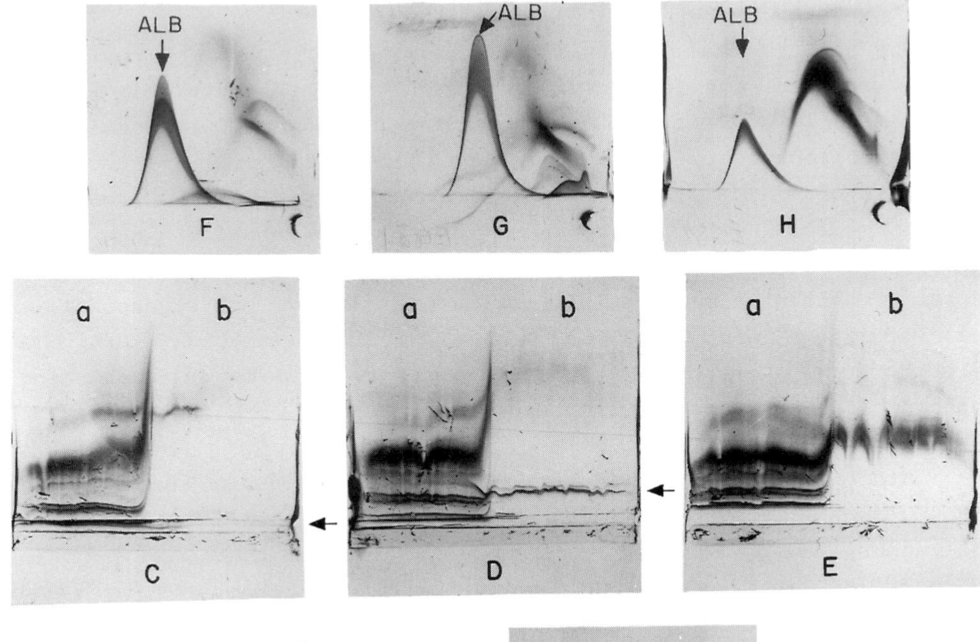

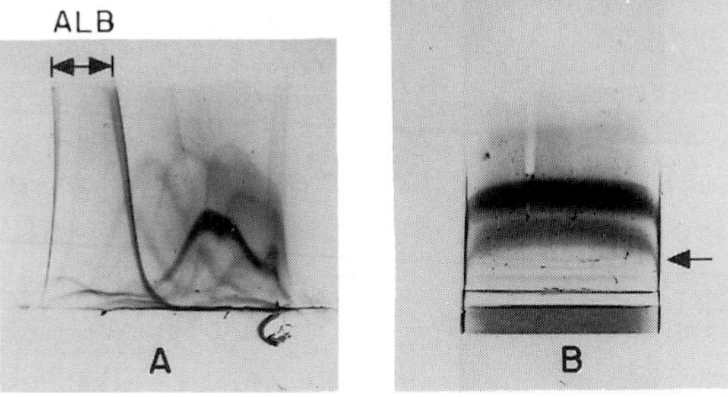

Figure 2. Immunoelectrophoresis of *Mya arenaria* antigens.

A. Two-dimensional electrophoresis. Ten µl of *M. arenaria* extract and 3 µl of bovine serum albumin (=BSA, 1 mg protein/ml) were electrophoresed in the first dimension (anode at the left) and electrophoresed against 0.1 ml antiserum to whole *M. arenaria* and 0.02 ml antiserum to BSA in the second dimension (anode at the top). The BSA reference peak (alb) is identified.

B. Line electrophoresis. 0.012 ml of *M. arenaria* antigen extract was electrophoresed into 0.12 ml antiserum to whole *M. arenaria*. Line 4 is marked with an arrow.

C. Split-line electrophoresis. 0.025 ml of *M. arenaria* extract was electrophoresed into (a) 0.12 ml antiserum to whole *M. arenaria* and (b) 0.12 ml antiserum to line 2. Line 2 is indicated with an arrow.

D. As in C, except antibody bed (b) contains 0.12 ml antiserum to line 4, not line 2. Line 4 is marked with an arrow.

E. As in C, except antibody bed (b) contains 0.12 ml antiserum to line 6. Line 6 is indicated with an arrow.

F. Two-dimensional electrophoresis. Fifteen µl of *M. arenaria* extract and 2 µl of BSA were electrophoresed in the first dimension and electrophoresed against 0.2 ml antiserum to BSA. The BSA-peak (alb) is identified.

G. As in F, except line-4 antiserum was used.

H. As in F, except line-6 antiserum was used.

modify those described in Caldwell *et al*. (1975 a & b) and Alexander and Kenny (1980).

Line immunoelectrophoresis produced 123 reference line patterns for *M. arenaria*. We cut out the lines designated 1, 4 and 6. On some slides the lines were not distinct, so the actual number of lines obtained for 1, 4 and 6 were: 121, 111 and 123, respectively.

The agarose and antigen-antibody precipitates, diluted with 2.5 ml of distilled water, were mixed with an equal volume of Freund's Complete Adjuvant and emulsified by repeated passage through a 20-ga needle. Using the procedures described in Alexander and Kenny (1980), we immunized each of two New Zealand white female rabbits (2.5-3.0 kg each) with approximately 2 ml of the antigen-agarose-adjuvant emulsion. Two weeks after the initial immunization, intramuscular injections were given to the rabbits with 1.0 ml antigen-agarose solution emulsified with 1.0 ml Freund's incomplete adjuvant per rabbit. Two weeks after the intramuscular injection, approximately 50 ml whole blood was obtained 3 times per week via the marginal ear vein. After two weeks whole blood was harvested by cardiac puncture. After overnight refrigeration, the antiserum separated from the clotted cells and was further purified by centrifugation (10,000 X g for 10 min). We noted only minor differences in antibody titre and specificity either between successive bleedings of the same rabbit or between replicate rabbits immunized with the same material. However, in other immunizations we have observed the antibody titre to drop significantly between successive bleedings.

Oligospecific but not monospecific antisera were produced to peaks 1, 4 and 6. Split-line immunoelectrophoresis (Figure 2C, D and E) clearly shows that these antisera recognize fewer antigens than the initial polyspecific antiserum. It also appears that an antiserum produced through the immunization of a line from line-immunoelectrophoresis often contains antibodies to antigens which form lines nearer the anode. Krøll (1981b) suggests that a reduced agarose concentration might alleviate this problem.

The two-dimensional profiles allow clear recognition of the antigens to which the oligospecific antisera react (Figure 2F, G and H). The line-6 antiserum produces a dense, broad, antigen-antibody precipitate. This antigen is found in several other molluscan species (e.g., *Macoma balthica*). With the high-titre oligospecific antisera to line 4 (Figure 2G), two-dimensional immunoelectrophoresis could be used to further purify the Peak-4 antigen to produce a monospecific antiserum. Purification of antigens with two-dimensional immunoelectrophoresis will be described in the next section.

Production of monospecific antisera to Corophium salmonis

We refined the purification procedure described in the previous section to produce a monospecific antiserum to the gammaridean amphipod *C. salmonis*, one of the most abundant taxa in the Skagit flats community. Counter immunoelectrophoresis was used to identify the antigen extracts from other members of the Skagit community that were cross-reactive with the polyspecific *C. salmonis* antiserum described in Feller *et al.* (1979) and Feller and Gallagher (1982).

We were able to obtain a very clear two-dimensional profile using the polyspecific antiserum (Figure 3A). We used tandem two-dimensional electrophoresis (Axelsen and Bock, 1972) to identify the antigens that were unique to *C. salmonis*. The antigen which formed Peak VI (Figure 3B) was unique to *C. salmonis*. Using the techniques described in the previous section, we immunized rabbits with Peaks II, III, V and VI. We immunized 2 rabbits with Peak-VI material and 1 rabbit each with the Peak-II, III and V material. We immunized the rabbits with the following numbers of peaks, obtained with two-dimensional immunoelectrophoresis:

Antigen	Intradermal injection (No. of peaks)	Intramuscular injection (No. of peaks)
Peak II	25	67
Peak III	27	34
Peak V	30	57
Peak VI	25	25

The Peak-III material used for the booster was contaminated with antigen-antibody precipitates from a second peak that precipitated near Peak III. As noted by Crowle *et al.* (1972), if the initial immunization is performed with pure antigen, slight contamination of the booster will not elicit significant antibody production to the contaminating antibody. The serum was obtained as described in the previous section.

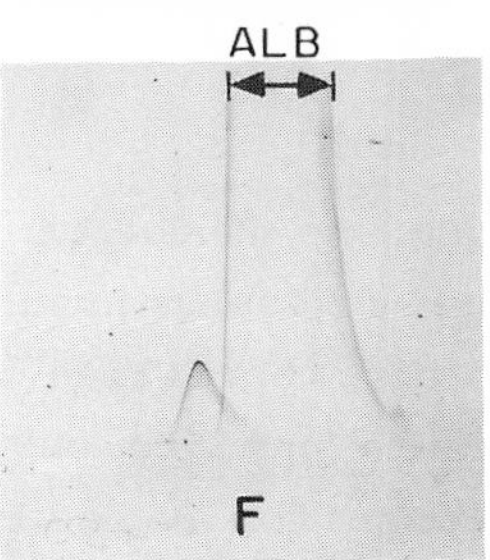

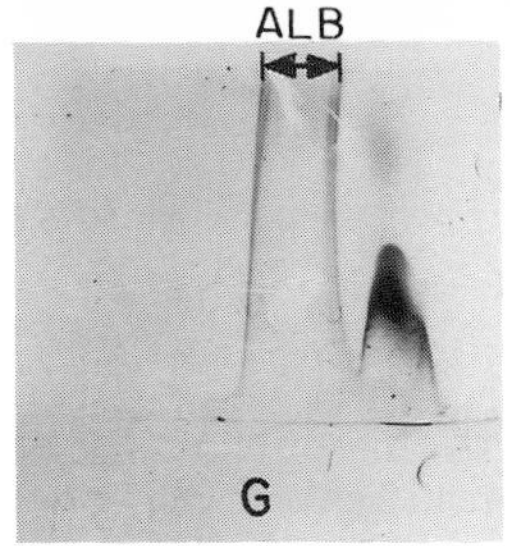

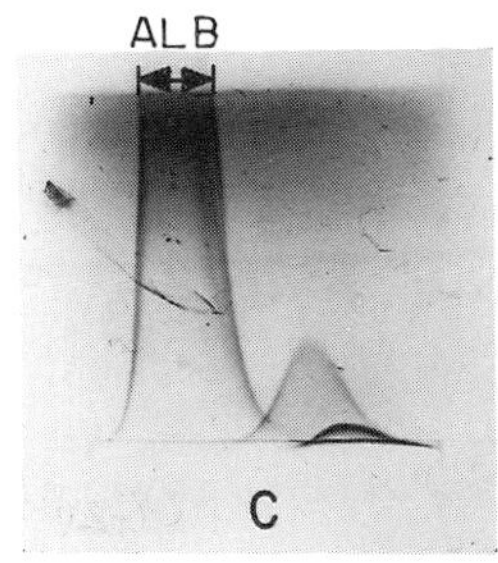

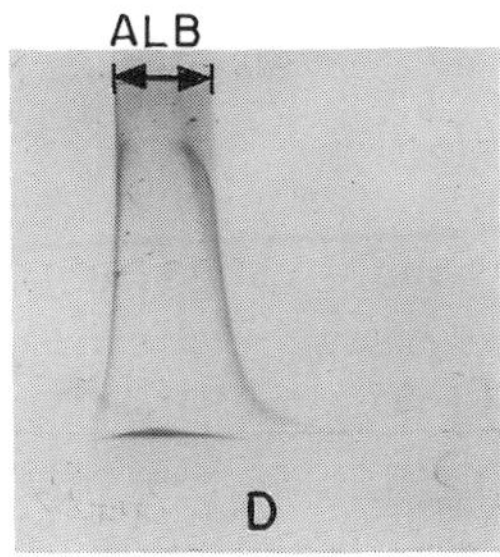

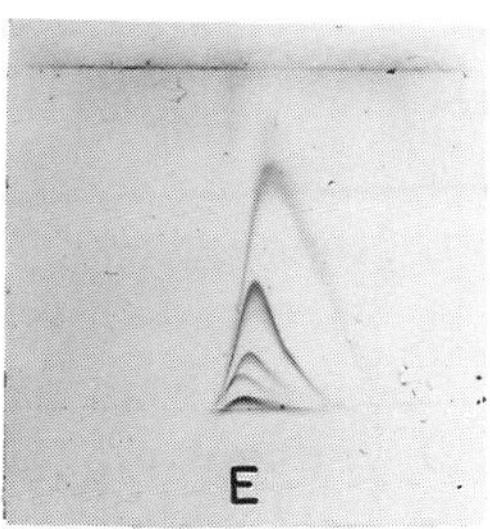

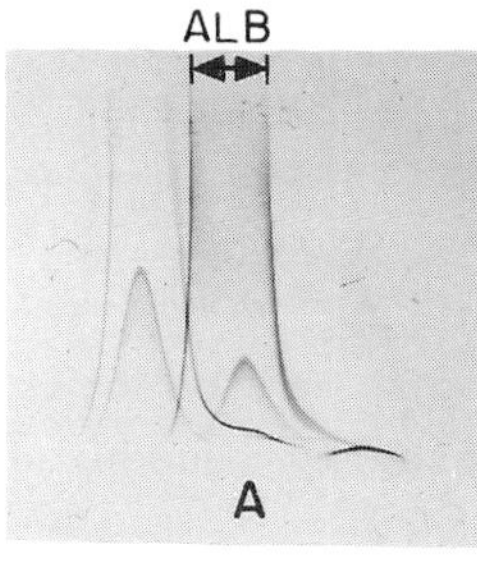

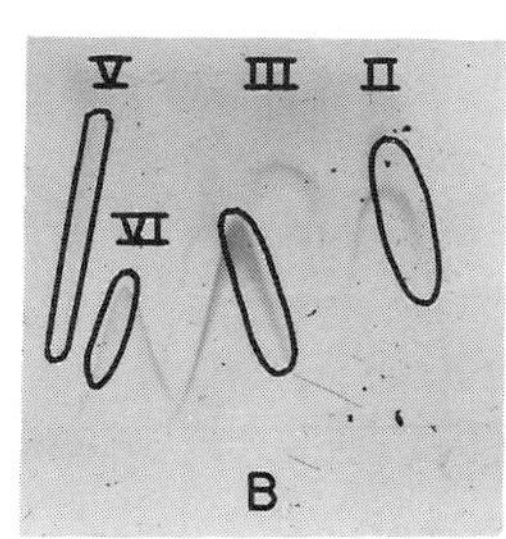

Figure 3. Two-dimensional immunoelectrophoresis of *Corophium salmonis* antigens.

A. Fifteen µl of *C. salmonis* extract and 2 µl of bovine serum albumin (=BSA, 1 mg protein/ml) were electrophoresed in the first dimension and electrophoresed against 0.2 ml antiserum to whole *C. salmonis* and 0.01 ml antiserum to BSA in the second dimension. The BSA reference peak (alb) is identified.

B. As in A, except BSA and anti-BSA were not included. The sections of the gel excised and used for immunizations are indicated.

C. As in A, except 0.28 ml of antiserum to Peak II were used. Two peaks in addition to the BSA peak are visible.

D. As in A, except 0.28 ml of antiserum to Peak III were used. Three peaks in addition to the BSA peak are visible.

E. Fifteen µl of *C. salmonis* extract were electrophoresed in the first dimension and electrophoresed against 0.28 ml of antiserum to Peak V. Each of the 6 antigens has an electrophoretic mobility greater than BSA, which is not shown in this slide.

F. Fifteen µl of *C. salmonis* extract and 3 µl of BSA were electrophoresed in the first dimension and against 0.14 ml antiserum to Peak VI and 0.01 ml antiserum to BSA in the second dimension. The antiserum to Peak VI is monospecific; only one antigen is detected.

G. Fifteen µl of an extract of *Calinectes sapidus*, the East Coast blue crab, and 3 1 of BSA were electrophoresed in the first dimension (anode at left) and electrophoresed into 0.14 ml of antiserum to Peak II and 0.01 ml antiserum to BSA in the second dimension (anode at top). One strongly cross-reacting antigen peak is visible.

Oligospecific antisera were produced to Peaks II (Figure 3C), III (Figure 3D) and V (Figure 3E). The antigens precipitated by each serum have very similar electrophoretic mobilities relative to BSA, as would be expected from the procedure used to purify the antigen. A monospecific antiserum was produced to the Peak-VI antigen (Figure 3F).

The Peak-II antiserum precipitates at least two antigens. One of these antigens is found in many crustaceans, including *Callinectes sapidus*, the East Coast blue crab (Figure 3G). This an antiserum would be worthless for detecting predator-prey links among Crustacea.

The Peak-III antiserum cross reacts with several other Crustacea. A fascinating feature of this antiserum is its specificity for a labile component of fresh *C. salmonis* extracts. This antigen is abundant in fresh *C. salmonis* extracts but disappears entirely from these extracts in a matter of hours. Crustacean hemolymph contains tyrosinase and phenoloxydase which oxidize phenols to quinones. The disappearance of the Peak-III antigen coincides with the discoloration of the extracts as the crustacean melanin is oxidized. Thus, this antiserum apparently recognizes an antigen that is readily oxidized.

The profile produced with the antiserum to Peak V (Figure 3E) clearly refutes the supposition that 1 peak equals one antigen. The antiserum to Peak V precipitates 6 different antigens, all possessing an electrophoretic mobility greater than that of BSA. These six highly mobile antigens were probably moving in a single band during the first-dimension electrophoresis.

The Peak-VI antiserum is monospecific because the antiserum precipitates only one antigen (Figure 3F). This antiserum is also taxon-specific for the Skagit flats community, where only *C. salmonis* has this antigen. However, the antiserum also reacts with extracts of *Corophium spinicorne* and *Corophium insidiosum* which occur allopatrically with *C. salmonis* at False Bay, an intertidal sandflat on San Juan Island on Puget Sound. We have assayed many other Crustacea from the Puget Sound Region and from the East Coast,including another gammarid amphipod (*Eogammarus confervicolus*, and to date, only members of the genus Corophium react with this antiserum.

Uses of a monospecific antiserum

First, the monospecific antiserum can be used to isolate and purify the protein required to make more monospecific antisera. As we have shown, 50 precipitin peaks are adequate to produce a high-titre, monospecific antiserum to Peak VI. With only 2 ml of monospecific antiserum and 2 ml of a concentrated extract of *C. salmonis* (requiring about 100 mg of freeze-dried *C. salmonis* = approximately 20 individuals), we can produce an additional 400 ml of monospecific serum to Peak VI. Thus, we can now produce another 400 ml of monospecific antiserum to a purified antigen without having to biochemically purify an antigen extract or produce additional cross-reactive polyspecific antisera.

Even with monospecific antisera, Ouchterlony's double diffusion assays are not sensitive enough to detect the presence of the trace amounts of *C. salmonis* proteins found typically in the guts of benthic predators. Of the immunoelectrophoretic techniques described in this paper, several are sensitive enough to detect these trace concentrations of protein. Counter immunoelectrophoresis is the most rapid and sensitive but cannot be used to quantify the amount of antigen present in an extract. Unfortunately, we have found that Peak-VI antigen, while a major antigen in *C. salmonis*, is not abundant enough to be measured quantitatively using the techniques described in this paper.

Serology, food-web studies and contact times

Counter immunoelectrophoresis can be used with a monospecific antiserum to produce source food webs (Cohen, 1978). In a source food web all predators of one type of prey are identified. Theoretically, the presence-absence data obtained by counter immunoelectrophoresis could be used to estimate quantitative predation rates if the contact times were known (Peterson and Bradley, 1978). The contact time is an estimate of the length of time a prey can be detected in the gut of a predator. Boreham and Ohiagu (1978) review several serological studies which estimated contact times. Davies (1969) and Davies *et al.* (1978; 1981) and Fichter and Stephens (1981) determined the serological contact times of prey in the guts of predators. Initially, we thought that the serological contact time would be easy to measure since antigens should be detectable immediately after ingestion and, after a fixed period in the gut, they would either be digested or expelled in a fecal pellet. However, the problem of determining contact times is not so simple. Different antigens found in a single prey have different contact times. For example, the *C. salmonis* Peak III antigen, which is easily oxidized, disappears shortly after ingestion, but other antigens persist much longer. The contact time is antigen specific.

The contact time is very difficult to estimate if only cross-reactive polyspecific antisera are available. The contact time should estimate the reactive period of only those antigens unique to the prey of interest. If the predator shares two antigens with the prey, the contact time would be the period during which antigens other than these two could be detected in the predator's gut.

A phenomenon noted with several predator-prey pairs further hinders the ready determination of contact times. We used a polyspecific antiserum to *Tanais* sp. and rocket-line immunoelectrophoresis to assay the gut contents of *Eogammarus confervicolus* that had been fed tanaids. Immediately after ingestion, we could not detect the antigens that were unique to *Tanais* sp.; we could detect them only after the tanaid had been in the gut of *E. confervicolus* for about an hour. Presumably the predator's enzymes solubilize the prey proteins, thus making them easier to detect with the serological assay. The number of prey antigens detected in the predator's gut increases with time and then decreases. The concentrations of tanaid antigens in the guts of *E. confervicolus* fed the same number of prey varies tremendously. Much of this variability may be due to sloppy feeding by *E. confervicolus*. As noted by Dagg (1974) amphipods are often sloppy feeders, losing much of the soluble organic matter present in prey. Tanaids are not ingested whole by *E. confervicolus*; they are held in the gnathopods and slowly chewed with the mandibles. The amount of time spent ingesting a single prey can vary greatly. Some *E. confervicolus* consume an individual tanaid in one or a few minutes, others take 15 min. Thus, the amount of protein ingested would be dependent on the size of the predator (larger predators consuming small prey more rapidly), the size of the prey and the handling time taken to consume a single prey. The number of prey present in a single predator's gut can vary substantially. When presented with an abundant supply of tanaids, some *E. confervicolus* consume 20 individuals before producing a single fecal pellet; others consume only 1 individual.

These studies indicate that contact times are difficult to determine even under carefully controlled laboratory conditions for one prey and one predator species. Moreover, a positive serological assay of a predator's gut could represent from one to at least 20 separate prey individuals, making the conversion of presence-absence data to numerical frequencies nearly impossible. In fact, contact times are not very useful in the analysis of source food webs, the types of food webs usually generated with serological assays. Contact times are used to convert presence-absence date to relative prey frequencies in predators' diets in both community food webs, in which all predator-prey links are identified, or sink food webs, in which all prey of a given predator are identified (*sensu* Cohen, 1978). Contact times are not very useful for analyzing source food webs, the type generated in most serological studies.

DISCUSSION

Serological assays have immense potential, but they are time-consuming to develop and to perfect. Is the payoff worth the investment of time and effort? In this discussion, we will evaluate food web studies, the usefulness of polyspecific antisera and the future of monospecific antisera. Finally, we will point out some recent technological advances in molecular immunology which may be applied eventually to the study of the ecology of marine benthos.

Despite a recent pulse of theoretical interest in food webs (Critchlow and Stearns, 1982; Pimm, 1982; Pimm, 1984), we conclude that community food-web studies, by themselves, do not divulge much about the processes controlling the abundances of infaunal populations. Paine (1980) reached a similar conclusion from his analysis of rocky intertidal food webs. Food webs, including those using serological assays, can be used to test only a few predictions from ecological theory. Cohen's (1978) observation that most food webs can be represented as interval graphs can be falsified using either a community food web or a sink food web, which is a subset of the entire community food web. Source food webs cannot be used to test Cohen's (1978) prediction. Furthermore, the connection between Cohen's (1978) prediction and an ecological theory which attempts to explain the distribution and abundance of animals is not readily apparent.

Optimal foraging theory, reviewed by Pyke *et al.* (1977), makes explicit predictions about the types and numbers of prey that should be eaten by a predator. As with Cohen's (1978) prediction, either community or sink food webs can be used to test predictions but source food webs cannot. Thus, the source food webs most easily revealed using serological assays, since only one antiserum is required, are not particularly useful for hypothesis testing.

Many simple hypotheses can be tested with serological assays. For example, Feller *et al.* (1979) document that meiofauna are often found in the guts of both surface-deposit feeders and epifaunal predators. These data refute McIntyre's (1969) prediction that the macrofauna do not feed on meiofaunal populations. Unfortunately, serological assays by themselves do not reveal whether predation alone determines the local abundances of meiofauna or even whether this trophic pathway represents a significant flow of energy from the meiofauna to the macrofauna. Food webs cannot show the strong competitive interactions which exist between prey species (Gallagher and Goodin, in preparation; Paine, 1980) nor can non-quantitative food webs weigh the relative importance of different trophic pathways in prey population dynamics. For example, *E. confervicolus*, an epifaunal gammarid amphipod in the Skagit community, consumes *Hobsonia florida*, harpacticoid copepods, *Manayunkia aestuarina* and tanaids. However, *E. confervicolus* can have rather unexpected effects on these prey populations. Manipulative field experiments reveal that *E confervicolus* reduces the abundances of harpacticoids in the spring (Gallagher *et al.*, in preparation). Surprisingly, the presence of *E. confervicolus* increases the survival of harpacticoid copepods in the fall because *E. confervicolus* inhibits the feeding activity of the large surface deposit feeders (e.g., *Pseudopolydora kempi japonica* and *Corophium salmonis*) which crop the harpacticoid copepod populations. These large surface deposit feeders are not abundant in the spring. Thus, despite the fact the *E. confervicolus* feeds on harpacticoid copepods in both fall and spring, the effect of this predator on the harpacticoid copepod populations differs seasonally.

Detailed studies of trophic interactions are most useful in addressing questions of community structure if predation has already been shown to be an important process controlling community structure. Detailed serological analyses on the importance of predation in a community should be conducted after, or concurrently with, rigorous tests of the other processes likely to affect benthic community structure. Serology is such a labor-intensive activity, that often relatively simple alternated hypotheses of observed patterns of community structure, such as competition for food, are left untested.

In the Skagit flats community, for example, predation on larvae and juveniles is not as important as habitat selection (Eckman, 1983, Gallagher *et al.*, 1983) and competition (Gallagher, 1983; Gallagher and Goodin, in preparation) in determining the local abundances of recently settled juveniles.

Once produced, monospecific antisera have several advantages over polyspecific antisera. They eliminate the extensive cross-reactions which confound assays using polyspecific, polyclonal, cross-reacting antisera (Feller and Gallagher, 1982). Cross-reactions are largely eliminated during the isolation of the unique antigen to produce the monospecific antiserum. Cross-reactions must be checked and re-checked continually if polyspecific, cross-reactive antisera are used for serological assays. Monospecific antisera can be used in the many sensitive and quantitative serological assays that require antisera which are not cross-reactive.

Monospecific antisera can produce detailed source food webs, documenting trophic links that are undetectable using other methods. In the Skagit community, a nemertean preys on *C. salmonis* by stunning the amphipod with its stylet and scouring the inside of the exoskeleton with its proboscis (E.D.G., personal observations). The nemertean does not ingest the exoskeleton of *C. salmonis*. Serological assays of extracts of these nemerteans disclose the presence of *C. salmonis* Peak VI antigen. Visual gut content analysis could not detect this trophic link.

Monospecific antisera have many marine ecological applications other than the study of predation. Serological assays, for example, can be used to estimate the abundances of marine ammonium-oxidizing bacteria in the water column (Ward and Perry, 1980; Ward, 1982). Serological assays using monospecific antisera may also prove useful in the systematics of some soft-bottom benthic taxa. The soft-bottom benthic polychaete formerly known as *Capitella capitata* has been shown to be a complex of at least six sibling species (Grassle and Grassle, 1974), distinguished by unique allozymes and reproductive modes (Grassle and Grassle, 1977). We have found that the most ubiquitous of these sibling species possesses antigens not found in the other sibling species (Gallagher and Grassle, in preparation). Monospecific antisera, produced to these unique antigens, could be used to quickly "key out" these sibling species.

Recently, Gallagher (with J.P.Grassle & D.D. Trueblood) adapted the techniques described in this paper to develop immunofluorescent assays to identify the larval dispersal stages of marine benthic invertebrates. Our ignorance regarding the factors controlling the distribution and abundance of marine invertebrate larvae has been one of the major gaps in our understanding the processes controlling benthic community structure. Serology offers a powerful tool for filling this lacuna in our knowledge.

The major impediments to the application of serology to benthic ecology and marine science are disappearing in the rush of technological advances occurring in the field of molecular immunology. A major problem has been obtaining sufficient quantities of pure antigen to develop a monospecific antiserum. The techniques described in this paper can be used to obtain a pure antigen from a mixture of antigens, but these techniques require two separate immunizations. Using crude antigen extracts and only a single immunization, the hybridoma technique (Köhler and Milstein, 1975) produces antisera which are not only monospecific but are monoclonal. All of the antisera described in this paper are polyclonal, they react with a variety of antigenic determinants on a molecle. Monoclonal antisera react with only a single antigenic determinant (epitope) on the molecule. Most of our research effort is devoted to producing species-specific serological assays. We regard the production of monoclonal antibodies much like playing high-stakes poker. The potential is high, but so are the initial costs and odds of obtaining monoclonal antibodies which are non-cross-reactive with phylogenetically related taxa. Biological oceanographers should carefully consider Lane and Koprowski's (1982, p. 200) arguments on the relative merits of polyclonal and monoclonal antisera before choosing to produce only monoclonal antibodies:

> *"A conventional antiserum raised in a rabbit against bovine serum albumin...contains a large number of different antibodies directed against discrete determinants or epitopes on the surface of the bovine albumin molecule. Some of these epitopes will also be present on a related molecule such as human serum albumin, and the two antigens will thus be cross-reactive. Other epitopes, on the other hand, will be present only on one or other albumin molecule. By appropriate absorption experiments polyclonal antisera can be used to distinguish the cross-reactive and the non-cross-reactive epitopes and are thus powerful tools for such purposes as determining phylogenetic relationships between species. Evidently a single monoclonal antibody will never yield this kind of information; and so complex are conventional antibody responses that it would be extremely difficult to mimic them with artificial cocktails' of monoclonal antibodies."*

There have been revolutionary advances in the serological techniques available to ecologists, but these elegant techniques will never eliminate the need for controlled field experiments. Without the field-manipulative experiments (Gallagher *et al.*, 1983; Gallagher and Goodin, in preparation), serologically detected prey in the guts of surface-deposit feeders (Feller *et al.*, 1979) and laboratory feeding studies might have led us to confirm the hypothesis that predation on recruits controls succession.

ACKNOWLEDGEMENTS

This work was supported by NSF Grant No. OCE 78-24687 to P. A. Jumars and NSF Grant No. OCE 83-00126 to J.P. Grassle & E.D. Gallagher. The text was greatly improved by comments from M.R. Landry, D.C. Miller, A. Schoener and R.F.L. Self.

REFERENCES

Alexander, A.G. and G.E. Kenny. 1980. Characterization of the strain-specific and common surface antigens of *Mycoplasma arginini*. Infection and Immunity 29: 442-451.

Axelsen, N.H. and E. Bock. 1972. Identification and quantification of antigens and antibodies by means of quantitative immunoelectrophoresis. A survey of methods. Journal of Immunological Methods. 1: 109-121.

Boreham, P.F.L. and C.E. Ohiagu. 1978. The use of serology in evaluating invertebrate prey-predator relationships: a review. Bulletin of Entomological Research. 68: 171-194.

Caldwell, H.D., C.C. Kuo and G.E. Kenny. 1975a. Antigenic analysis of Chlamydiae by two-dimensional immunoelectrophoresis I. Antigenic heterogeneity between *C. trachomatis* and *C. psittacti*. The Journal of Immunology 115: 963-968.

Caldwell, H.D., C.C. Kuo and G.E. Kenny. 1975b. Antigenic analysis of Chlamydiae by two-dimensional immunoelectrophoresis. II. A Trachoma-LGV specific antigen. The Journal of Immunology. 115: 969-975.

Critctlow, R.E. and S.C. Stearns. 1982. The structure of food webs. American Naturalist. 120: 478-479.

Cohen, J.E. 1978. Food webs and niche space. Princeton University Press, Princeton.

Crowle, A.J. 1973. Immunodiffusion, 2nd Edition, p. 1-373. Academic Press Inc., New York.

Crowle, A.J., G.J. Revis and K. Jarrett. 1972. Preparatory electroimmunodiffusion for making precipitins to selected native antigens. Immunological Communications. 1: 325-336.

Dagg, M.J. 1974. Loss of prey body contents during feeding by an aquatic predator. Ecology. 55: 903-906.

Davies, R.W. 1969. The production of antisera for detecting specific triclad antigens in the gut contents of predators. Oikos. 20: 248-260.

Davies R.W., F.J. Wrona and R.P. Everett. 1978. A serological study of prey selection by *Nephelopsis obscura* Verrill (Hirudinoidea). Canadian Journal of Zoology. 56: 587-591.

Davies, R.W., F.J. Wrona, L. Linton and J. Wilkialis. 1981. Inter- and intra specific analyses of the food niches of two sympatric species of *Eprobellidae* (Hirudinoidea) in Alberta, Canada. Oikos. 37: 105-111.

Eckman, J.E. 1983. Hydrodynamic processes affecting benthic recruitment. Limnology and Oceanography. 28: 241-257.

Feller, R. J. 1984a. Dietary immunoassay of *Ilyanassa obsoleta*, the Eastern mud snail. Biological Bulletin. 166: 96-102.

Feller, R. J. 1984b. Serological tracers of meiofaunal food webs. Hydrobiologia. 118: 119-125.

Feller, R.J., G.L. Taghon, E.D. Gallagher, G.E. Kenny and P.A. Jumars. 1979. Immunological methods for food web analysis in a soft-bottom benthic community. Marine Biology. 54: 61-74.

Feller, R.J. and E.D. Gallagher. 1982. Antigenic similarities among estuarine soft-bottom benthic taxa. Oecologia. 52: 305-310.

Feller, R.J., G. Zagursky and E. A. Day. 1985. Deep-sea food web analysis using cross-reacting antisera. Deep-Sea Research. 32: 485-497.

Fichter, B.L. and W.P. Stephen. 1981. Time related decay in prey antigens ingested by the predator *Podius maculiventris* (Hemiptera, Pentatomidae) as detected by ELISA. Oecologia. 51: 404-407.

Gallagher, E.D. 1983. The mechanisms of benthic succession. Ph.D. Dissertation, University of Washington, Seattle, WA. 263 pp.

Gallagher, E.D., P.A. Jumars and P. Goodin. Competition among the pioneers in soft-bottom benthic succession. In preparation.

Gallagher, E.D., P.A. Jumars and D.D. Trueblood. 1983. Facilitation of soft-bottom benthic succession by tube-builders. Ecology. 64: 1200-1216.

Grassle, J.F. and J.P. Grassle. 1974. Opportunistic life histories and genetic systems in marine benthic polychaetes. Journal of Marine Research 32: 253-284.

Grassle, J.R. and J.P. Grassle. 1977. Temporal adaptation in sibling species of *Capitella*. IN: Ecology of marine benthos, B.C. Coull, Ed., University of South Carolina, Columbia, South Carolina. pp. 177-189.

Kenny, G.E. and H.M. Foy. 1975. Detection and quantitation of circulating polysaccharide in pneumococcal pneumonia by immunoelectroosmophoresis (counterelectrophoresis) and rocket electrophoresis. IN: Microbiology 1975, D. Schlessinger, Ed. American Society for Microbiology, Washington, D.C.pp. 92-102.

Köhler, G. and C. Milstein. 1975. Continuous cultures of fused cells secreting antibody of predefined specificity. Nature. 256: 495-497.

Krøll, J. 1973a. Line immunoelectrophoresis. Scandinavian Journal of Immunology. 2 (supplement no. 1): 61-67.

Krøll, J. 1973b. Rocket-line immunoelectrophoresis. IN: Scandinavian Journal of Immunology. 2 (Supplement no. 1): pp. 83-87.

Krøll, J. 1981a. Quantitation of antigens by densitometric scanning of immunoelectrophoretic precipitates. Methods in Enzymology. 73: 370-383.

Krøll, J. 1981b. Production of specific antisera by immunization with precipitin lines. Methods in Enzymology. 73: 52-57.

Lane, D. and H. Koprowski. 1982. Molecular recognition and the future of monoclonal antibodies. Nature. 296: 200-202.

Laurell, C-B. and E.J. McKay. 1981. Electroimmunoassay. Methods in Enzymology. 73: 339-370.

McIntyre, A.D. 1969. Ecology of marine meiobenthos. Biological Reviews 44: 245-290.

Ouchterlony, O. 1968. Diffusion-in-gel methods for immunological analysis. Ann Arbor Science Publishers, Ann Arbor, Michigan.

Oudin, J. 1980. Immunochemical analysis by antigen-antibody precipitation in gels. Methods in Enzymology. 70: 166-198.

Paine, R.T. 1980. Food webs: Linkage, interaction strength and community infrastructure. Journal of Animal Ecology. 49: 667-685.

Peterson, C.H. and B.P. Bradley. 1978. Estimating the diet of a sluggish predator from field observations. Journal of the Fisheries Research Board Canada 35: 136-141.

Pimm, S.L. 1982. Food webs. Chapman and Hall, London.

Pimm, S.L. 1984. Food chains and return times. IN: Ecological communities: conceptual issues and the evidence, D.R. Strong, D. Simberloff, L.G. Abele, and A.B. Thistle, (eds.). Princeton University Press, Princeton, N.J. pp. 397-412.

Pyke, G.H., H.R. Pulliam and E.L. Charnov. 1977. Optimal foraging: a selective review of theory and tests. Quarterly Review of Biology 51: 137-154.

Thirkill, C.E. and G.E. Kenny. 1974. Serological comparison of five arginine-utilizing *Mycoplasma* species by two-dimensional immunoelectrophoresis. Infection and Immunity. 10: 624-632.

Thirkill, C.E. and G.E. Kenny. 1975. Antigenic analysis of three strains of *Mycoplasma arginini* by two-dimensional immunoelectrophoresis. The Journal of Immunology. 114: 1107-1111.

Vaitukaitis, J.L. 1981. Production of antisera with small doses of immunogen: multiple intradermal injections. Methods in Enzymology. 73: 46-52.

Vaitukaitis, J., J.B. Robbins, E. Nieschlag and G.T. Ross. 1971. A method for producing specific antisera with small doses of immunogen. Journal of Clinical Endocrinology. 33: 988-991.

Ward, B.B. 1982. Oceanic distribution of ammonium-oxidizing bacteria determined by immunofluorescent assay. Journal of Marine Research 40: 1155-1172.

Ward, B.B. and M.J. Perry. 1980. Immunofluorescent assay for the marine ammonium-oxidizing bacterium *Nitrosococcus oceanus*. Applied and Environmental Microbiology 39: 913-918.

PREPARATION AND USE OF MONOSPECIFIC ANTISERA TO *CAPITELLA* SP. I.

Eugene D. Gallagher
Environmental Sciences Program
University of Massachusetts, Boston
Boston, MA 02125

J.P. Grassle
Marine Biological Laboratory
Woods Hole, MA 02543

D. D. Trueblood
Environmental Sciences Program
University of Massachusetts, Boston
Boston, MA 02125

M.G. Gleason
Department of Zoology
University of California
Berkeley, CA 94720

ABSTRACT

The opportunistic polychaetes formerly known as *Capitella capitata* are the archetypal pioneers in soft-bottom benthic succession. *Capitella* is now known to be comprised of several morphologically indistinguishable but reproductively isolated sibling species. *Capitella* sp. I is widely distributed on the east coast of the United States and has been used in a number of experimental and toxicological studies. We used three immunoelectrophoretic techniques (i.e., 2-dimensional, Krøll's line, and Laurell's rocket electrophoresis) to identify and purify a dominant antigen in this species. To study the role of *Capitella* species in field experiments, we developed a monospecific antiserum to this major antigen. We plan to use this monospecific typing serum and FITC-immunocytochemical staining to identify the formalin-preserved larvae and juveniles of *Capitella* sp. I. This immunocytochemical staining technique will enable us to test processes controlling the distribution and abundance of *Capitella* sp. I in the field.

CELL AFFINITY CHROMATOGRAPHY

FOR A MARINE NITRIFYING BACTERIUM

U. Heyman
Institute of Limnology
University of Uppsala
Uppsala, Sweden

B. Heyman
Department of Immunology
Uppsala University Biomedical Center
Uppsala, Sweden

Bess B. Ward
Institute of Marine Resources
Scripps Institution of Oceanography
University of California, San Diego
La Jolla, CA 92093

INTRODUCTION

Difficulty in the identification and separation of species of bacteria remains a major obstacle to the understanding of microbial processes in aquatic environments. A routine, widely applicable method for obtaining total cell counts by means of acridine orange or other DNA-staining fluorochromes has been available for several years (Hobbie *et al.*, 1977; Porter and Feig, 1980; Paul, 1982) and recently the use of immunofluorescent staining has made counts on separate species possible (Ward and Perry, 1980; Campbell *et al.*, 1983; Dahle and Laake, 1982: Ward and Carlucci, 1985). Species specific antibodies can also be used in radioimmunoassay to enumerate bacteria (Benbough and Martin, 1976). Although these methods give reasonable estimates of abundance of cells of a particular strain, methods used to estimate metabolic activity of cells of individual strains are still inadequate. Microautoradiography (Fliermans and Schmidt, 1975: Ward, 1984) and ETS-activity (Baker and Mills, 1982) in combination with immunofluorescence have been applied with some success.

Affinity chromatography and flow cytometry both represent technical advances which may provide solutions to the above obstacles. Herein, we describe an immunosorbent chromatographic method which should be applicable to many experimental approaches with diverse microorganisms, including bacteria, phytoplankton and microflagellates. This method is designed to separate individual cells of a unique subpopulation from all other cells in an environmental sample, to yield a pure, or highly enriched, collection of cells of one particular type. The purified cells are then ready for further experimentation, or analysis of experimental manipulations carried out prior to the separation. For example, species specific parameters such as cellular composition (DNA, RNA, protein, particular enzymes) and metabolic activity with regard to radiolabelled precursors or substrates, could be determined independently on a defined subset of the natural population. We have used the method for marine bacteria, but the results have implications for other cells in the size range up to tens of microns.

The separation method we describe is based on the bonding between antibody and antigen, which is usually strong (Davis *et al.*, 1973) and is capable of binding a cell to a solid surface while liquid flows by. This specific binding property has been exploited in the use of immunosorbent columns (cell affinity chromatography) by which different kinds of lymphocytes have been separated from each other (Chess *et al.*, 1974; Manderino *et al.*, 1978; Duarte *et al.*, 1982). The same principle can be used to separate cells of different bacterial species. Antibodies (specific to one strain of bacteria) are immobilized on the solid phase of a gel filtration column. Cells of the homologous species passing through the column stick to the bound antibodies; the cells can subsequently be eluted from the column and examined, counted, concentrated or used for measurement of other species specific parameters (see above). Incubation of a sample with isotopically labelled substrates or metabolic precursors prior to the separation would allow measures of the metabolic activity of the species, independently of all other microorganisms present in the population. Lymphocytes remain viable after separation on affinity columns and can be used for further experiments (Chess *et al.*, 1974; Manderino *et al.*, 1978). If this were possible for separated bacteria, the potential for diverse applications of the separation approach is increased, although many applications are possible with preserved cells.

In the present work, we have tried to separate *Nitrosomonas* sp. (marine) by passing cells over columns to which anti-*Nitrosomonas* antibodies had been bound. Our purpose was to evaluate the potential of cell affinity chromatography in aquatic microbial research, to roughly characterize the properties of the columns and to elucidate the important variables in the necessary experimental procedures.

METHODS

Antibodies

Immunoglobulin G (IgG) antibodies in serum from rabbits immunized with *Nitrosomonas* sp. (marine) (Ward, 1982) were purified on protein A-sepharose (Miller and Stone, 1978). After elution from the protein A column, the antibodies were concentrated and dialyzed against phosphate buffered saline (PBS; pH 8.0) by ultrafiltration. Protein concentration was determined from the absorbance at 280 nm, assuming 1 mg ml^{-1} of antibody had an absorbance of 1.5 (10 mm cuvettes) (Hanson and Phillips, 1981).

Preparation of immunosorbent columns

The standard columns were prepared from cyanogenbromide activated Sepharose 6MB macrobeads (Pharmacia trademark) but we also evaluated the use of Sepahrose 4B and glass beads. The volume of the macrobead and glass bead columns was 5 ml and for the smaller bead size Sepahrose 4B, 1 to 2 ml. The Sepharose beads were swelled, rinsed and coated with antibody according to the instructions provided by the manufacturer. Two concentrations of antibody solutions (4 mg ml^{-1} and 1 mg ml^{-1}) were used to obtain two different densities of antibody coating on the columns (high and low). Remaining binding sites on the beads were inactivated by incubation with 0.2 M glycine. This procedure resulted in the covalent binding of IgG antibodies to the Sepharose beads; thus the columns could be used repeatedly. Columns were stored at 4°C with sodium azide to inhibit degradation of the antibodies.

The glass beads (200-300 µm diameter) were prepared for use as antibody supports by acid washing (1:1 concentrated $HCl:HNO_3$), followed by thorough rinsing with deionized water and PBS (Wigzell, 1976). The beads were coated with antibodies by incubating in purified IgG overnight at 4°C. Nonspecific adsorption (by bacteria to the beads) was inhibited by coating the beads with normal serum by incubation for 1 hr at 4°C following the incubation with IgG. In the case of glass beads, the antibodies simply adsorb onto the beads rather than binding permanently, and subsequent elution of attached cells also resulted in desorption of the antibodies. These columns thus had to be recoated for each separation, requiring the destructive use of much antiserum.

Bacterial Cultures

Scintillation counting of small volumes was much easier and faster than cell counting by acridine orange or immunofluorescence so radioactively labelled cells were used routinely to allow rapid, reliable quantification of the contents of fractions eluted from the columns. However, we were able to convert the scintillation counts to cell counts by knowing the cell concentration of the cell suspension added to the columns, and computing dpm per cell.

Nitrosomonas sp. (marine) was grown in W/2 medium (Ward, 1982) in a stirred, aerated (sterile air) 4-liter carboy in the dark at 18°C. Cells were grown in the presence of 0.13 µCi ml^{-1} $NaH^{14}CO_3$ for 3 weeks with daily pH adjustment. When the cell density reached 1.5×10^5 cells ml^{-1}, and the culture was in exponential phase, it was killed by adding formalin (final concentration 2% v/v). The preserved cells were washed in several columns of PBS and resuspended in PBS or PBS + 0.1% gelatin (see below) immediately before use, and used at a final concentration of 1 to 10 times over the original culture concentration. Activity per cell in the resuspended culture was initially 0.011 dpm per cell, and declined with time. Dpm per cell was reassayed regularly.

A pure culture of an unidentified marine heterotrophic bacterium was used as a nonspecific control. The culture, NB3, isolated as a heterotrophic contaminant from an enrichment of a marine nitrite-oxidizing bacterium, was chosen because it is an elliptical rod of dimensions similar to *Nitrosomonas* (0.5 x 1.2 µm) when both are grown under their respective optimal culture conditions. Two hundred and fifty ml of CP medium (Carlucci and Pramer, 1957) were inoculated from a one day old tube of NB3, and a tritiated amino acid mixture was added (final activity, 0.36 µCi/ml). The culture was preserved 6 days later at which time it was turbid (3.5×10^7 cells ml^{-1}) and sufficiently labelled. Preserved cells were washed and resuspended in PBS as described above, resulting in a dilution of 1 to 5 times the original culture concentration. Activity per cell in the resuspended heterotrophic cell suspension was initially 0.001 dpm per cell; activity per cell was reassayed at the time of each experiment and had decreased to half the original level in 6 months.

Chromatography Procedure

Several variations on the standard procedure, described here, were performed to test the variables influencing column performance, and are described below with results. Regulation of flow rate through the columns was obtained by connecting the outflow to a peristaltic pump (Harvard Apparatus Co.). After thorough rinsing with PBS (pH 8.0), the columns were coated with gelatin (Difco; 1% in PBS, pH 8.0) to minimize the nonspecific attachment of cells. Coating was performed by recirculating the gelatin solution from the pump to the top of the column at a slow (0.3 ml min^{-1}) flow rate for 30 min. Excess gelatin solution was drained off the column and 1 to 5 ml of cell suspension were added to the top of the column. The cells were allowed to recirculate in a similar manner at 0.3 ml min^{-1} for 30 min. Unattached cells were removed by rinsing with 100 ml of PBS at 4 ml min^{-1}. Cells remaining on the column at this point were considered to be antibody-bound. Attached cells were eluted by adding citrate buffer (pH 3.0) or 2 M $MgCl_2$ (5 ml flowed through the column and another 10 ml retained in the column) and turning the column end over end 20 times. Elution was accomplished with three such end over end washes and the last eluted cells were removed by rinsing with 20 ml PBS. A last PBS rinse with 1% NaN_3 was added to the column for storage. If the standard procedure, or various permutations thereof used during test procedures, resulted in less than 100% recovery of cells originally added during the experiment, elution procedures were repeated until no more counts could be recovered before the columns were used in further experiments.

One to five milliliters of each fraction eluted from the column was dissolved in 10 ml of Betaphase (WestChem) and counted on a Beckman LS 100C Scintillation counter. Counting efficiency was determined using an external standard ratio quench curve. Cells eluted from the column were inspected, and sometimes counted, by epifluorescent microscopy using immunofluorescent (Ward, 1982) or acridine orange (Hobbie *et al.*, 1977) staining procedures.

RESULTS

Tests with pure cultures

When *Nitrosomonas* sp. (marine) was added to a column, a reproducible fraction (33%; standard procedure, Table 1) of the cells attached to the antibody coated beads and could not be removed by simple PBS rinses. Most of the attached cells could be eluted with

Table 1. Standard column procedures. Percent recovery of bound cells = (number of cells eluted/number of cells bound) x 100; percent recovery of total cells added = total cells recovered in rinses plus elution as a percent of cells originally added. 1 thru 3: Comparison of results from high (4 mg ml^{-1}) and low (1 mg 1 ml^{-1}) antibody density columns under various elution conditions. 4: Effect of using seawater instead of phosphate buffer to suspend cells. 5: Heterotroph control.

Cell Affinity Chromatography. Standard Procedure

Experimental Conditions	Antibody Density (mg ml^{-1})	Cells added (10^6 ml^{-1})	Percent bound	Percent eluted	Percent Recovery of Bound cells	Percent Recovery of total cells added
1 suspension in PBS elution in PBS	1	9.5	22.1	15.7	71.0	93.6
+ mixing	4	9.5	38.1	24.9	65.4	86.8
2 suspension in PBS elution in citrate	1	8.0	57.8	44.1	76.3	86.6
+ mixing	4	8.0	33.3	23.4	70.3	90.4
3 suspension in PBS elution in 2M $MgCl_2$	1	7.8	34.5	33.0	95.7	98.6
+ mixing	4	7.8	28.3	24.9	88.0	96.6
Average of 1 through 3	1	9.1	38.1	30.9	81.1	92.9
	4	9.1	33.2	24.4	74.5	91.3
4 suspension in seawater elution in citrate	1	9.6	18.6	9.8	52.7	91.1
+ mixing	4	9.6	34.3	16.8	49.0	82.5
5 heterotroph control suspension in PBS elution in citrate + mixing	4	14.0	0.2	0.2	100.0	100.0

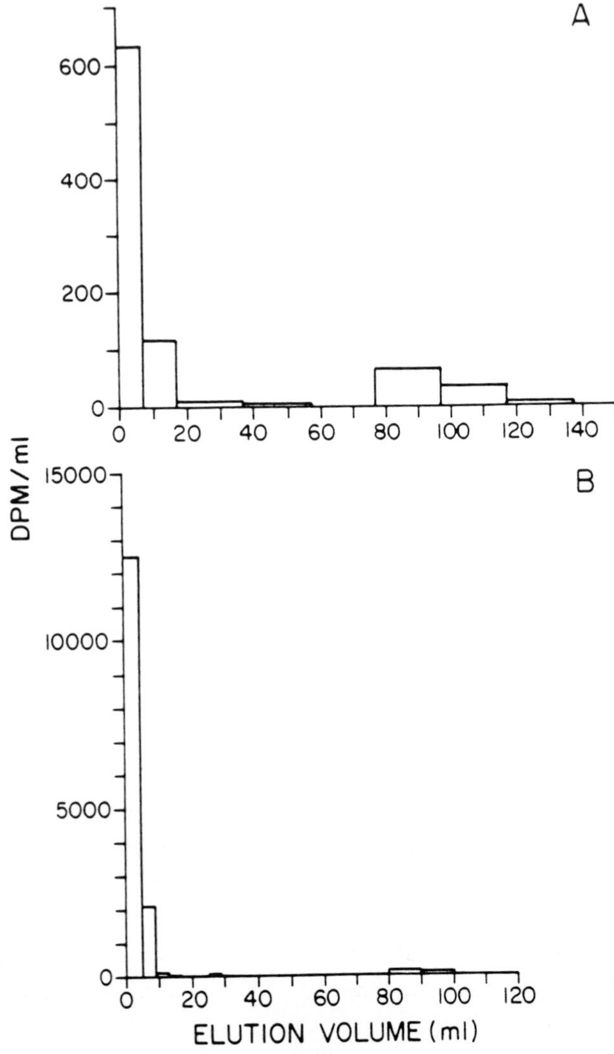

Figure 1. A. Results of homologous cell binding experiment with standard procedure. $^{14}CO_2$-labelled *Nitrosomonas* cells suspended in PBS added, recirculated 30 min., eluted with $MgCl_2$ and end over end mixing. B. Results of control experiment with standard procedure. ^{3}H-amino acid labelled heterotroph cells suspended in PBS added, recirculated 30 min., eluted with citrate and end over end mixing.

citrate buffer or $MgCl_2$ and end over end mixing of the column (Figure 1A) Cells eluted from the column intact and could be visualized by immunofluorescence or acridine orange staining. Eluted cells appeared indistinguishable from cells sampled directly from a culture.

The same procedure was used in a control experiment using the labelled culture of heterotrophic bacteria. Eighty-four percent of the counts were recovered in the first 9 ml of PBS rinse and >99% were recovered in the total 100 ml PBS rinses (Figure 1B). A very small amount of activity (0.21% of the total added) was recovered in the elution fractions, indicating a minor amount of nonspecific attachment of cells (Table 1).

Separation of cell mixture

A mixture of labelled NB3 cells (tritium labelled) and *Nitrosomonas* cells (^{14}C labelled) was run using the standard procedure described above, but using a column with slightly lower antibody density (0.8 mg ml^{-1}). Capture efficiency of this column for *Nitrosomonas* was lower than usual: 10 to 13% of added counts bound and could be eluted by the standard protocol. In the experiment shown (Figure 2), 10% of the added *Nitrosomonas* cells bound, while 2% of the heterotroph cells in the mixture also bound. *Nitrosomonas* constituted 22.3% of the cell mixture added to the column and was preferentially retained, so that despite the low binding efficiency, the eluted cell mixture was enriched relative to the original mixture and contained 48% *Nitrosomonas* cells (Table 2). Factors responsible for low binding efficiency are discussed below.

Table 2. Results of Mixed Cell Experiment

	Cell concentration (10^6 ml^{-1})		
	NB3	*Nitrosomonas*	Percent *Nitrosomonas*
Initial mixture	4.09	1.21	22.3
Eluted mixture	0.097	0.09	48.1

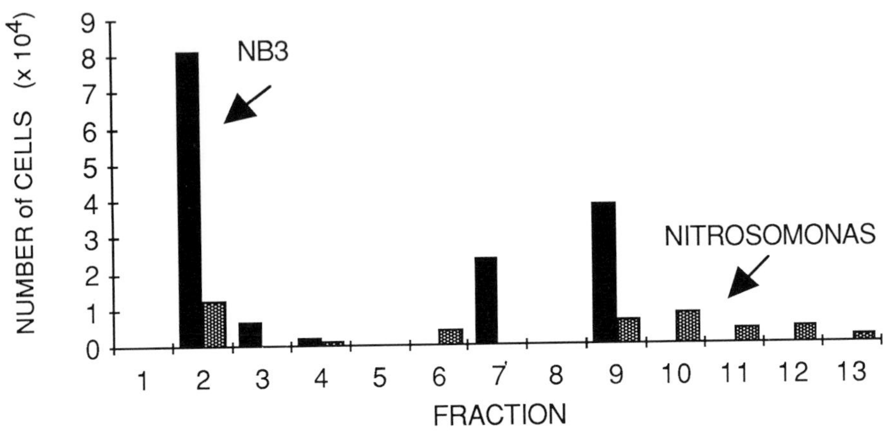

Figure 2. Separation of cell mixture on column of antibody density 0.8 mg ml^{-1} protein using standard procedure. Initial cell mixture was 4.09 x 10^6 cell ml^{-1} NB3 and 1.21 x 10^6 cell ml^{-1} *Nitrosomonas* (22.8% *Nitrosomonas*). Bound cells (9.8% of added *Nitrosomonas* cells) were eluted in fractions 9-13, which contained 48.1% *Nitrosomonas*.

Optimization of protocol

The degree of attachment of cells was dependent on the length of time the cells were in contact with the column during the cell addition step prior to rinsing. A constant flow (through recirculation) resulted in more attached cells than one slow passage (Table 3A). Longer contact times (>30 min) or slower recirculation speed (0.2 ml min^{-1}) resulted in the attachment of a larger fraction of the added cells and, more importantly, attached cells stuck more firmly so that the normal elution procedure was insufficient to remove all attached cells (Table 3A).

Simply changing the pH (from 8.0 in PBS to 3.0 in citrate) was not sufficient to cause elution of the antibody-bound cells; some sort of mechanical mixing of the columns was necessary. Indeed, 60% of the bound cells could be loosened with no pH change by turning the columns end over end with PBS (Table 1). Most complete elution was obtained with either citrate buffer (pH 3.0) or 2 M $MgCl_2$ combined with end over end mixing of the columns. The possibility of eluting

Table 3. Effects of column variables on cell binding and elution efficiency. See Table 1 caption for definitions. A. Effect of varying contact time (all experiments with sepharose 6MB; antibody density 4 mg ml^{-1}). B. Use of serum to control nonspecific binding.

Results of varying experimental conditions

A. Varying contact time and flow rate during cell addition

Contact time (min)	Flow rate (ml min^{-1})	Recirculation	Cells added (10^6 ml^{-1})	Percent bound	Percent eluted	Percent Recovery of bound cells	Percent Recovery of total cells added
30	gravity	-	4.9	19.4	11.2	57.7	91.8
30	0.5	+	11.1	11.2	11.2	100.0	104.1
30	0.2	+	9.3	35.6	17.2	48.3	81.7
30	0.2	+ (wide tubing)	9.8	41.9	27.4	65.4	85.5
60	0.5	+	46.0	36.6	24.4	66.7	87.9
60	0.5	+	1.2	60.9	36.7	60.3	75.8

B. Use of serum to reduce nonspecific binding

glass beads without antibody	with serum	99.5
	without serum	43.0
Sepharose-4 without antibody	without serum	73.0
Sepharose-4 with antibody (1 mg ml^{-1})	without serum	84.0
Sepharose-4 with antibody (4 mg ml^{-1})	without serum	79.0
Sepharose-4 without antibody	with serum	84.9
Sepharose-4 with antibody (4 mg ml^{-1})	with serum	83.2

cells by means of mixing alone implies that the flow rate of the PBS rinses must be kept fairly low and smooth. The effect of physical forces on the elution of the columns was also evident when the columns accidentally went dry and activity in the following fraction increased (data now shown).

The capacity of the columns was high and some 10^5 to 10^6 cells could easily be attached to a 5 ml column of Sepharose 6MB beads. We could not detect any increased relative attachment with decreasing numbers of added cells and, in order to avoid high counts of uneluted cells, did not try to find the maximum capacity of the columns. In the standard experiments, the columns were coated with gelatin or normal serum prior to the addition of cells (gelatin was preferred due to its lower cost). This procedure had a marked effect in decreasing the nonspecific attachment of cells, both in glass bead and sepharose columns (results for serum in Table 3B). The inclusion of gelatin in the PBS used to prepare cell suspensions and to rinse the column after cell addition had the same effect of "lubricating" the columns.

The density of antibodies on the beads affected the performance of the columns; a high density did not always result in more cells attaching, but it did cause firmer bonding of cells to the beads, which made the elution more difficult (compare high and low columns in Table 1).

When filtered seawater was used as buffer instead of PBS, the binding efficiency of the columns was reduced (Table 1). This result is expected since the higher salinity tends to weaken the antigen-antibody complex (cf. the use of $MgCl_2$ for elution).

The use of smaller sepharose beads (Sepharose 4B) as column material resulted in columns that were harder to rinse and also harder to elute. Nonspecific attachment of cells was also more troublesome with smaller beads. Nonspecific attachment was negligible with glass beads precoated with normal serum, but the binding efficiency of glass bead columns was very low. This effect was probably due to the fact that adsorption alone could not produce a high enough density of antibodies on the bead surfaces to bind cells (data not shown).

The sepharose columns were used many times over a period of several months; the efficiency decreased eventually and finally new columns had to be made. We have no direct data on the absolute lifetime of a column under conditions of routine use and storage; our observations are that the columns are fairly sturdy and capable of withstanding long periods at room temperature under relatively unfavorable conditions (low pH, high salinity).

DISCUSSION

IgG coated sepharose macrobead columns used in our standard procedure bound about 2×10^5 *Nitrosomonas* cells per ml of column volume. In a typical experiment, 60 to 70% of the added cells ran through the column and were eluted in the first fractions without binding (first two fractions, Figure 1A). About 30% of added cells bound specifically to the antibodies and could be eluted in later fractions. Although this is a lower than desired capture efficiency for *Nitrosomonas*, it demonstrates preferential binding for the homologous cells. When heterologous cells, NB3, were added to the column, 99% of the added cells eluted immediately without binding.

Preliminary trials separating cell mixtures demonstrated that passage through the column once could enrich the concentration of *Nitrosomonas* in the initial cell mixture by at least two-fold. A greater enrichment might be obtained using a column with higher binding efficiency. Further improvements are necessary before application to natural populations is feasible.

Regardless of the *Nitrosomonas* concentration added, more than half of the cells passed through the column without binding (Table 2A). Since the relative number of attached cells increased by only two-fold when the number of cells added decreased by five-fold, the inefficiency of the selective binding may be due to inadequate presentation of the cells to the antibodies rather than saturation of the binding sites by bacteria. The increased attachment when the contact time was increased also implies orientation of the cells in relation to the antibody-coated surfaces as an important variable influencing attachment. Furthermore, the low density column adsorbed almost as many cells as the high density column, although the difference in antibody density was a factor of 4. In similar experiments with lymphocytes (Pharmacia Fine Chemicals AB) relatively more cells attached, and attachment increased with contact time. Thus it is likely that the smaller size of the bacteria decreases the probability of adequate presentation.

Increasing the contact time (i.e., lengthening the recirculation time during the cell addition step) in an effort to allow more cells to adsorb to the column had the adverse effect of increasing the nonelutable fraction of cells (Table 2A). The number of antigen-antibody reactions per cell increases with increasing contact time, so that attached cells become more firmly bound to the beads with time. Achieving optimum efficiency of the columns requires consideration of antibody density, flow rate and contact time.

The polyclonal antibodies used here probably contained antibodies that differed in affinity as well as the number of binding sites (surface antigens per surface area of the target cell). This diversity of antibodies, and the possibility that the density of cell surface binding sites for different antibodies may differ among cells in an apparently homogeneous population, may also be a factor in the relatively low efficiency of our columns. The use of monoclonal antibodies has the potential to greatly increase the efficiency, since the column matrix could thereby be coated with antibodies of optimal affinity. Also antibody response to column variables would then be uniform.

The columns and standard procedure used here are a compromise among these variables (antibody inhomogeneity, antibody density, flow rate, contact time), which was designed to separate out a small but highly enriched subpopulation of cells (e.g., *Nitrosomonas*) from a mixed population (e.g., a natural marine bacterial assemblage). The lower antibody density performed best for this purpose because it allowed more efficient elution of attached cells. The small size of bacteria seemed to result in a presentation problem but size may also affect the efficiency of elution. As long as a particle bound to a bead in the column extends only within the region of laminar flow of liquid past the bead surface, both the force on the particle and the size of the bound area will be proportional to the square of the radius of the bead (cf. Rouse, 1938). If, however, a particle extends into the region of turbulent flow, the shear will increase considerably, as will the force per unit area bound between particle and bead. In the latter case, elution of particles from the bead surfaces should be much easier; i.e., require lower flow rates. The Reynolds number for the beads when the columns are turned end over end will be well above 0.5 (7-20) and turbulent flow is thus likely (Hutchinson, 1967). Experiments with lymphocytes (Manderino et al., 1978) showed that quantitative elution is much easier for these larger cells than for the bacteria used in the present study.

The cell suspensions added to the columns in this study represent concentrations of *Nitrosomonas* about 1000 times higher than their concentration in natural seawater (Ward and Carlucci, 1985). Current work with cell suspensions at natural concentrations indicates that high concentration may not be necessary for successful separation. Natural samples could be concentrated if necessary by centrifugation or filtration before separation with the column; this might be preferable to direct separation of natural samples because smaller columns (i.e., smaller quantities of antibodies) could be used. Also, column efficiency is reduced in seawater compared to PBS, so the concentration step could be combined with resuspension in PBS.

Nitrosomonas is much less abundant in seawater than many other kinds of microorganisms of ecological interest. Thus, it may be a good test organism, but the use of cell affinity chromatography with many different kinds of cells now seems possible. Efficiency of removal of cells from an unknown mixed population could be estimated by spiking the sample with labelled cells. The purity of the eluted cell fraction can be checked by standard immunofluorescent enumeration and acridine orange counts on the same fraction, as was done in this study. The potential of affinity chromatography to enrich desired cell type in natural samples for autecological study of, for example, nutrient status, cell cycle distribution or metabolic activity with regard to isotopically labelled substrates, is matched only by flow cytometry. Neither flow cytometry nor affinity chromatography (as described here) performs optimally on very small cells. All the results presented here were obtained using the same batch of preserved cells to maintain reproducibility of column results independently of possible variations in cell characteristics. Thus far, our column experiments have used only preserved cells; we do not know if live cells will retain viability upon passage through the column. Because many experimental manipulations can be performed prior to separation, use of preserved samples does not greatly constrain the applicability of the method.

There are two main areas where further developments of the method might result in significant improvements. One is the use of monoclonal antibodies (discussed above). The other area is improvement of the column matrix. Sandstrom and Watz (1984) used a thin plastic tube coated with antibodies against *Fransiscella tularensis*. The unconcentrated sample was allowed to flow through the tube and the number of cells in the column effluent was estimated by an enzyme linked immunosorbent assay (ELISA) technique with an efficiency which allowed detection of 100 cells/ml. In this setup, antibodies were simply adsorbed onto the plastic, and therefore the tube had to be recoated before each experiment. It may be possible to manufacture a long thin tube with antibodies bound covalently to the inside of the tube, since there are several materials that can be cyanogenbromide activated.

ACKNOWLEDGEMENTS

This research was supported by the National Science Foundation (grant no. OCE83-16608 to BBW) and was performed while UH and BH were visitors to the Institute of Marine Resources, University of California at San Diego, and Scripps Clinic and Research Foundation, respectively.

REFERENCES

Baker, K. H. and A.L. Mills. 1982. Determination of the number of respiring *Thiobacillus ferrooxidans* cells in water samples by using combined fluorescent antibody-2-(p-iodophenyl) -3-(p-nitrophenyl)-5- phenyltetrazolium chloride staining. Appl. Environ. Microbiol. 43: 338-344.

Benbough, J. E. and K.L. Martin. 1976. An indirect radiolabelled antibody staining technique for the rapid detection and identification of bacteria. Jour. Appl. Bact. 41: 47-58.

Carlucci, A.F. and D. Pramer. 1957. Factors influencing the plate method for determining abundance of bacteria in sea water. Proc. Soc. Exp. Biol. Med. 96: 392-394.

Campbell, L., E.J. Carpenter and V.J. Iacono. 1983. Identification and enumeration of marine chroococcoid cyanobacteria by immunofluorescence. Appl. Environ. Microbiol. 46: 553-559.

Chess, L., R.P. MacDermott and S. F. Schlossman. 1974. Immunologic functions of isolated human lymphocyte subpopulations. I. Quantitative isolation of human T and B cells and response to mitogens. J. Immunol. 113: 1113-1121.

Dahle, A.B. and M. Laske. 1982. Diversity dynamics of marine bacteria studied by immunofluorescent staining on membrane filters. Appl. Environ. Microbiol. 43: 169-176.

Davis, B.D., R. Dulbecco, H.N. Eisen, H.S. Ginsberg and W.B. Wood. 1973. Microbiology, 2nd ed. Harper and Row Publishers, Inc., Hagerstown, MD.

Duarte, A.J.S., C.B. Carpenter and T.B. Strom. 1982. Expression of T cell differentiation antigens and Ia on rat cytotoxic T lymphocytes. Jour. Immunol. 128: 580

Fliermans, C.B. and E.L. Schmidt. 1975. Autoradiography and immunofluorescence combined for autecological study of single cell activity with *Nitrobacter* as a model system. Appl. Microbiol. 30: 676-684.

Hanson, R.S. and J.A. Phillips. 1981. Chemical composition. IN: Manual of Methods for General Bacteriology. P. Gerhardt, (editor in chief). Amer. Soc. Microbiol., Washington, D.C. pp. 328-364.

Hobbie, J.E., R.J. Daley and S. Jasper. 1977. Use of Nuclepore filters for counting bacteria by fluorescence microscopy. Appli. Environ. Microbiol. 33: 1225-1228.

Hutchinson, G.E. 1967. A Treatise on Limnology. Vol. 2. John Wiley & Sons, Inc.

Manderino, G.L., G.T. Gooch and A.B. Stavitsky. 1978. Preparation, characterization and functions of rabbit lymph node cell populations. I. Preparation of KLH primed T and B memory cells with anti-Fab' affinity columns. Cell. Immunol., 41: 264-275.

Miller, T.J. and H.O Stone. 1978. The rapid isolation of ribonuclease-free immunoglobulin G by protein A-sepharose affinity chromatography. J. Immunol. Methods. 24: 111-125.

Paul, J.H. 1982. Use of Hoechst dyes 33258 and 33342 for enumeration of attached and planktonic bacteria. Appl. Environ. Microbiol. 43: 939-944.

Porter, K.G. and Y.S. Feig. 1980. The use of DAPI for identifying and counting aquatic microflora. Limnol. Oceanogr. 25: 943-948.

Pharmacia Fine Chemicals AB. 1980. Cell Affinity Chromatography: Principles and Methods. Pharmacia Fine Chemicals, Uppsala, Sweden.

Rouse, H. 1938. Fluid mechanics for hydraulic engineers. Dover Publications, Inc. New York.

Sandstrom, G. and H.W. Watz. 1984. The duct ELISA: A new technique to identify low numbers of *Fransiscella tularensis*. FOA Report C 40202-133. National Defense Research Institute, S90182 Umea, Sweden.

Ward, B.B. 1982. Oceanic distribution of ammonium-oxidizing bacteria determined by immunofluorescent assay. Jour. Mar. Res. 40: 1155-1172.

Ward, B.B. 1984. Combined autoradiography and immunofluorescence for estimation of single cell activity by ammonium-oxidizing bacteria. Limnol. Oceanogr. 29: 402-410.

Ward, B.B. and A.F. Carlucci. 1985. Marine ammonia- and nitrite-oxidizing bacteria: Serological diversity determined by immunofluorescence in culture and in the environment. Appl. Environ. Microbiol. 50: 194-201.

Ward, B.B. and M.J. Perry. 1980. Immunofluorescent assay for the marine ammonium-oxidizing bacterium *Nitrosococcus oceanus*. Appl. Environ. Microbiol. 39: 913-918.

Wigzell, H. 1976. Specific affinity fractionation of lymphocytes using glass or plastic bead columns. Scand. Jour. Immunol. 5(suppl. 5): 23-30.

APPLICATION OF IMMUNOBLOTTING FOR DIETARY ANALYSIS

Gregory Zagursky and Robert J. Feller
Department of Biology
Belle W. Baruch Institute for
Marine Biology and Coastal Research
University of South Carolina
Columbia, SC 29208

INTRODUCTION

Information on trophic interactions within a community is important for answering questions concerning energy flow and the effects of predation on community structure. Conventional visual analysis of stomach contents is often complicated by the differential digestion of prey, mastication of prey, ingestion of only fluids from prey, and small sizes of both predators and prey. Serological methods have been utilized to alleviate some of these difficulties. In most cases polyspecific antisera produced against whole-organism extracts of suspected prey items have been utilized to detect prey antigens with immunoprecipitation techniques. Passive immunodiffusion tests have been utilized extensively because of their speed and relative simplicity (e.g., Young, 1980; Adams, 1981; Giles and Phillips, 1985; Feller, 1986). Increased sensitivity and specificity of serological analysis have been achieved by utilizing immunoelectrophoretic techniques (Healy and Cross, 1975; Grisley and Boyle, 1985). These immunoassays have proven adequate in situations for which the diversity of potential prey is low (so most serological cross-reactions can be accounted for) and for predators having a large quantity of stomach content material.

The development of techniques for blotting proteins onto nitrocellulose and enzyme-linked immunosorbent assays (ELISA) has permitted production of antisera for specific prey and detection of low concentrations of prey protein in predators with small gut volumes (Fichter and Stephen, 1984; Theilacker *et al.*, 1986). We are utilizing ELISA techniques to develop monospecific tracers for detecting small prey (e.g., meiofauna, larvae, or particles of vascular plant detritus) in a highly diverse salt marsh community. Our goal is to isolate those prey proteins which are monospecific, immunogenic, and detectable for extended periods in a predator's digestive tract. To test the feasibility of this approach, we have examined the digestive degradation of three different types of prey ingested in the laboratory by grass shrimp, *Palaemonetes pugio*.

MATERIALS AND METHODS

Prey organisms

Artemia salina nauplii were hatched in the lab from cysts (Biomarine, P.O. Box 5, Hawthorne, CA 90250). Cultures were filtered through a 64 µm sieve to retain 24 hr-old individuals and rinsed with distilled water before use.

Thalassiosira weissflogii from the CCMP - Bigelow Laboratory, were cultured in 20 l batch cultures using f/2 medium (Guillard, 1975) and were in the log-phase of growth for feeding experiments.

Spartina alterniflora detritus was prepared from standing dead plant material which was rinsed and incubated at room temperature in f/2 medium for one month. The material was then ground, wet sieved to a size range of 63-250 µm and dried at 55°C for 24 hrs. Detritus for a feeding experiment was soaked in filtered seawater for 24 hrs before use.

Grass shrimp culture

Grass shrimp were collected with a seine from salt marsh creeks in North Inlet, SC, and kept in 75 l aquaria on a diet of all three prey organisms *ad libitum*.

Feeding experiments

Four different feeding experiments were conducted. In three of the experiments single prey items were utilized - brine shrimp (*Artemia*) nauplii, a diatom (*Thalassiosira weissflogii*), and laboratory-prepared marsh cordgrass detritus (*Spartina alterniflora*). The fourth experiment utilized a mixture of all three prey types. In all cases prey were present in sufficient quantities to permit maximum ingestion.

For each experiment groups of 15 grass shrimp (carapace length 11-17 mm) were placed in nylon windowscreen cages which were suspended in 38 l aquaria. The cages facilitated removal of animals during the experiment and reduced the potential for ingestion of feces. The grass shrimp were previously starved for 48 hr at which time unfed shrimp were removed for the control samples and then prey were added to the remaining shrimp. Feeding was permitted for 30 min before the cage was lifted out of the aquarium, rinsed in clean seawater, and placed into a new aquarium without food. At this time the t = 0 sample was taken and individual shrimp were then collected and frozen on dry ice at t = 0.25, 0.5, 1, 2, 4, 6, 8, 12, and 24 hrs. All experiments were performed at 25°C with 30 o/oo seawater in dim, indirect light.

Antigen preparation

Protein extracts from the three prey items were prepared by homogenizing whole organisms in TES (N - tris [hydroxymethyl] methyl -2- aminoethanesulphonic acid) saline, pH = 7.3. Grass shrimp extracts were prepared in a similar manner, except the digestive tract was removed before homogenization. The homogenate was then centrifuged at 700 x g and the supernate stored frozen at -20 °C for use in antiserum production and as antigenic standards. The standards were prepared by acetone precipitating the soluble proteins, centrifugation at 15600 x g, and lyophilization of the protein pellet.

Antiserum production

Solubilized proteins of known concentration were used to produce antisera in New Zealand white female rabbits following the protocol of Feller *et al.* (1979) by Cocalico Biologicals, Inc., P.O. Box 265, Reamstown, PA 17567. The antiserum to *Spartina alterniflora* detritus was produced to 6 month aged freeze-dried material provided by K.R. Tenore (see Tenore (1981) for preparation). Thus the detritus antiserum was not produced with exactly the same material used in the feeding experiment; however, use of micro-Ochterlony, double-diffusion immunoassay tests has demonstrated that this antiserum produces the strongest cross-reactions with other ages (3 months - 1 year) of laboratory prepared *S. alterniflora* detritus. The other antisera, however, were produced from material used in the feeding experiments.

Gut contents sample preparation

The entire digestive tract of the control and time-series shrimp was removed and individually homogenized in 0.2 ml TES with glass beads added to facilitate grinding. The solubilized proteins were precipitated with acetone, centrifuged (15600 x g), and the protein pellet lyophilized.

Polyacrylamide gel electrophoresis (PAGE)

For each experiment a polyacrylamide electrophoretic separation of the proteins for *P. pugio*, the prey item(s), and the control and time-series gut content samples was performed. The electrophoresis followed the methods of Laemmli (1970) for a discontinuous system resolving proteins denatured with SDS.

Each gel had a resolving gel with an acrylamide concentration of 10% T, 2.7% C and a stacking gel of 4% T, 2.7% C. Both gels had a Tris-HCl buffer (resolving gel - 0.375 M Tris-HCl, pH = 8.8; stacking gel - 0.125 M Tris-HCl, pH = 6.8), and the tank buffer was 0.02 M Tris, 0.192 M glycine at pH = 8.3. All buffers contained 0.1% SDS.

The protein pellets were denatured in SDS extraction buffer (2% SDS, 5% B-mercaptoethanol, 1 mM PMSF, 10 mM EDTA, 50 mM Tris-HCl, pH = 7.0) by boiling for 3-5 min. A 10 µl sample of extracted protein was mixed with 10 µl of tracking dye (10% glycerol, 0.001% bromophenol blue), boiled for 30 sec, and the entire 20 µl sample loaded in a lane on the polyacrylamide gel for electrophoresis at constant current (30 mA) until the solvent front reached the bottom of the 14 cm gel.

Enzyme-linked immunosorbent assay (ELISA)

Resolved proteins from the polyacrylamide gel were transferred to nitrocellulose paper (NC) following Towbin *et al.* (1979). Transfers were performed in buffer (25 mM Tris, 192 mM glycine, 20% v/v methanol, pH = 8.3) at constant current (100 mA) for 13 hrs after which the current was increased to 200 mA for an additional 2 hrs.

The transferred proteins were assayed at room temperature with the Bio-Rad Immuno-Blot (GAR-HRP) Assay Kit (Cat. #170-6502 Bio-Rad, 2200 Wright Ave., Richmond, CA 94804). The NC was blocked with 3% gelatin in Tris buffer saline (TBS = 20 mM Tris-HCl, 500 mM NaCl, pH = 7.5) for 1 hr and then incubated 2 hrs in the appropriate primary antiserum at a dilution of 1:50 v/v with TBS. This was followed by two 10 min washes in TBS-Tween buffer (20 mM Tris-HCl, 500 mM NaCl; 0.05% Tween-20, pH = 7.5). Next the NC was transferred to a 1:2000 dilution of goat antirabbit horseradish peroxidase antiserum for a 1 hr incubation. After incubation with the second antiserum there were two more 10 min washes of the NC in TBS-Tween buffer and then the horseradish peroxidase was visualized by a 30 min incubation of the NC in 0.015% H_2O_2 - HRP development solution (4-chloro-1-naphthol). Antigen-antiserum conjugates attained a purple-blue coloration.

For each of the three feeding experiments offering grass shrimp a single prey organism, the primary antiserum was that of the prey. For the feeding experiment in which all three prey were available simultaneously, three separate PAGE-ELISA's were performed, each using one of the three prey antisera as the primary antiserum.

RESULTS AND DISCUSSION

Several observations arose from examining the six immunoblots from the feeding experiments (Figures 1-6). In each case the primary antiserum cross-reacted most heavily with its homologous antigenic proteins. Cross-reactions between proteins of the predatory grass shrimp and the polyspecific antisera were present in all cases but were most prominent when the primary antiserum was for *Artemia*. This was expected, since closely-related taxa share many immunogenic components (Feller and Gallagher, 1982).

The number of bands detected in the homologous cross-reactions appears to be related to the structural complexity of the prey involved, with *Artemia* producing the greatest number of bands and detritus the fewest. Also, there was a high background staining level due to incomplete blocking of the nitrocellulose by gelatin. Use of a pre-bleed control antisera at the same dilution (blood taken from the rabbit before immunization) might lower the background staining considerably.

Almost none of the high molecular weight proteins (> 100,000 daltons) were efficiently transferred to the nitrocellulose by this method. After the immunoblot transfer, some of the polyacrylamide gels were silver stained to check transfer efficiency. In all cases higher molecular weight proteins remained in the gel while almost all of the other proteins were completely transferred. This inefficient transfer can be attributed to the use of long-term (overnight) low-power transfers. Ideal conditions call for a constant current of 2.5 A for 5 hrs, but a power supply of this rating, unfortunately, was not available.

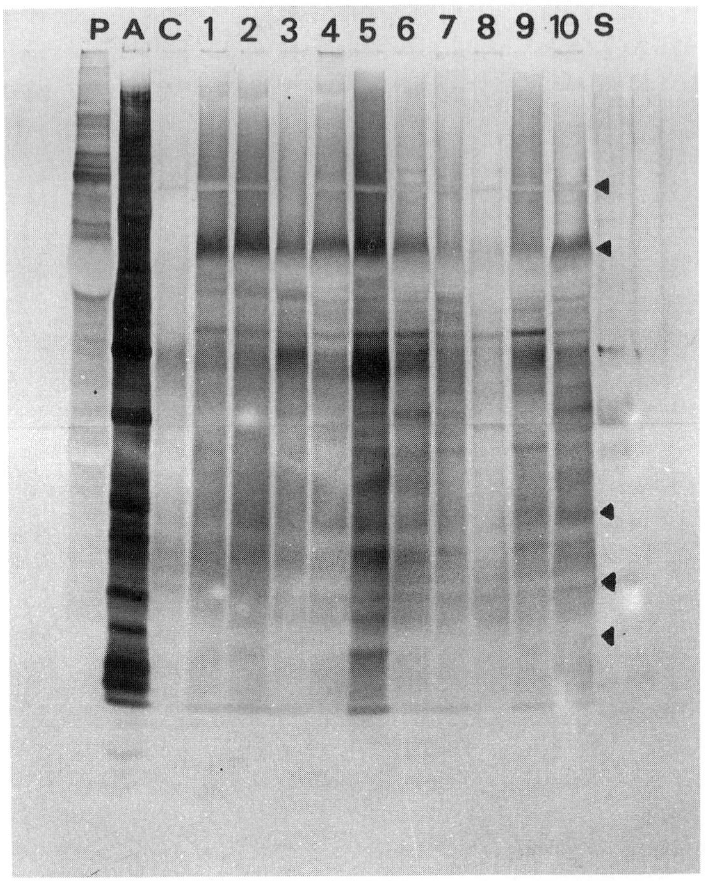

Figure 1. Immunoblot for single prey experiment using *Artemia salina* nauplii probed with *Artemia salina* nauplii antiserum. Lane legend: P = *Palaemonetes pugio*, A = *Artemia salina* nauplii, T = *Thalassiosira weissflogii*, D = *Spartina alterniflora* detritus, C = control (unfed) gut contents. Lane 1 contains grass shrimp gut contents collected at t = 0 (after 30 min with prey). Lanes 2-10 contain gut contents from time-series samples collected at t = 0.25, 0.5, 1, 2, 4, 6, 8, 12 and 24 hr respectively. S = molecular weight standards. Photographic reproduction may not reveal some bands which were clearly visible on the original immunoblots.

Adjacent lanes do not necessarily exhibit a temporal decrease in the number of protein bands (i.e., the same bands appear, disappear, and reappear in successive lanes), nor is there a decrease in staining intensity through time as one might predict (Figure 1). We attribute this to our having loaded equivalent volumes of gut material, though not necessarily equal amounts of protein, to each

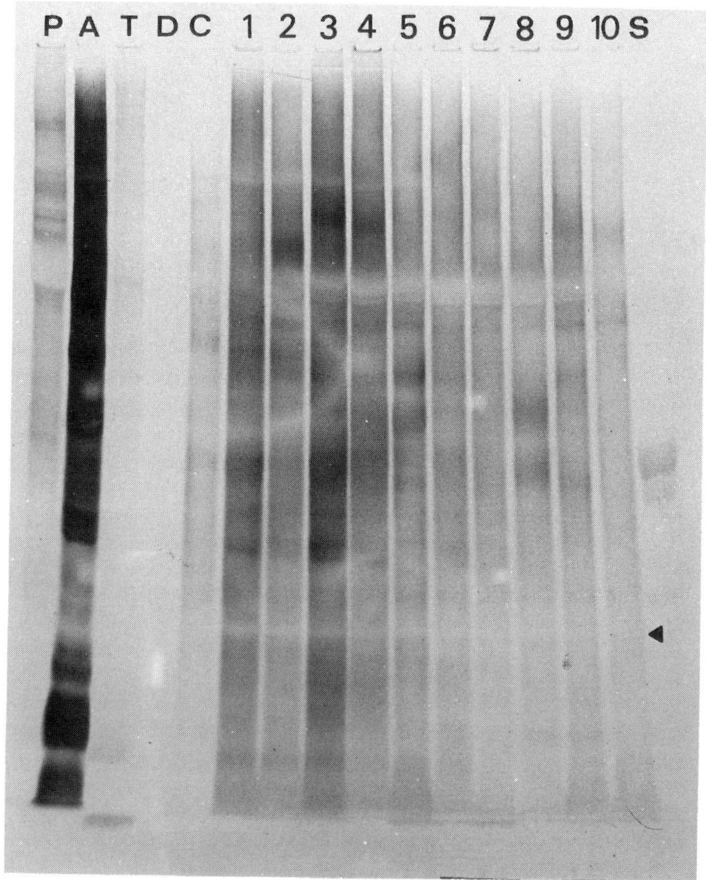

Figure 2. Immunoblot for multi=prey experiment probed with *Artemia salina* nauplii antiserum. (See Figure 1 for lane legend)

lane. In a concurrent study to examine animal-to-animal variability, a single feeding experiment with *Artemia* as prey was performed utilizing the same protocol as previously described, except the only samples taken were groups of 5 shrimp each at t = 1, 6, 12 hrs. The individual gut content proteins were separated on a polyacrylamide gel and stained with Coomassie Brilliant Blue R. The number of protein bands resolved was the same for each gut content but the staining intensity within a time interval was related to the size of the individual shrimp (i.e., larger shrimp had larger gut volumes and therefore, with TES dilution, slightly higher concentrations and greater total amounts of each protein). Future investigations should utilize predators of equal size, or gut protein concentration should be standardized.

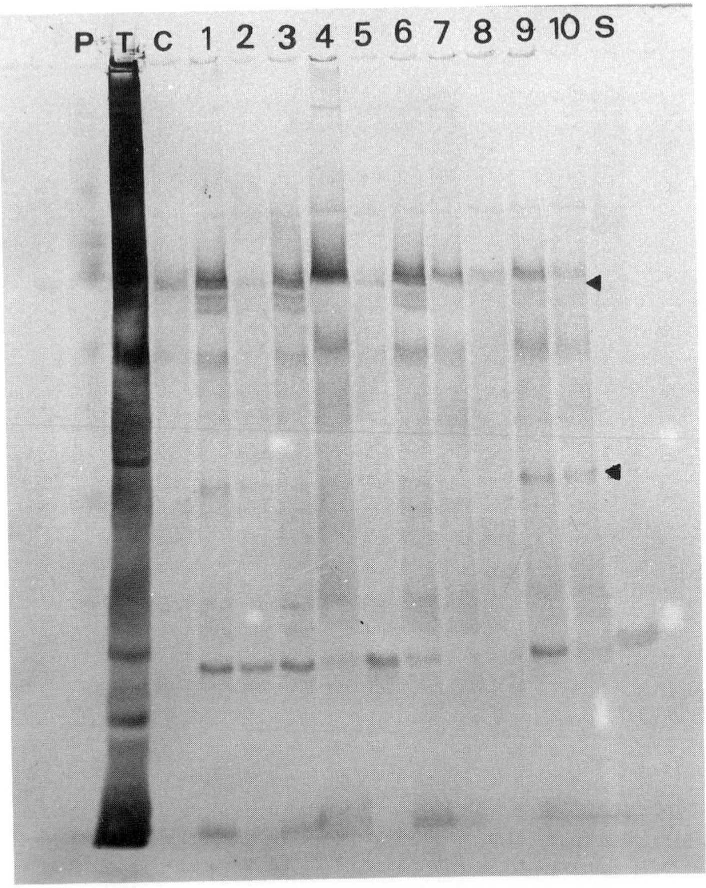

Figure 3. Immunoblot for single prey experiment using *Thalassiosira weissflogii* probed with *Thalossiosira weissflogii* antiserum. (See Figure 1 for lane legend)

Despite the large number of cross-reactions and high background staining levels, several distinct prey antigens persisted in the grass shrimp guts for the full 24 hrs post-ingestion. The experiments in which *Artemia* antiserum was the primary antiserum for the immunoblot produced very different results for the single and mixed prey experiments. When presented as a single food source, at least 5 distinct *Artemia* antigen bands persisted for 24 hrs (Figure 1). When fed in a mix with the other prey, only one *Artemia* antigen band persisted; none of the other visible *Artemia* proteins showed

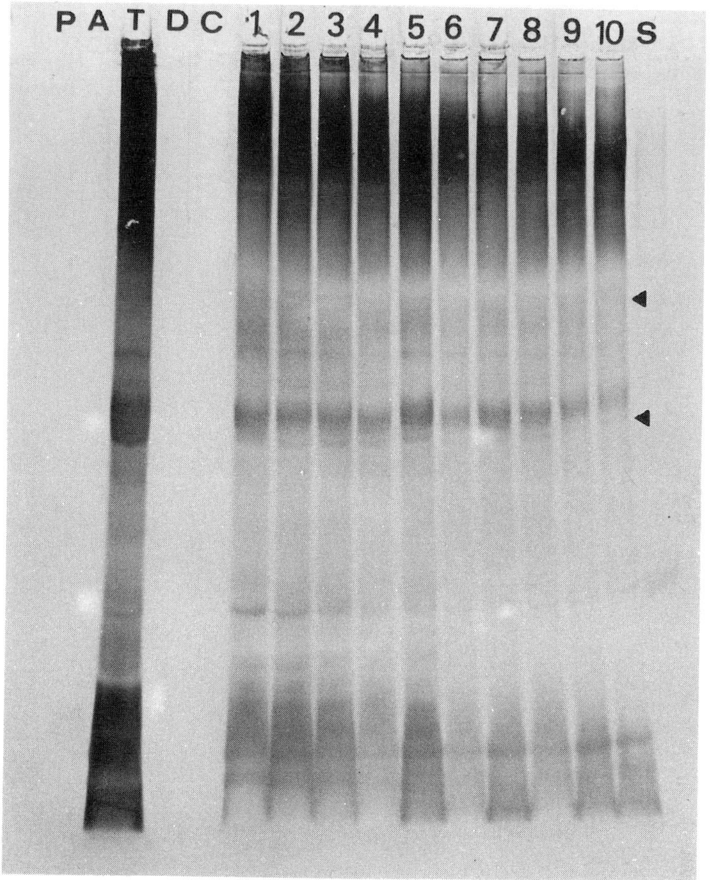

Figure 4. Immunoblot for multi-prey experiment probed with *Thalassiosira weissflogii* antiserum. (See Figure 1 for lane legend)

lane-to-lane consistency (Figure 2). This may be attributed to differences in grass shrimp size, differential digestion of proteins in mixed prey cases, or possibly some other physiological mechanism.

The *T. weissflogii* experiments both exhibited very similar results. Although the mixed prey immunoblot had a high degree of background staining, the pattern of antigen bands was very similar in both feeding situations. Two antigen bands of similar motility persisted throughout both experiments (Figures 3 and 4).

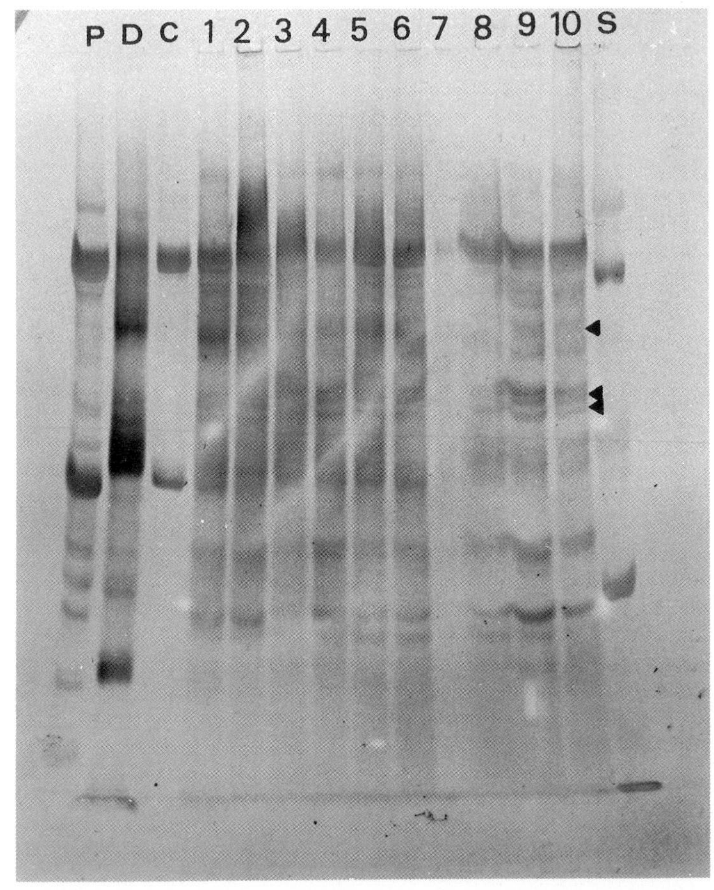

Figure 5. Immunoblot for single prey experiment using *Spartina alterniflora* detritus probed with *Spartina alterniflora* detritus antiserum. (See Figure 1 for lane legend)

Results for the experiments in which detritus was present as prey were encouraging since the detritus antiserum was not produced against the same material used in the feeding experiments. Two antigens of similar motility are visible in both the single and multi-prey experiments (Figures 5 and 6). An additional antigen was also visible in the single prey experiment (Figure 5). We chose to use antiserum to 6-month old *S. alterniflora* because cross-reacted with antigens from other ages of *S. alterniflora* detritus and gave positive results for the presence of detritus in the gut contents of field-collected animals (Zagursky and Feller, 1985).

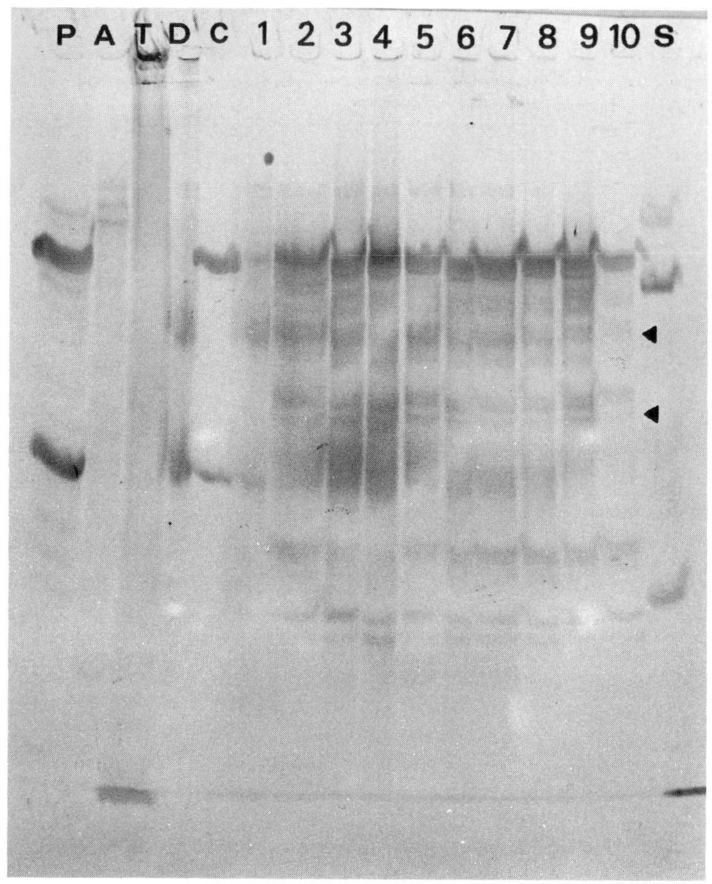

Figure 6. Immunoblot for multi-prey experiment probed with *Spartina alterniflora* detritus antiserum. (See Figure 1 for lane legend)

For future experiments we plan to improve the immunoblots, increase band resolution, and decrease lane-to-lane variability. A stronger power supply will be utilized to increase the transfer efficiency of high molecular weight proteins. To reduce background staining, we will try goat IgG as the blocking agent. We would like to try gradient SDS-polyacrylamide gels to increase antigen resolution. A preliminary study indicates that better separation will be achieved with a 10%-20% acrylamide gradient. A final improvement will be to standardize the amount of gut contents protein applied to each lane, since the temporal variability of protein bands may be an artifact of the amount of protein applied to a lane rather than a breakdown of proteins during the time course of digestion.

The immunoblotting technique has advantages over other serological methods (e.g., rocket electrophoresis, counterimmuno-electrophoresis) because of its greater sensitivity (picogram levels), increased resolution of individual proteins, and only very small samples are required. This latter advantage is particularly important when dealing with small predators. The increased sensitivity of immunoblotting is not without some tradeoff. Higher sensitivity increases the chances of getting a false positive result due to cross-reactions or from secondary prey (prey eaten by prey).

In conclusion, we feel that the use of immunoblotting should prove useful for detecting and isolating persistent prey antigens in a predator's gut. These methods may lead to the production of monospecific antisera which can be utilized as tracers for modeling the flow of materials and/or energy in a community or for expanding our knowledge on the modes of digestion, prey selection and frequency of feeding by predators. Also, since the assays are quantitative, it is possible to quantify predator diet with certain assumptions regarding predator feeding behavior (Feller and Ferguson, this volume).

ACKNOWLEDGEMENTS

This work was supported by NSF grants OCE 81-10148 and OCE 85-21345 (Biological Oceanography Program) and grants from the South Carolina Sea Grant Consortium. This is contribution number 660 from the Belle W. Baruch Institute.

REFERENCES

Adams, J. 1981. Serological analysis of the diet of *Bdellocephala punctata*, a freshwater triclad. Oikos. 36: 99-106.

Feller, R.J. 1986. Immunological detection of *Mercenaria mercenaria* in a predator and preparation of size-class specific antibodies. The Veliger. 28: 341-347.

Feller, R.J. and E.D. Gallagher. 1982. Antigenic similarities among estuarine soft-bottom benthic taxa. Oecologia. 52: 305-310.

Feller, R.J., G.L. Taghon, E.D. Gallagher, G.E. Kenny and P.A. Jumars. 1979. Immunological methods for food web analysis in a soft-bottom benthic community. Mar. Biol. 54: 61-74.

Fichter, B.L. and W.P. Stephen. 1984. Time-related decay of prey antigens ingested by arboreal spiders as detected by ELISA. Environmental Entomology. 13: 1583-1587.

Giles, N. and R.S. Phillips. 1985. A note on the production and use of antisera for the detection of part-digested sticklebacks in predator stomach-content samples. J. Fish. Biol. 27: 827-829.

Grisley, M.S. and P.R. Boyle. 1985. A new application of serological techniques to gut content analysis. J. Exp. Mar. Biol. Ecol. 90: 1-9.

Guillard, R.R.L. 1975. Culture of phytoplankton for feeding marine invertebrates. IN: Culture of Marine Invertebrate Animals. W.L. Smith and M.H. Canley. (eds.). Plenum Publishing Co. NY. pp. 29-60.

Healy, J.A. and T.F. Cross. 1975. Immunoelectroosmophoresis for serological identification of predators of the sheep tick *Ixodes ricinus*. Oikos. 26: 97-101.

Laemmli, U.K. 1970. Cleavage of structural proteins during the assembly of the head of bacteriophage T4. Nature. 227: 680-685.

Tenore, K.R. 1981. Organic nitrogen and caloric content of detritus. I. Utilization by the deposit-feeding polychaete *Capitella capitata*. Estuarine Coastal Shelf Sci. 12: 39-47.

Theilacker, G.H., A.S. Kimball and J.S Trimmer. 1986. Use of an ELISPOT immunoassay to detect euphausiid predation on larval anchovy. Mar. Ecol. Prog. Ser. 30: 127-131.

Towbin, H., T. Staehelin and J. Gordon. 1979. Electrophoretic transfer of proteins from polyacrylamide gels to nitrocellulose sheets: Procedure and some applications. Proc. Natl. Acad. Sci. 76: 4350-4354.

Young, J.O. 1980. A serological investigation of the diet of *Helobdella stagnalis* (Hirudinea:Glossiphonudae) in British lakes. J. Zool. 192: 467-488.

Zagursky, G. and R.J. Feller. 1985. Macrophyte detritus in the winter diet of the estuarine mysid, *Neomysis americana*. Estuaries. 8: 355-362.

TWO-DIMENSIONAL ELECTROPHORESIS WITH SILVER STAINING: A SENSITIVE METHOD FOR ANALYZING COMPLEX PROTEIN MIXTURES

Marcia F. Goldfarb
Anatek-EP
999 Forest Avenue
Portland, ME 04103

INTRODUCTION

About 10 years ago, O'Farrell combined two electrophoretic methods which separated proteins by different parameters to produce a greatly enhanced separation of proteins in serum. This idea was developed by Anderson and Anderson (1978a) at Argonne Laboratory, and they have called their methodology the Iso-Dalt system (Anderson *et al.*, 1978a; Anderson *et al.*, 1978b).

METHODS

The first separation is an isoelectric focus. It is done in a 1.5 mm cylindrical polyacrylamide gel to which ampholytes have been added. Ampholytes are synthetic polypeptides which form a pH gradient when placed in an electric field. The wide range ampholyte mixture gives a gradient from pH 3-10. However, addition of narrow range (2-pH units) mixtures (i.e., 4-6, 5-7) can flatten the gradient and expand the separation for a pH area of interest. The sample is solubilized and placed on top of the gel, which is usually the basic end. At the end of the focus, the gels are extruded, equilibrated and placed in a -80°C freezer, until ready to be used. A typical focus is 10,000 volthours (hrs. run X voltage). The Iso gels are poured and run as a set of 20. The Iso unit is purchased from Electronucleonics, Oak Ridge, TN.

The second separation is by size. The Iso is placed on a 10-20% gradient vertical slab gel. This gel contains Tris-HCl and SDS, and is run in a Tris, glycine, SDS buffer. Molecular weight markers can be run down one side of the gel. The run is complete 30 minutes after a Bromophenol blue dye front from the Iso exits the bottom of the gel. This run is typically 6-7 hours. The Dalt gels are poured as a set of 10 in our lab and run 1-4 per time. The Dalt unit is a Hoefer SE 600. Tanks which run 20 Dalts per time can be purchased from Electronucleonics.

The Dalt gel at the end of the run is fixed and stained. Stain can be Coomassie or silver. The new silver staining techniques increase detectability 25 to 200 times. Our lab used a $AgNO_3$ method from Merril *et al.* (1982).

For identification of specific spots, the Western Blot technique is utilized (Symington, 1984). The completed Dalt gel is placed on a membrane (nitrocellulose or nylon) in a cassette. The cassette is placed in a tank (Hoefer Transphor) and the proteins are electrophoresed out of the gel onto the membrane. The proteins are then identified with probes. The most common probe is antibody. This is visualized with a second antibody which is enzyme linked.

RESULTS AND DISCUSSION

Our laboratory is involved in a 2-D analysis of the protein fraction of human milk. Casein is the major protein in milk and is the nutrient. There are, at least, 100 other proteins which have received the general name of whey proteins. Some of these are unique to milk. Many are serum proteins; enzymes, protease inhibitors, or immunoglobulins. Two-D gives a visual image of the proteins, and concentrations and relationships can be observed under varying circumstances. This study has already identified a diurnal difference in the concentration of the casein fraction.

Figure 1 shows a 2-D separation of human milk with some spot patterns identified. Many proteins give spot patterns, because they are polymorphic and the variants are clearly separated by this method. Ten additional areas have been identified using the Western Blot technique. The goal of the project is to draw a 2-D map of human milk proteins. Maps of human serum, urine and certain cell lines already exist and identifications can be made by referring to these maps.

Two-dimensional electrophoresis is a powerful analytical tool for several reasons. Each point on the gel has 2 coordinates, an isoelectric point and a molecular size, which help in identification. The separation is very sensitive and differentiation of genetic variants is easily accomplished. It is the only method which can screen complex protein mixtures and pick up aberrant situations without knowing beforehand they exist. While there still may be overlapping spots, 2-D gives separation far greater than other existing electrophoretic methods.

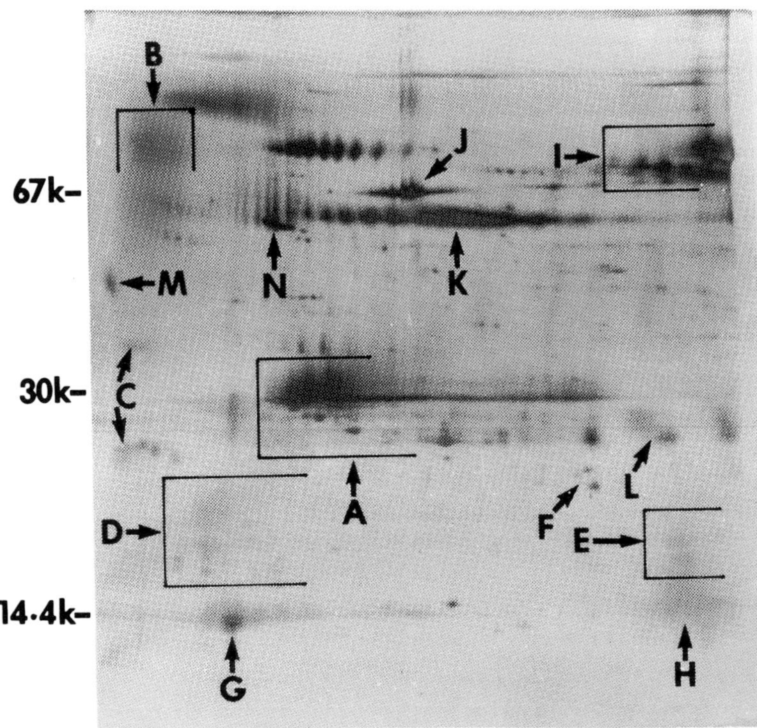

Figure 1. Two-dimensional electrophoresis pattern of a human milk sample collected one month after parturition. Acid end of isoelectric focusing gel is to the left, Identification: A. B-casein. This consists of a large streaked area and four major spots that can be individually identified. The spots start at approximately 25K and go up diagonally, running into the streak. This is blue with silver stain, which clearly sets it off from the many other proteins in this area. One of the other major molecules in this area is the immunoglobulin light chains. B. α-casein-micellar form. C. α-casein D. Casein - a large heterogenous area. E. Casein isomers with very basic pI. F. Casein - approximate pI 6.0, 24K. These spots are usually coordinated with presence of β-casein. G. α-lactalbumin H. lysozyme I. lactoferrin J. albumin K. sIgA - heavy chain L. Ig - light chains M. α_1- acidglycoprotein N. α_1 antitrypsin.

REFERENCES

Anderson, N.G. and N.L. Anderson. 1978a. Analytical Techniques for cell fractions. XXI. Two-dimensional analysis of serum and tissue proteins: Multiple isoelectric focusing. Anal. Biochem. 85: 331-340.

Anderson, N.L. and N.G. Anderson. 1978b. Analytical techniques for cell fractions. XXII. Two-dimensional analysis of serum and tissue proteins: Multiple gradient-slab electrophoresis. Ibid., pp 334-354.

Merril, C.R., D. Goldman, and M.L. Van Keuren. 1982. Simplified silver protein detection and image enhancement methods in polyacrylamide gels. Electrophoresis 3: 17-23.

Symington, J. 1984. Electrophoretic transfer of proteins from two-dimensional gels to sheets and their detection. IN: Two dimensional gel electrophoresis of proteins. Celis, Bravo. (ed.). Academic Press. pp 127-168.

Clinical Chemistry. 1982. Part II, 28(4): pp 737-1092.

Clinical Chemistry. 1984. Part I, 30(2): pp 1897-2108.

BREVETOXINS AND BINDING: SODIUM CHANNELS VERSUS ANTIBODIES

Daniel G. Baden
University of Miami
Rosenstiel School of Marine and Atmospheric Science
4600 Rickenbacker Causeway
Miami, Florida 33149

Thomas J. Mende
University of Miami
School of Medicine
PO Box 016129
Miami, Florida 33101

Alina M. Szmant
University of Miami
Rosenstiel School of Marine and Atmospheric Science
4600 Rickenbacker Causeway
Miami, Florida 33149

INTRODUCTION

The marine dinoflagellate *Ptychodiscus brevis* is the dinoflagellate responsible for Florida's red tides (Steidinger and Joyce, 1973). The organism produces potent toxins, both *in situ* (Gervais and MacLean, 1985; Pierce *et al.*, 1985) and in the laboratory (Shimizu *et al.*, 1986; reviewed in Baden, 1983). To date, seven potent toxins have been described (Shimizu *et al.*, 1986; Figure 1), all of which are based on two multi-ring polyether carbon backbones.

The polyether toxins affect neuronal function in the nanomolar to picomolar concentrations ranges (Catterall and Risk, 1980; Baden *et al.*, 1984a; Gallagher and Shinnick-Gallagher, 1980), suggesting specific loci of action exist in excitable membranes (Catterall and Risk, 1980). These loci have recently been shown to be associated with the voltage-sensitive sodium channel (Poli *et al.*, 1985, 1986; Sharkey *et al.*, 1986), and binding constants for tritium-labeled brevetoxin in a synaptosomal assay system are in the nanomolar concentration range, in good agreement with its dose-response effects in model nerve (Huang *et al.*, 1984) and neuromuscular preparations (Baden *et al.*, 1984a).

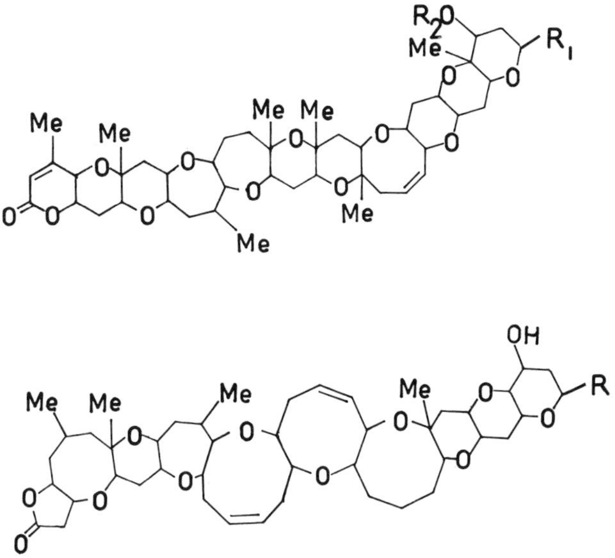

Figure 1. The brevetoxins. Upper panel. PbTx-2 [R_2=H, R_1=$CH_2C(=CH_2)CHO$]; PbTx-3 [R_2=H,R_1=$CH_2C(=CH_2)CH_2OH$]; PbTx-5 [R_2=Ac, R_1=$CH_2C(=CH_2)CHO$]; PbTx-6 [R_2=H, R_1=$CH_2C(=CH_2)CHO$], 27,28 epoxide; PbTx-8 [R_2=H, $R_1CH_2COCH_2Cl$].
Lower panel. PbTx-1[R=CHO]; PbTx-7 [R=CH_2OH]. PbTx-4 structure is unknown. Redrawn from Poli *et al.*, 1986.

We have also developed antibodies to one subset of the brevetoxins, using PbTx-3 covalently bound to bovine serum albumin as immunogen (Baden *et al.*, 1984b). The antibodies raised in goats against this toxin-BSA conjugate bind both PbTx-2 and PbTx-3 with equal affinity and avidity (Baden *et al.*, 1984b), and using tritiated PbTx-3 as probe, immunoassays can detect and quantitate each of these toxins from *P. brevis* cell extracts (Baden *et al.*, 1985). The antibodies cross-react with, and hence can also quantitate, polyether toxins such as ciguatoxin (Hokama *et al.*, 1984; Baden *et al.*, 1985) and other polyethers derived from toxic dinoflagellates (Baden *et al.*, 1985).

This paper describes continuing work related to both types of binding assays, and is a preliminary comparison of brevetoxin binding in each system.

MATERIALS AND METHODS

Purification of Toxins

Toxins were purified from laboratory cultures by solvent extraction and a combination of column and thin-layer chromatography as previously described (Baden *et al.*, 1981; Baden and Mende, 1982). The final step of purification was a reverse phase high pressure liquid chromatography step utilizing a C-18 column and an isocratic elution employing 85% methanol/15% water solvent. Eluted fractions were monitored using ultraviolet absorbance at 215 nm, and were collected separately (Figure 2). Each was dried under reduced pressure on a rotary evaporator, redissolved in minimal acetone, dried on aluminum weighing pans, and quantitated utilizing a Cahn gram electrobalance.

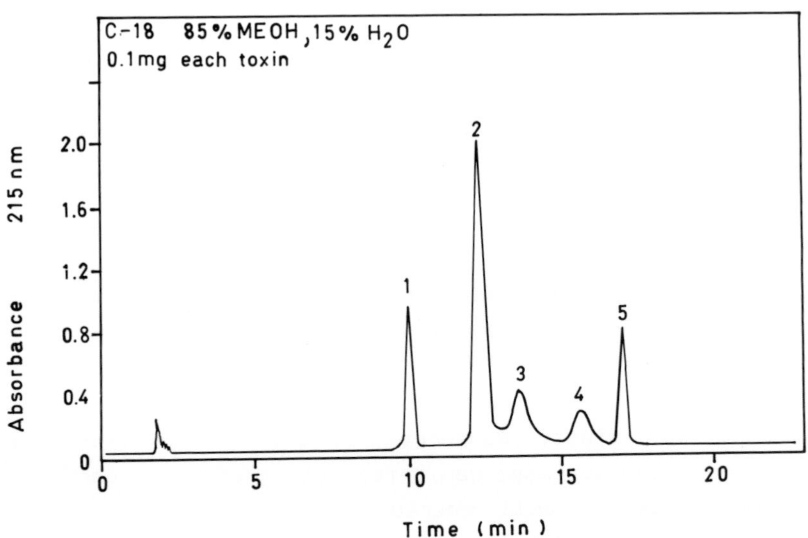

Figure 2. HPLC tracing of brevetoxin separations. Conditions: 100 µg of each toxin mixed together and injected in 100 µl volume; C-18 reverse phase column 1.5 x 25 cm.; flow rate 1.4 ml/min. 850 psi, 85% methanol/15% water; ultraviolet detection at 215 nm. 2.0 AUFS. Peak 1= PbTx-3, Peak 2= PbTx-2, Peak 3= mixture of PbTx-5 and 6, Peak 4= PbTx-7, Peak 5= PbTx-1. PbTx-8 is not present in cultures.

Assays

Both immunoassay (Baden *et al.*, 1984b) and synaptosomal (Poli *et al.*, 1986) binding assays have been described previously. Assays in both cases were performed at 4°C. Tritiated probe (Figure 3) was prepared using tritiated sodium borohydride in acetonitrile solution, reducing the aldehyde function in PbTx-2 to the corresponding primary alcohol function in PbTx-3. The reduction mixture was subjected to reverse phase HPLC as described for Figure 2, collecting the single peak corresponding to PbTx-3. The reduced product possesses one covalently-bound tritium atom per PbTx-3 molecule.

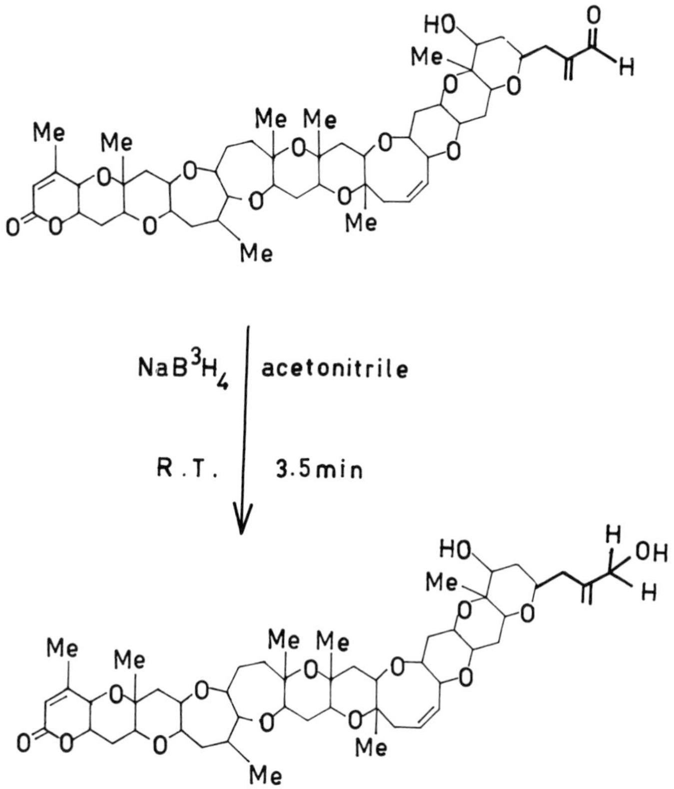

NaB^3H_4 acetonitrile

R.T. 3.5min

Figure 3. Chemical reduction of PbTx-2 to PbTx-3 using sodium borotritiide. Conditions: equimolar reducing equivalents with PbTx-2; room temperature (22°C); reaction time 3.5 minutes. Following reaction, remaining reductant is destroyed with acetone. Mixture is dried, chromatographed on thin layer silica gel plates using 70% ethyl acetate, 30% petroleum ether. Active fraction is further purified on HPLC as described in Figure 2, taking peak 1. Specific activities ranged from 10-15 Ci/mmole, or 25% of the specific activity of the reductant.

Unlabeled toxins were prepared in each respective binding medium, to yield final concentrations of potential competitor toxins ranging from 1-1000 nM.

RESULTS AND DISCUSSION

Identity of Toxic Fractions

Shimizu *et al.* (1986) has described six distinct toxic compounds which are isolated from laboratory cultures of *Ptychodiscus brevis*. Four of the seven are based on the polyether backbone structure shown in the upper structure of Figure 1. The remaining two are based on the lower structure of Figure 1. Our previous publications have shown that peaks 1 and 2 of Figure 2 correspond to toxins PbTx-3 and PbTx-2, respectively. In support of this, reduction of peak 2 (PbTx-2) with sodium borohydride results in peak 1 (PbTx-3). Likewise, peak 5 is known to be PbTx-1 based on its migration in this reverse phase system, from its characteristics on isolation (obtained from the mother liquor of PbTx-2 crystallizations), and by its reduction with sodium borohydride to yield peak 4. By analogy with the PbTx-2 reduction to PbTx-3 situation, peak 4 is known to be PbTx-7. Upon closer analysis using 70% aqueous methanol as solvent, peak 3 was discovered to be a mixture of two compounds, most likely the 27,28 epoxide of PbTx-2 and acetoPbTx-2. This accounts for all the toxins isolated with these polyether backbones. Their absolute identities shall be ascertained using Fourier Transform Infrared Spectrometry.

Binding Assays

Both binding assays are based on tritiated PbTx-3, prepared from chemical reduction of PbTx-2 using sodium borotritiide. Using equimolar amounts of reducing equivalents and toxin, the overall yield of PbTx-3 ranges from 40-70%; the remainder of the toxin PbTx-2 being either doubly reduced (the methylene function in addition to the aldehyde), or not reduced. The specific activity of the labeled toxin is 10-15 Ci/mmole, or 25% that of the reducing agent.

Figure 4 is an illustration of the binding assays. As already stated, each is based on specific binding of labeled toxin to its binding component. In the case of radioimmunoassay, the binding component is antibody directed against PbTx-3; in the synaptosomal assay, it is a portion of the voltage-dependent sodium channel which specifically associates with toxin. The binding component of the former assay is in soluble form. The binding component in the latter assay is associated with a suspended tissue component.

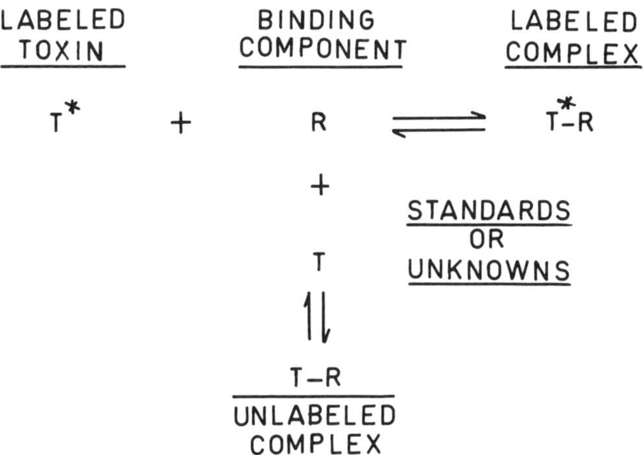

Figure 4. Competitive displacement binding assays. This technique requires some form of a label (whether it be radioactive, fluorescent, enzyme-linked, or some other "tagging" method) be attached to the toxin (T*) being assayed. Using fixed and stooiciometric amounts of both labeled toxin (T*) and binding component (R), an equilibrium is established between free toxin (T*) and bound toxin (T*-R). Standard curves are developed by adding known amounts of unlabeled toxin (T), and determining the <u>decreased</u> amount of (T*-R) present and each (T) concentration. Toxin concentrations in unknowns are then determined by comparison with standard curves. Results are expressed in [T*] equivalents. (Note: A true quantitative measurement is possible only if the dissociation constants of T*-R and unknown T-R are equivalent).

Following association of toxin with the binding component, a mechanical step is required to separate bound from unbound radioactive toxin. In both cases, we used Eppendorf microcentrifugation. In the case of the synaptosomal assay, the tissue preparation forms a compact pellet. When the supernatant solution is aspirated, and the pellet is rapidly washed, bound radioactivity, which remains associated with the tissue pellet, may be counted.

In a soluble binding assay such as radioimmunoassay, the addition of dextran-coated charcoal serves to adsorb unbound toxin. Subsequent microcentrifugation isolates unbound toxin (now adsorbed to charcoal) in a pellet fraction, and soluble antibody-toxin conjugate remains in the supernatant solution. Thus, in this case, the soluble fraction is counted.

By adding known quantities of competitor toxins, standard curves may be developed; and once standard curves have been developed, unknowns may be evaluated. Standard curves for each of the toxins separated by HPLC are shown in Figures 5 and 6. Note there is a dissimilarity in the types of displacement curves one obtains with each assay; the resulting curves are much less uniform with respect to the immunoassay, and all toxins seemed to displace tritiated PbTx-3 in a similar manner using synaptosomes.

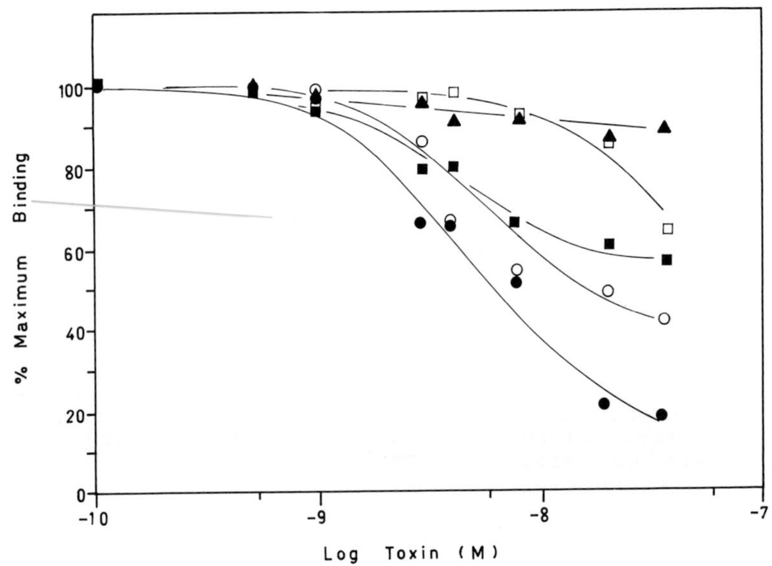

Figure 5. Displacement of tritiated PbTx-3 by competitor brevetoxins in radioimmunoassay. Labeled toxin, immunoglobulin, and increasing concentrations of unlabeled potential competitor toxins were mixed and incubated as previously described (Baden *et al.*, 1984). (●) Peak 1; (○) Peak 2; (■) Peak 3; (□) Peak 4; (▲) Peak 5.

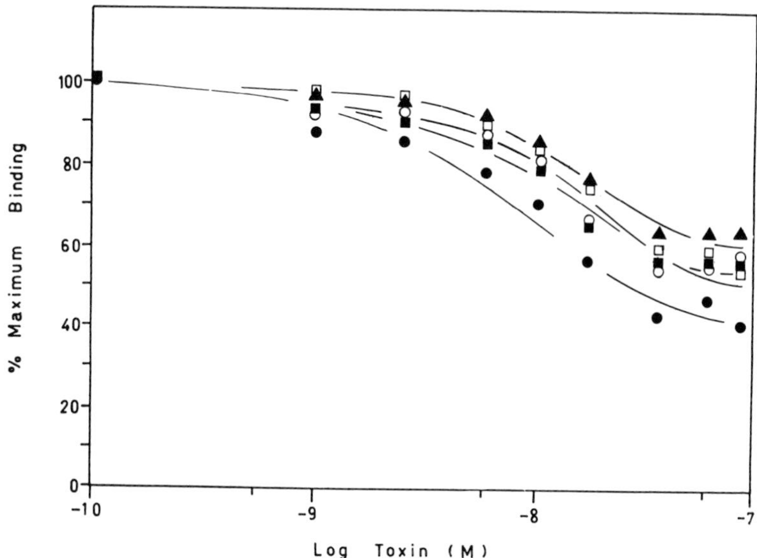

Figure 6. Displacement of tritiated PbTx-3 by competitor brevetoxins in the synaptosomal binding assay. Labeled toxin, synaptosomes, and increasing concentrations of unlabeled potential competitor toxins were mixed and incubated as previously described (Poli *et al.*, 1986). (●) Peak 1; (o) Peak 2; (■) Peak 3; (□) Peak 4; (▲) Peak 5.

We feel it is important to point out that each of these assays depends entirely upon the measurement of tritiated toxin binding (tritiated toxin of the structural backbone type illustrated in the upper structure of Figure 1). In the case of the immunoassay, the antibodies were raised using protein-linked PbTx-3 and binding affinities would be expected to reflect similarities in structure to PbTx-3. Thus, it is not surprising that toxins of the structural class illustrated in the lower panel of Figure 1 do not displace well in radioimmunoassay (Figure 5). Hence, in correlation with Hall and Shimizu (1985), immunoassay may better be classified as an analysis (for measurement of concentrations of PbTx-3-like molecules) than an assay. One would expect that antibodies directed specifically against PbTx-7 would detect and quantitate it and PbTx-1 better than PbTx-2,3,5 and 6. These experiments are progressing.

The molecular pharmacological synaptosome assay, on the other extreme, is a true assay and measures "composite potency". Since binding is the pharmacologically-significant event in the onset of toxicity, binding affinity should closely parallel potency. Using a single highly potent (i.e. binds with some high affinity) radioactive toxin, any toxin which binds to the same site could be measured by competitive displacement, and intuitively the degree of displacement should correlate with potency relative to the labeled probe. This evaluation is currently progressing. The sodium channel assay is particularly appealing, for by using labeled saxitoxin (or tetrodotoxin) instead of labeled brevetoxin, one could measure composite potency of unknown gonyautoxin preparations (See papers in Anderson *et al.*, 1985).

It remains to be seen if the diarrheic shellfish toxins or the pectenotoxins (Yasumoto, 1985) interact with similar sites in excitable membranes. Being polyether toxins, they may also interact with the brevetoxin binding site in excitable tissues. If they do, the labeled brevetoxin probe may be of use in detecting these materials, just as it is useful in the detection of ciguatoxin in some fish tissues (Baden *et al.*, 1985).

Regardless, for purposes of toxicological significance and as a general method for assessing potential lethality, this assay has great potential (Hall and Shimizu, 1985). If a sodium channel preparation can be stabilized and reduced to a reproducible form, it could become useful for measurement of many excitable membrane perturbing agents.

ACKNOWLEDGEMENTS

The technical assistance of Laurie Roszell is gratefully acknowledged. This work was supported in part by the US Army Medical Research Acquisition Activity, Contract No. DAMD17-85-C-5171, and by grant No. ES 02651, National Institutes of Health, DHHS.

REFERENCES

Anderson, D.M., A.W. White and D.G. Baden. 1985. Eds. Toxic Dinoflagellates. Elsevier Science Publishers, NY. 561 pp.

Baden, D.G. 1983. Marine food-borne dinoflagellate toxins. Int. Rev. Cytol. 80: pp. 99-150.

Baden, D.G. and T.J. Mende. 1982. Toxicity of two toxins from the Florida red tide dinoflagelate *Ptychodiscus brevis*. Toxicon. 20: 457-461.

Baden D.G., T.J. Mende and L. Brand. 1985. Cross-reactivity in immunoassays directed against toxins isolated from *Ptychodiscus brevis*. IN: Toxic Dinoflagellates. D.M Anderson, A.W. White and D.G. Baden (eds.). Elsevier Science Publishers, NY. pp. 363-368.

Baden, D.G., G.M. Bikhazi, S.J. Decker, F.F. Foldes and I. Leung, 1984a. Neuromuscular blocking action of two brevetoxins from the Florida red tide organism *Ptychodiscus brevis*. Toxicon. 22: pp. 75-84.

Baden, D.G., T.J. Mende, W. Lichter and L. Wellham. 1981. Crystallization and toxicology of T34: A major toxin from Florida's red tide organism (*Ptychodiscus brevis*). Toxicon. 19: pp. 455-462.

Baden, D.G., T.J. Mende, J. Walling and D.R. Schultz. 1984b. Specific antibodies directed against toxins of *Ptychodiscus brevis* (Florida's red tide dinoflagellate). Toxicon. 22: pp. 783-789.

Catterall, W.A. and M.A. Risk. 1980. Toxin T46 from *Ptychodiscus brevis* enhances activation of voltage-sensitive sodium channels by veratridine. Mol. Pharmacol. 19: pp. 345-348.

Gallagher, J.P. and P. Shinnick-Gallagher. 1980. Effect of *Gymnodinium breve* toxin in the rat phrenic nerve diaphragm preparation. Br. J. Pharmacol. 69: pp. 367-374.

Gervais, A.J. and J.L. MacLean. 1985. Management. IN: Toxic Dinoflagellates. D.M. Anderson, A.W. White and D.G. Baden (eds.). Elsevier Science Publishers, NY. pp. 530-533.

Hall, S. and Y. Shimizu. 1985. Toxin analysis and assay methods. IN: Toxic Dinoflagellates. D.M. Anderson, A.W. White and D.G. Baden, (eds.). Elsevier Science Publishers, NY. pp. 545-548.

Hokama, Y., L.H. Kimura, M.A. Abad, L. Yokochi, P.J. Scheuer, M. Nukina, T. Yasumoto, D.G. Baden and Y. Shimizu. 1984. An enzyme immunoassay for the detection of ciguatoxin and competitive inhibition by related natural polyether toxins. IN: Seafood Toxins, ACS Symposium Series. E.P. Ragelis (ed.). American Chemical Society, Washington, D.C., 262: pp 307-320.

Huang, J.M.C., C.H. Wu and D.G. Baden. 1984. Depolarizing action of a red tide dinoflagellate brevetoxin on axonal membranes. J. Pharmacol. exp. Therapeut. 229: pp. 615-621.

Pierce, R. 1986. Red tide (*Ptychodiscus brevis*) aerosol toxins: A review of current knowledge. Toxicon. 24: 955-966.

Pierce, R., R.C. Brown and J.R. Kucklick. 1985. Analysis of *Ptychodiscus brevis* toxins by reverse phase HPLC. IN: Toxic Dinoflagellates, D.M. Anderson, A.W. White and D.G. Baden. (eds.). Elsevier Science Publishers, NY. pp. 309-314.

Poli, M.A., T.J. Mende and D.G. Baden. 1985. Characterization of the *Ptychodiscus brevis* polyether binding component in excitable membranes. IN: Toxic Dinoflagellates, D.M. Anderson, A.W. White and D.G. Baden. (eds.). Elsevier Science Publishers, NY. pp. 357-362.

Poli, M.A., T.J. Mende and D.G. Baden. 1986. Brevetoxins, unique activators of voltage-sensitive sodium channels, bind to specific sites in rat brain synaptosomes. Mol. Pharmacol. 30: 129-135.

Sharkey, R.G., E. Jover, E., F. Courand, D.G. Baden and W.A. Catterall. 1986. Allosteric modulation of neurotoxin binding to voltage-sensitive sodium channels by *Ptychodiscus brevis* toxin 2. Mol. Pharmacol. in press.

Shimizu, Y., H.N. Chou, H. Bando, G. VanDuyne and J.C. Cardy. 1986. Structure of brevetoxin-a (GB-1), the most potent toxin in the Florida red tide organism *Ptychodiscus brevis*. J. Am. Chem. Soc. 108: pp. 514-515.

Steidinger, K.A. and E.A. Joyce. 1973. Florida red tides. Florida Department of Natural Resources Laboratory Education Series, 17: pp. 1-26.

Yasumoto, T. 1985. Recent progress in the chemistry of dinoflagellate toxins. IN: Toxic Dinoflagellates. D.M. Anderson, A.W. White and D.G. Baden. (eds.). Elsevier Science Publishers, NY. pp.259-270.

PREPARATION OF MODEL HAPTENS TO EXPRESS COMMON EPITOPES OF PSP TOXINS

Patrick E. Guire, Peter H. Duquette
Richard A. Amos, Jerome C. Behrens
Mark W. Josephson and Ross P. Chambers

Bio Metric Systems, Inc
9932 W 74th St.
Eden Prairie, MN 55344

INTRODUCTION

Massive dinoflagellate blooms in coastal waters are the cause of the phenomenon known as "red tide" (Boyer *et al.*, 1979). Unfortunately, these organisms produce an extremely potent group of tetrahydropurine-based toxins known as the paralytic shellfish poisons (PSP) (Hall *et al.*, 1984). This group of compounds serves as a significant potential risk to humans through the ingestion of shellfish which bioaccumulate the PSP.

The mode of action of the PSP is based on their ability to act as potent blocking agents of the voltage-sensitive sodium ion channels in neuromuscular membranes (Agnew *et al.*, 1978; Catterall, 1980; Kao, 1981). The disruption of ion passage prevents signal transmission in the neuron, resulting in paralysis and death. Since extraordinarily small amounts of total PSP are needed for a lethal dose (<1 mg) in humans, there is a significant need for convenient analytical methods for field and/or laboratory evaluation of potential exposure to PSP.

In the past a variety of assays have been utilized for the assessment of PSP but none of the tests are rapid enough for routine testing in the laboratory or field. The most widely utilized laboratory tests have been bioassays in animals with the current test for shellfish toxins being based upon a time-of-death mouse assay (Association of Official Analytical Chemists, 1984). The advantage of the assay is that it indicates total toxicity; but has the following distinct disadvantages: 1) need to maintain a mouse colony with mice in the 19-22 gram weight range, 2) sensitivity dependent on mouse strain, 3) possibility for underestimating marginally toxic clams, 4) subsequent time of death determination, 5) laborious to perform, and 6) relatively expensive (Shimizu, 1979). Consequently, a variety of assay methods have been developed as alternatives to the mouse assay. These include the following: a) fluorimetric and colorimetric methods based upon hydrogen peroxidase oxidation (Bates and Rapoport, 1975; Bates *et al.*, 1978; Gershey *et al.*, 1979; McFarren *et al.*, 1958) and b) high performance liquid chromatography methods (Rubinson, 1982; Sullivan *et al.*, 1982; Sullivan *et al.*, 1985). Each of the assays has a serious disadvantage which has prevented it from replacing the mouse assay. The peroxide oxidation methods require following a strict protocol (Bates *et al.*, 1978) and the neosaxitoxin (NEO) sub-group of the PSP is not efficiently detected (Bose *et al.*, 1979). The HPLC methods require extensive sample preparation and the need for expensive equipment and trained personnel. Furthermore, the HPLC methods for detection of the PSP toxins require the use of post-column derivatization since they do not give a significant UV/VIS absorption band.

An attractive alternative to the established assay methods would be an analysis based on anti-PSP antibodies. Several previous attempts have been made to prepare such antibodies but the results have been marginal at best because of relatively poor anti-PSP antisera and corresponding low sensitivities (Chu *et al.*, 1985; Johnson *et al.*, 1964; Johnson *et al.*, 1966; Koehn *et al.*, 1981). Our studies have led to successful production and characterization of an anti-PSP antisera using a stable saxitoxinol (STX-ol) - bovine serum albumin (BSA) conjugate (Carlson *et al.*, 1984). This antisera has been used in the development of an STX radioimmunoassay (Carlson *et al.*, 1984) and has been evaluated for use in anti-STX therapy (Davio, 1985).

GOALS

While the anti-STX-ol antisera has proven to be effective in therapeutic and analytical protocols for STX and closely related PSP, it does have the disadvantage that it does not cross-react with the NEO subgroup of the PSP and is only a weak competitor against the PSP binding protein. It has been our primary goal to produce anti- PSP antibodies which will be useful for both the analysis of the PSP and for prophylaxis/therapy of PSP poisoning. Analytical use of the antibody will require high cross-reactivity so that the antibody and the assay will be readily usable for the determination of total PSP. Therapeutic use of the antibody will require high PSP affinity so

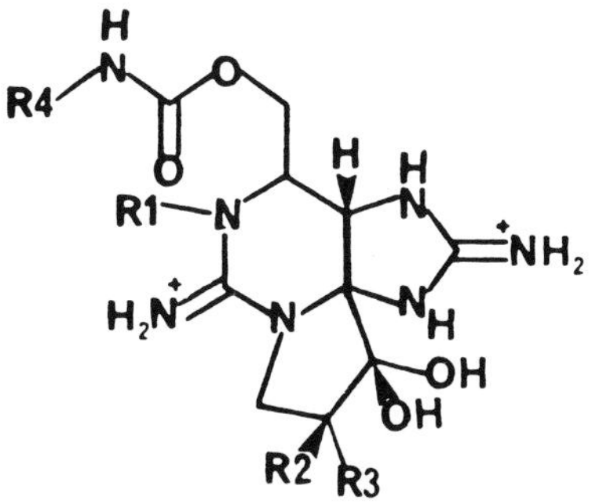

Compound	R1	R2	R3	R4
IA-- Saxitoxin (STX)	H	H	H	H
IB - Gonyautoxin2 (GTX2)	H	H	OSO_3	H
IC - Gonyautoxin3 (GTX3)	H	OSO_3	H	H
ID - Neosaxitoxin (NEO)	OH	H	H	H
IE - Gonyautoxin1 (GTX1)	OH	H	OSO_3	H
IF - Gonyautoxin4 (GTX4)	OH	OSO_3	H	H
IG - Saxitoxinol (STXOL)	H	H	H	H

Figure 1. Structures of the Paralytic Shellfish Poisons

that the antibody can effectively compete with the PSP binding protein for the toxin, thus protecting the individual from the toxic effects of the PSP.

The generation of a single antibody to the entire class of PSP is made difficult by the significant differences in structures throughout the series (Figure 1). One approach to this problem is to focus on the portion of the molecule which is common to all members of the family, utilizing that structure fragment as a hapten for antibody production. In this manner, the opportunity exists for antibody production which would exhibit good cross-reactivity toward all the PSP. In keeping with this approach the focus of our research has been on the use of the five-membered guanidinium ring species (1) as the hapten of choice. In addition to this compound, the six membered guanidinium ring system has also been a target because of the similarity in the synthetic route and because it is an integral part of each member of the PSP family.

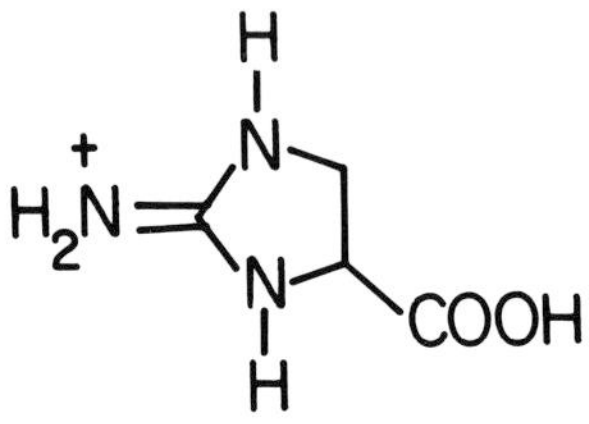

RESULTS AND DISCUSSION

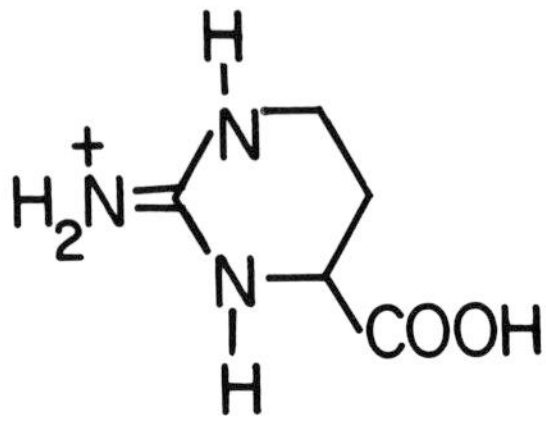

Synthesis

The synthesis of hapten (1) is outlined in Figure 2. S,S-Dimethyl-N-tosyliminodithiocarbonimidate (3) was easily prepared by reaction of p-toluenesulfonamide with carbon disulfide in a concentrated aqueous NaOH solution with a DMF co-solvent. The resulting anion was then trapped by alkylation with methyl iodide, giving a 78% yield of 3 after recrystallization (Gompper *et al.*, 1966).

The tosylated cyclic precursor 5 was prepared by first neutralizing 2,3-diaminopropionic acid monohydrochloride with two equivalents of aqueous NaOH and then refluxing with 3 in aqueous ethanol solvent for a period of 24 hours. The recovered product was recrystallized from water in 87% yield.

Finally, the desired guanidinium compound was prepared by detosylation of 5 using HBr and phenol (Snyder *et al.*, 1952a; Snyder *et al.*, 1952b). For this reaction, 5 was refluxed 2 hours in a mixture of the HBr and phenol, yielding a rather complex mixture of products whose purification proved quite difficult. However, selective removal of organic soluble impurities by extraction, followed by purification on an ion-exchange column gave the desired guanidinium compound 1 in an 88% yield. This material was thus suitable for coupling to protein through the carboxylic acid group, work to be discussed under the section on immunogen preparation.

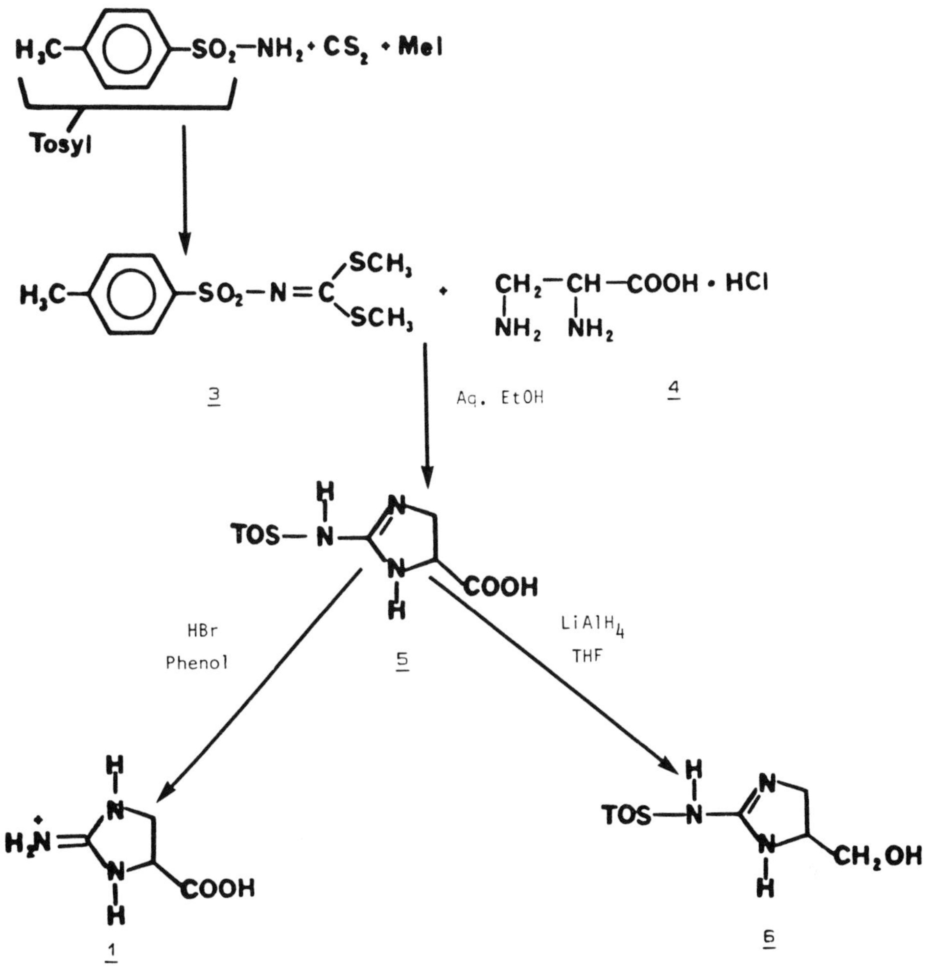

Figure 2. Synthetic Route to the Five-Membered Guanidinium Ring System

A second approach for coupling the five-membered ring to protein would proceed via a 1,1'-carbonyldiimidazole (CDI) coupling of the alcohol analog, obtainable by reduction of the acid group already prepared. Toward this goal, 6 has been prepared by reduction of 5 using lithium aluminum hydride (LAH) as the reducing agent in a THF solvent. The desired alcohol 6 was isolated in a 93% yield. Detosylation using the HBr/phenol technique has been attempted but product isolation and characterization is not complete at this time. The selective reduction of the carboxyl group in 1 is an alternative route to the alcohol that is also being investigated.

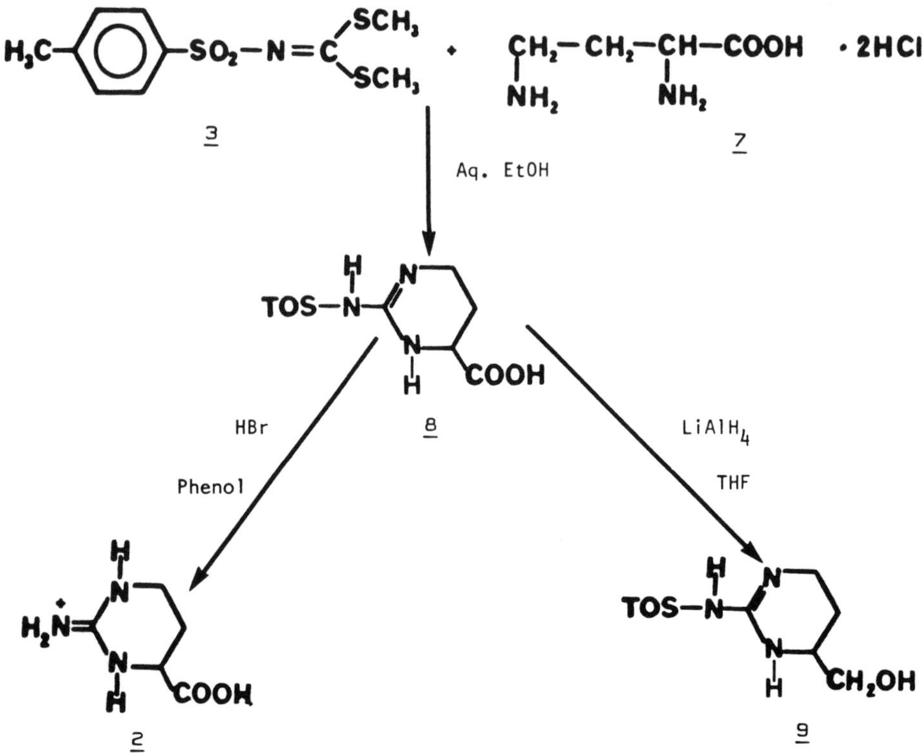

Figure 3. Synthetic Route to the Six-Membered Guanidinium Ring System

Using a similar synthetic strategy, the six-membered guanidinium compound 2 has been prepared as outlined in Figure 3. The (S) - (+)-2,4-diaminobutyric acid dihydrochloride (7) was first neutralized with three equivalents of aqueous NaOH and then refluxed 22 hours with 3 in an aqueous ethanol solvent. The resulting acid, 8, was isolated in an 85% yield after recrystallization from water. Detosylation of 8 was accomplished using HBr/phenol reflux (2 hours), giving a complex mixture of products. The final product was isolated by extraction of organic soluble impurities from an aqueous solution of impure 2, followed by ion-exchange chromatography to give a 93% yield of 2. Further recrystallization of that product gave a 54% yield of pure white crystals (Rodricks *et al.*, 1971). The corresponding alcohol derivative 9 has also been prepared in tosyl protected form by a LAH reduction in THF solvent, providing an 85% yield of a crystalline white solid. No effort has been made to provide the alcohol in the free guanidinium form but the chemistry being developed for the five-membered ring should be equally applicable to this material.

Immunogen Preparation

Having in hand the model compounds, the investigations have turned to coupling of the guanidinium acids (1 and 2) with protein for antibody production. The approaches considered for immunogen preparation were: 1) coupling of the N-ethoxycarbonyl-2-ethoxy-1, 2-dihydroquinoline (EEDQ) activated ester of 1 directly to bovine serum albumin (BSA); 2) coupling of the N-hydroxysuccinimide (NHS) or N-hydroxysulfosuccinimide (sulfo NHS) activated ester to protein; 3)preparation of 4-aminobutyric acid (GABA) or 3,3'-diamino-dipropylamine (DADPA) derivatives of 1 and coupling those analogs to BSA by either of the above activation methods; and 4) direct coupling of 1 by the use of 1-ethyl-3-(3-dimethylaminopropyl) carbodiimide (EDC).

Each of the above approaches has been pursued with varying success. Direct coupling of 1 to BSA with EDC failed to yield any of the desired immunogen. Examination of the reaction product indicated that only intramolecular cross-linking of the protein had resulted. The preparation of derivatives of 1 using the GABA and DADPA spacers has proven difficult and only small amounts of the desired materials could be isolated by preparative silica gel chromatography plates. Attempts were made to couple the GABA analog to BSA using the EEDQ and sulfo NHS activated esters and to directly couple the DADPA analog to BSA using water soluble EDC. Neither of these analogs could be coupled to BSA in sufficient yield to be useful as immunogens.

The most successful approach has involved the coupling of the sulfoNOS activated ester of 1 to BSA and modified BSA (described below). This activated ester was prepared by first preparing the free base of 1 using an Amberlite IRA-400 (OH form) ion exchange column. The free base was derivatized with sulfoNHS by reaction in DMF at pH 8 to 9 using dicyclohexylcarbodiimide (DCC) as the dehydrating agent. The resulting activated ester was then incubated with BSA or modified BSA in a pH 9 bicarbonate buffer. The resulting immunogen was dialyzed extensively against PBS and concentrated by ultrafiltration. Preliminary evidence indicated that 18-20 moles of hapten/mole of protein had been incorporated but available analysis techniques make this number less than certain. To accurately determine the loading of the protein, compound 1 is currently being prepared in radiolabeled form using the chemistry described previously, and will be utilized in the above coupling reactions.

To further enhance the loading of BSA with hapten, three modified BSA preparations have been produced: 1) direct coupling of the DADPA to the BSA carboxylic acid groups using EDC activation; 2) acetic anhydride addition to BSA to acetylate the free amines on BSA, followed by DADPA coupling to the remaining carboxylic acid groups using EDC activation; and 3) addition of succinic anhydride to BSA with subsequent coupling of DADPA to all carboxylic acid groups to give an increased number of amine sites available for coupling to the activated ester of 1. Evaluation of these modified BSA proteins is currently underway.

Polyclonal Antibody Production

Using immunogens prepared by the sulfoNHS coupling with BSA and modified BSA, efforts are currently underway to raise antibodies in rabbits. For each immunogen five rabbits were injected intradermally in multiple sites (40 sites - 0.1 mg/rabbit) along the back with an emulsion containing a 50% mixture of complete and incomplete Freund's Adjuvant. The rabbits were given subcutaneous booster injections (0.2 mg/rabbit in incomplete Freund's Adjuvant) at three week intervals. Evaluation of antibody response is in progress.

SUMMARY

The development of polyclonal antibodies having good cross-reactivity against the entire family of the paralytic shellfish poisons has been the focus of much research. One approach to this problem is the design of small hapten molecules which express the epitopes common to all the PSP toxins. Since the guanidinium ring systems are integral parts of all of the toxin molecules, synthetic routes to both the five- and the six-membered ring systems have been developed. In both cases, a carboxylic acid group is attached to the ring system for coupling to carrier protein via an activated ester route. The coupling of the five-membered guanidinium ring to a variety of carrier proteins has been investigated and the antibody response to these conjugates in rabbits is currently being evaluated.

REFERENCES

Agnew, W.S., S.R. Levinson, J.S. Brabson and M.A. Raftery. 1978. Purification of the Tetrodotoxin-Binding Component Associated with the Voltage-Sensitive Sodium Channel from *Electrophorus electric* vs Electroplax Membranes. Proc. Natl. Acad. Sci. USA. 75: 2606.

Association of Official Analytical Chemists. 1984. IN: Official Methods of Analysis. S. Williams (ed.). 14th Edition, Washington, D.C. p. 344.

Bates, H.A., R. Kostriken and H. Rapoport. 1978. A Chemical Assay for Saxitoxin. Improvements and Modifications. J. Agric. Food Chem. 26: 252.

Bates, H.A. and H. Rapoport. 1975. A Chemical Assay for Saxitoxin, the Paralytic Shellfish Poison. J. Agric. Food Chem. 23: 237.

Bose, R.J. and J.E. Reid. 1979. Evidence for the Heterogeneity of Paralytic Shellfish Toxin in Clam and Mussel Samples Gathered near Prince Rupert, British Columbia. IN: Toxic Dinoflagellate Blooms. D.L. Taylor and H.H. Seliger. (eds.). Elsevier North Holland, Inc., New York, NY. pp. 399-402.

Boyer, G.L., C.F. Wichmann, J. Mosser, E.J. Schantz and H.K. Schnoes. 1979. Toxins Isolated from Bay of Fundy Scallops. IN: Toxic Dinoflagellate Blooms. D.L. Taylor and H.H. Seliger. (eds.). Elsevier North Holland, Inc., New York, NY. pp. 373-376.

Carlson, R.E., M.L. Lever, B.W. Lee and P.E. Guire. 1984. Development of Immunoassays for Paralytic Shellfish Poisoning: A Radioimmunoassay for Saxitoxin. IN: Seafood Toxin. E.P. Ragelis (ed.). ACS Symposium Series, Washington D.C. 262: 181.

Catterall, W.A. 1980. Neurotoxins that Act on Voltage-Sensitive Sodium Channels in Excitable Membranes. Ann. Rev. Pharmacol. Toxicol. 20: 15.

Chu, F.S. and T.S.L. Fan. 1985. Indirect Enzyme-Linked Immunosorbent Assay for Saxitoxin in Shellfish. J. Assoc. Off. Anal. Chem. 68: 13.

Davio, S.R. 1985. Neutralization of Saxitoxin by Anti-Saxitoxin Rabbit Serum. Toxicon 23: 669.

Gershey, R.M., R.A. Neve, D.L. Musgrave and P.B. Reichardt. 1979. A Colorimetric Method for Determination of Saxitoxin. J. Fish Res. Board Can. 34: 559.

Gompper, R. and W. Hagele. 1966. N-Sulfonyl-iminodithiokohlensaureester und N-Sulfonyl-dithiourethane. Chem. Ber. 99: 2885.

Hall, S. and P.B. Reichardt. 1984. Cryptic Paralytic Shellfish Toxins. IN: Seafood Toxins. E.P. Ragelis (ed.). ACS symposium Series, Washington D.C. 262: 113-122.

Johnson, H.M., P.A. Frey, R. Angelotti, J.E. Campbell and K.H. Lewis. 1964. Haptenic Properties of Paralytic Shellfish Poison Conjugated to Proteins by Formaldehyde Treatment. 1964. Proc. Soc. Exp. Biol. Med. 117: 425.

Johnson, H.M. and G. Mulberry. 1966. Paralytic Shellfish Poison: Serological Assay by Passive Haemagglutination and Bentonite Flocculations. Nature 211: 747.

Kao, C.Y. 1981. Tetrodotoxin, Saxitoxin, Chiriquitoxin: New Perspectives on Ionic Channels. Fed. Proc. 40: 30.

Koehn, R.E., V.E. Ghazarossian, E.J. Schantz, H.K. Schnoes and F.M. Strong. 1981. Derivatives of Saxitoxin. Bioorg. Chem. 10: 412.

McFarren, E.F., E.J. Schantz, J.E. Campbell and K.H. Lewis. 1958. Chemical Determination of Paralytic Shellfish Poison in Clams. J. Assoc. Off. Agric. Chem. 41: 168.

Rodricks, J.V. and H. Rapoport. 1971. Synthesis of Cyclic Guanidines. J. Org. Chem. 36: 46.

Rubinson, K.A. 1982. HPLC Separation and Comparative Toxicity of Saxitoxin and its Reaction Products. Biochem. Biophys. Acta. 687: 315.

Shimizu, Y. and E. Ragelis. 1979. Alternatives to the Mouse Assay. IN: Toxic Dinoflagellate Blooms. D.L. Taylor and H.H. Seliger (eds.). Elsevier North Holland, Inc., New York, NY. pp. 453-455.

Snyder, H.R. and H.C. Geller. 1952. Cleavage of Sulfonamides with Aqueous Hydrobromic Acid and Phenol. II. J. Am. Chem. Soc. 74: 4864.

Snyder, H. R. and R.E. Heckert. 1952. A Method for the Rapid Cleavage of Sulfonamides. J. Am. Chem. Soc. 74: 2006.

Sullivan, J.J. and W.T. Iwaoka. 1982. High Pressure Liquid Chromatographic Determination of Toxins Associated with Paralytic Shellfish Poisoning. J. Assoc. Off. Anal. Chem. 66: 297.

Sullivan, J.J., M.M. Wekell and L.L. Kentala. 1985. Application of HPLC for the Determination of PSP Toxins in Shellfish. J. Food Sci. 50: 26.

MONOCLONAL ANTIBODIES (MONOABS) TO CIGUATOXIN AND RELATED POLYETHERS

Y. Hokama, S.A.A. Honda, M.N. Kobayashi,
L.K. Nakagawa, L.K. Shirai and J.T. Miyahara

Department of Pathology
University of Hawaii, Manoa
Honolulu, Hawaii 96822

INTRODUCTION

An immunological approach for the examination of low molecular weight lipid polyether marine toxins in fish tissues was initiated in 1977 with the development of a modified radioimmunoassay (RIA) using anti-ciguatoxin antibody prepared in sheep (Hokama *et al.*, 1977). Subsequently, a solid phase enzyme immunoassay (EIA) using the same sheep anti-CTX was established (Hokama *et al.*, 1983; Hokama *et al.*, 1984). The EIA procedure permitted the evaluation of cross-reactivity between ciguatoxin (CTX), okadiac acid (OA), brevetoxin (PbTX), maitotoxin (MTX) and monensin by competitive EIA using these purified toxins (Hokama *et al.*, 1984). In 1984, Baden *et al.*, developed a competitive radioimmunoassay for assessment of PbTX. More recently (Baden *et al.*, 1985), using this same RIA procedure and goat anti-PbTX the cross-reactivity of PbTX and CTX was demonstrated.

The cross-reactivity between CTX and related polyethers shown for the pure compounds (Baden *et al.*, 1985; Hokama *et al.*, 1984) may also occur in detection in fish tissues, which may pose a problem in attempts to screen fishes for the presence of CTX. The less toxic OA and polyether ionophores (Hokama *et al.*, 1984; Hokama *et al.*, 1985b; Westley, 1975) may contribute to increasing the number of false positives in the test for screening CTX specifically.

In attempts to circumvent this high cross-reactivity shown with the sheep anti-CTX (Hokama *et al.*, 1983; Hokama *et al.*, 1984), we have employed hybridoma technology for the preparation of monoclonal antibodies to CTX and OA (Hokama *et al.*, 1985b). The results of this study are presented in this report.

MATERIALS AND METHODS

Purified Polyether Toxins

Purified okadaic acid, isolated from Halichondria (Tachibana *et al.*, 1981), and ciguatoxin, isolated from Gymnothorax livers (Nukina *et al.*, 1984; Scheuer *et al.*, 1967; Tachibana *et al.*, 1986) were obtained through the courtesy of Professor P.J. Scheuer of the Department of Chemistry, University of Hawaii, Honolulu, Hawaii.

Preparation of Monoclonal Antibody to Okadaic Acid

Okadiac acid (0.5 mg) was conjugated to 10 mg human serum albumin (HSA) in 7.5 ml 0.05 M phosphate buffered saline (PBS) by the carbodiimide procedure (Hokama *et al.*, 1985a).

Three BALB/c mice, 10 weeks of age, were injected intraperitoneally once a week for three consecutive weeks with 0.10 ml of the okadaic acid-HSA conjugate and a booster was given on the 4th or 6th week, 3 days prior to being sacrificed. Each mouse received a total of 50 µg OA/1 mg HSA and 0.2 ml Freund's Complete Adjuvant (FCA) subcutaneously for each injection of 0.1 ml conjugate.

The non-immunoglobulin synthesizing mouse myeloma cells selected for fusion were those previously reported by Kearney *et al.*, (1978) designated PBX63-Ag8.65B. These cells were grown in Dulbecco's Modified Eagles Medium (DMEM) supplemented with 10% fetal calf serum (FCS). Myeloma cells in the logarithmic growth phase were used for the fusion.

The method of Köhler and Milstein (1975) as described in detail by Schreier *et al.*, (1980) was used for the fusion of the okadaic acid-HSA sensitized BALB/c spleen cells with myeloma cells. The selection of hybridomas, limiting dilution methods, etc. were those previously reported (Schreier *et al.*, 1980).

MonoAb to CTX was prepared in the same manner as for OA, except that a much lower concentration of CTX was conjugated to human serum albumin. Approximately 1-10 µg of CTX was conjugated to 0.1-1.0 mg of HSA. MonoAbs from hybridomas were designated 10C3 and 5C8 for monoAb-OA and monoAb-CTX, respectively.

Sheep Anti-CTX-horseradish Peroxidase

The sheep anti-CTX used in this study was prepared previously (Hokama *et al.*, 1977; Hokama *et al.*, 1983) and coupled to horseradish peroxidase (HRP) according to the method of Voller *et al.* (1978).

Ammonium Sulfate Preparation of MonoAb

Hybridoma culture supernatants, ascites and sheep anti-ciguatoxin serum were fractionated by ammonium sulfate precipitation. One hundred ml of culture supernatants were treated with 30 g of solid ammonium sulfate. The salt was allowed to dissolve at room temperature with gentle stirring. The resulting cloudy solution was placed in the cold room at 4°C overnight and the residue collected by centrifugation at 3000 rpm for 15 min. The precipitate was washed twice with a solution of 50% saturated ammonium sulfate, then collected by centrifugation and dissolved in 10 ml of distilled water and placed into a dialysis membrane. The latter was previously boiled in distilled water containing a few crystals of dissolved EDTA. The residue in the dialysis tubing was then immersed in 1 liter of 0.01 M PBS, pH 8.0 buffer. The dialysis buffer was changed 3 times at 6 hr intervals.

Preparation of coated beads

The beads from Falcon Assay Screening Test (FAST) plates (Beckton-Dickinson, Oxnard, California) were coated with an oil-base, high-gloss enamel paint by dipping the entire plate cover containing beads row by row into a trough containing the paint. The excess paint was removed by tapping and blotting gently twice onto a paper towel. The beads were air-dried overnight.

Enzyme Immunoassay for the Selection of Monoclonal Antibody from the Hybridomas

The procedure for the direct detection of the specific antibody from the hybridoma culture supernatants, Protein A fractions and BALB/c ascites of injected hybridomas was carried out as follows: (a) Enamel paint coated beads of FAST plates were coated with either purified OA (50 ng/ml of methanol) or purified CTX (50 ng/ml of methanol), then washed in Tris-buffer containing 5% FCS; (b) the coated beads were then immersed in 200 µl of ammonium sulfate fractions to be tested (culture supernatants and ascites) in a 96 well microwell plate for 30 min; (c) the beads were removed and washed three times in 10 nM Tris-buffer, pH 8.0, containing 0.05%

Tween 20 and 0.1% merthiolate; (d) the beads were then placed in a solution of goat anti-mouse immunoglobulin horseradish peroxidase antiserum at 37° C for 60 min; (e) the beads were then washed three times in 0.05 M phosphate, 0.14 M NaCl buffer, pH 7.4; (f) then the beads were immersed in wells containing 150 µl HRP substrate for 30 min at room temperature and the subsequent color reaction (yellow-orange) stopped with 150 µl 1 M sulfuric acid. The final color reaction was read at 492 nm in the Multiscan Titertek (Flow Laboratories, Inc., McLean, Virginia). The HRP substrate, (step f), consisting of 4% phenylenediamine in distilled water, was prepared 15 min prior to the end of step (c) and mixed with an equal part of 0.3% hydrogen peroxide. This 1:1 solution mixture was further diluted 1 part to 9 parts with citrate-phosphate buffer (17 nM citric acid, 65 nM phosphate, pH 6.3, with 0.1% merthiolate). The sequence and concept of the test procedure are presented in Figure 1.

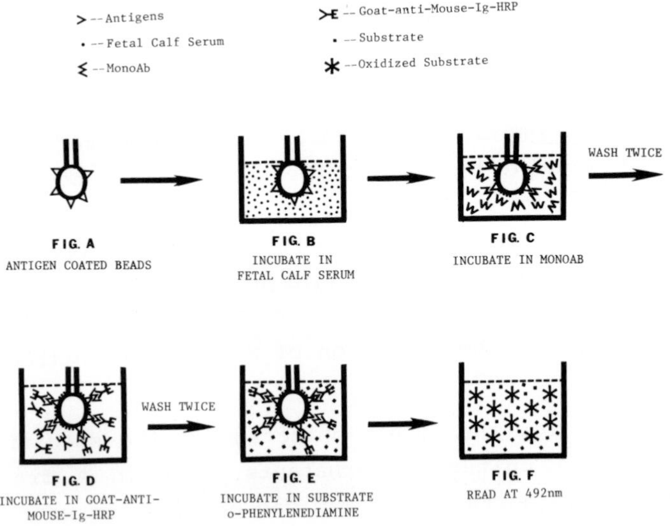

Figure 1. Illustration of the coated bead EIA concept and the sequence in which the test procedure is carried out (see methods section for details of the test).

Assessment of MonoAb to CTX and OA

The procedure for assessing cross-reactivity of monoAb-CTX (5C8) and monoAb-OA (10C3) was similar to that described in the preceding paragraphs for selection of the monoclones except as follows: (1) Various concentrations of CTX, OA, ionomycin and A23187 were used to coat the painted beads. Concentrations from 0.5 to 20 ng/ml of methanol of toxins or ionophores were examined. (2) MonoAb-CTX and monoAb-OA were used in 2 to 10 µg protein/ml concentrations in buffer. The MonoAbs were partially purified by ammonium sulfate precipitation as described in the earlier paragraph. (3) Sheep anti-CTX conjugated to horseradish peroxidase was used at 10 µg protein/ml and used directly against the coated beads without the use of a secondary antibody. The substrate used for the sheep anti-CTX-HRP was also 4% phenylenediamine in distilled water containing hydrogen peroxide. (4) The OD was examined at 492 nm.

The results are means of triplicate determinations of OD at 492 nm.

RESULTS

MonoAb-CTX at 10 µg protein/ml reacted with CTX (10 ng/ml) to give an O.D. of 1.24 at 492 nm, while OA, ionomycin and A23187 showed 0.18, 0.06 and 0.09 O.D. at 492 nm, respectively. These represented 16%, 2.0% and 9.5% respectively, for OA, ionomycin and A23187 reactivity with MonoAb-CTX. Thus, a greater activity of monoAb-CTX to homologous CTX was demonstrated than for the other polyethers, when all were present at the same concentrations (Figure 2).

Figure 3 summarizes the reactions between monoAb-CTX (5C8) at 2 different dilutions, 1:10 and 1:50, against various concentrations of purified CTX and OA. Optimum binding of monoAb-CTX occurred at 10 ng/ml CTX for both 1:10 and 1:50 dilutions of the monoAb-CTX. Little or no reactions were shown for monoAb-CTX against purified OA at 10 ng/ml. However, monoAb-CTX at both dilutions reacted with OA at 5 ng/ml.

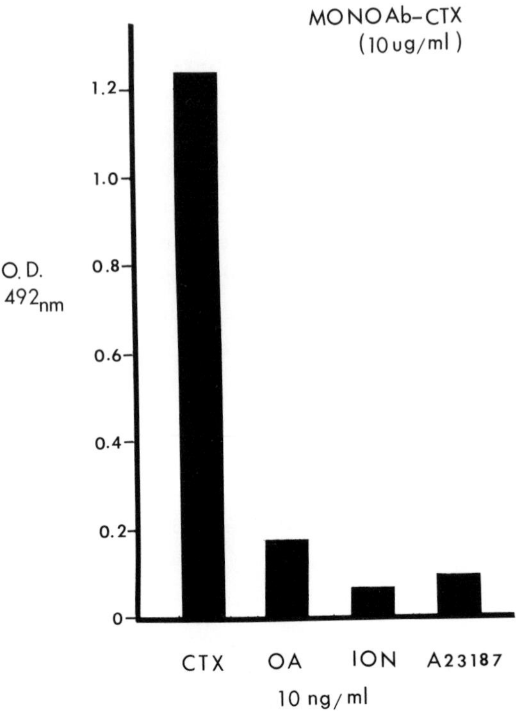

Figure 2. Examination of monoAb-CTX (5C8) with a concentration of 10 μg protein/ml against ciguatoxin (CTX), okadaic acid (OA), ionomycin (ION) and ionophore A23187 at 10 ng/ml of methanol. Optical density (OD) read at 492 nm.

MonoAb-OA

On the other hand, monoAb-OA (10C3) at 2 μg/ml concentration reacted with its homologous purified OA, at 5 ng/ml concentration to give an O.D. at 492 nm of 0.68 which is 3.4 times greater than that of CTX which gave an O.D. of 0.20 at 492 nm. Therefore, the CTX reaction was approximately 30% of that of OA. Little or no reaction was demonstrated between ionomycin (0.15%) and A23187 (3.3%) with monoAb-OA (Figure 4).

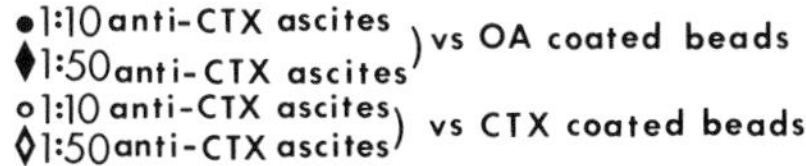

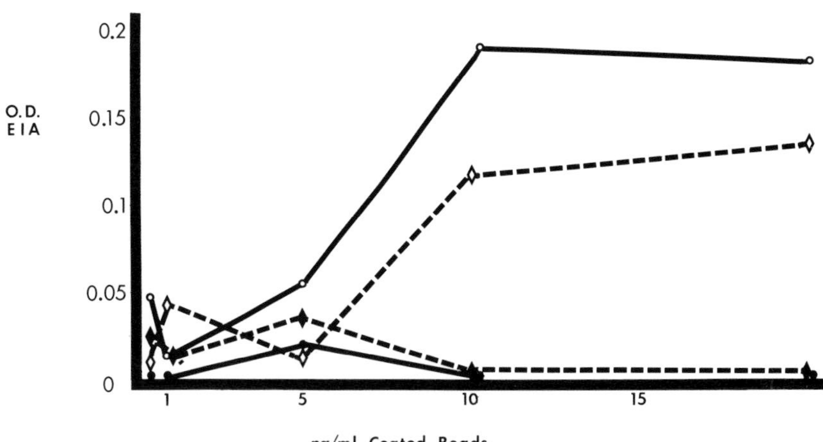

Figure 3. Examination of monoAb-CTX at two dilutions, 1:10 and 1:50, against coated beads containing various concentrations of ciguatoxin (CTX), okadaic acid (OA), and ionophores. Optical density (OD) read at 492 nm.

MonoAb-OA at 2 µg Ab protein/ml examined with varying concentrations of CTX or OA ranging from 0.5 to 20.0 ng/ml in methanol is shown in Figure 5. At the optimum reaction, 5 ng/ml toxins, monoAb-OA reacted with OA (homologous system) to give an O.D. at 492 nm of 0.64 in contrast to CTX at 0.20 (30.0% cross-reactivity). A significant decrease in reactivity occurred at 10 ng/ml and greater. This represents a classical precipitin or Ab-Ag complex pattern.

Sheep anti-CTX-HRP

Sheep Ab-CTX-HRP (10 g/ml) reacted equally with both CTX and OA at concentrations of 1 and 10 ng/ml in methanol as shown in Figure 6. A slightly greater activity with OA was observed. This suggests the close relationship of the antigenic determinants present in CTX and OA, both polyethers of dinoflagellate origin.

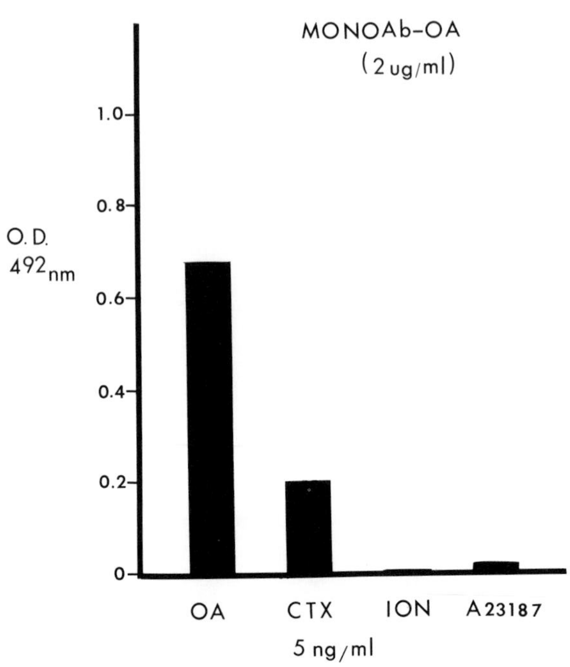

Figure 4. Examination of monoAb-OA (10C3) at 2 µg protein/ml concentration tested with 5 ng/ml toxins and ionophores; okadaic acid (OA), ciguatoxin (CTX), ionomycin (ION) and ionophore A23187. Optical density (OD) read at 492 nm.

DISCUSSION

A polyclonal sheep anti-CTX-HRP prepared in 1977 and in 1983 (Hokama *et al.*, 1977; Hokama *et al.*, 1983) has been employed for the assessment of ciguatoxin and related polyethers by radioimmunoassay, enzyme immunoassay and more recently a simple stick-EIA test (Hokama *et al.*, 1977; Hokama *et al.*, 1983; Hokama *et al.*, 1984; Hokama, 1985; Kimura *et al.*, 1982a; Kimura *et al.*, 1982b). This polyclonal antibody demonstrated an equal affinity for both ciguatoxin and okadiac acid as was shown previously by EIA (Hokama *et al.*, 1984), and confirmed in this study (Figure 6). This ability of sheep anti-CTX-HRP to react with CTX, OA and other polyethers obviously accounted for the so-called false positives reported in previous RIA, EIA and stick-EIA studies (Hokama *et al.*, 1977; Hokama, 1985; Kimura *et al.*, 1982a; Kimura *et al.*, 1982b). However, in these same studies no <u>false negatives</u> were reported.

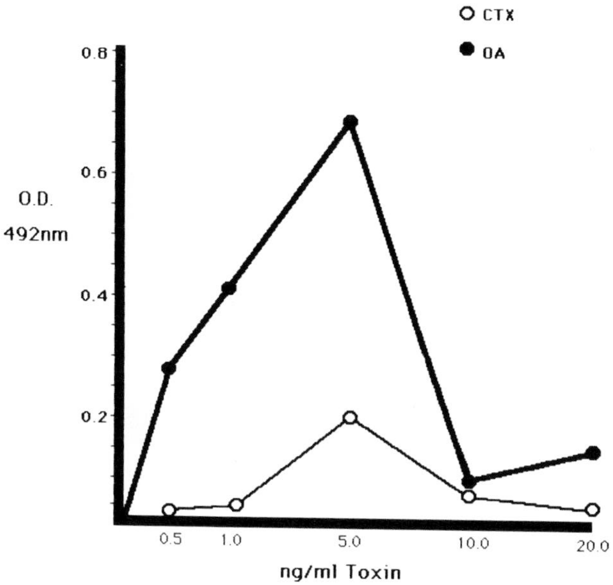

Figure 5. Examination of monoAb-OA (10C3, 4/4) at 1:5000 dilution (2 μg/protein/ml) against OA and CTX coated beads containing various concentrations of the toxins (1,5,10, and 20 ng/ml of methanol). Optical density (OD) read at 492 nm.

In this study the preparation of monoclonal antibodies to CTX and OA has been presented. A preliminary report on monoAb to okadiac acid was presented previously (Hokama *et al.*, 1985b). The examination of this monoAb-OA (10C3) confirms our previous observations of its moderate cross-reactivity with CTX (Hokama *et al.*, 1984). MonoAb-CTX (5C8) appears to react specifically with CTX. It showed some cross-reactivity with OA (16%), but little or none with ionophores (polyethers), ionomycin and A23187 (Westley, 1975). This cross-reactivity of CTX and OA with monoAb-CTX (5C8) is not surprising since it has been suggested that part of the structure of CTX is very similar to OA (Tachibana *et al.*, 1986).

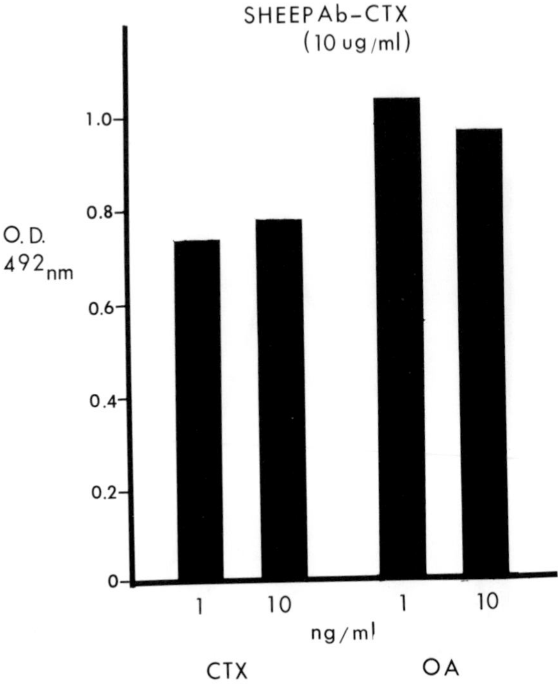

Figure 6. Sheep anti-CTX (sheepAb-CTX) at 10 µg protein/ml against ciguatoxin (CTX) and okadaic acid (OA) at 1 and 10 ng/ml of methanol. Optical density (OD) read at 492 nm.

These monoAbs are presently being employed to examine toxic fishes (implicated in ciguatera poisoning), pre-screening of fishes caught in areas suspected of ciguatoxin containing fishes and in routine examination of reef fishes. The simplified stick enzyme immunoassay (Hokama, 1985) is being used for this study.

The method of using oil-base paint coated beads for the solid phase enzyme immunoassay in this report is an extension of an earlier study (Hokama *et al.*, 1985b). This report verifies the usefulness of the oil-base paint coated beads to extract and enhance the adherence of lipid ciguatoxin and related polyethers to the surface of the beads as postulated in Figure 1A.

It is hoped to further clone and examine monoAbs to CTX and other polyethers to enhance the specificity for each toxin, since the relative toxicity of each of the polyethers vary greatly. Furthermore, recent observations in our laboratory suggests that ciguatoxin in herbivores may differ slightly from that found in carnivores and that alteration of the structure of ciguatoxin(s) may occur through metabolism in passage through different fish species.

ACKNOWLEDGEMENTS

This study has been supported in part by the Division of Aquatic Resources, Department of Land and Natural Resources, State of Hawaii.

REFERENCES

Baden, D.G., T.J. Mende and L.E. Brand. 1985. Cross-reactivity in immunoassays against toxins isolated from *Ptychodiscus brevis*. IN: Toxic Dinoflagellates, Anderson, D.M., A.W. White and D.G. Baden (eds.). Elsevier Science Publ. Co., Inc., NY. pp 363-368.

Baden, D.G., T.J. Mende, J. Walling and D.R. Schultz. 1984. Specific antibodies directed against toxins of *Ptychodiscus brevis* (Florida's red tide dinoflagellate). Toxicon 22: 783-790.

Hokama, Y. 1985. A rapid simplified enzyme immunoassay stick test for the detection of ciguatoxin and related polyethers from fish tissues. Toxicon. 23: 939-946.

Hokama, Y, M.A. Abad and L.H. Kimura. 1983. A rapid enzyme immunoassay for the detection of ciguatoxin in contaminated fish tissues. Toxicon 21: 817-824.

Hokama, Y, A.H. Banner, and D. Boyland. 1977. A radioimmunoassay for the detection of ciguatoxin. Toxicon 15: 317-325.

Hokama, Y, L.H. Kimura, M.A. Abad, L. Yokochi, R.J. Scheuer, M. Nukina, T. Yasumoto, D.G. Baden and Y. Shimizu. 1984. An enzyme immunoassay for the detection of ciguatoxin and competitive inhibition by related polyether toxins. IN: Seafood Toxins, ACS Symposium Series 262, E.P. Ragelis (ed.). Washington, D.C., Am. Chem. Soc., pp 307-230.

Hokama, Y, L.H. Kimura and J.T. Miyahara. 1985a. Immunological approaches to understanding marine toxins. IN: Aquaculture, Public Health Regulatory Management Aspects. C.E. Kimble (ed.). Zarcon Press, Zarcon Corp., Silver Spring, MD. pp 80-96.

Hokama, Y., A.M. Osugi, S.A.A. Honda and M. Matsuo. 1985b. Monoclonal antibodies in the detection of ciguatoxin and related toxic polyethers in fish tissues by a rapid stick test. IN: Proceed., 5th Int. Coral Reef Cong., Tahiti, French Polynesia. 4: 449-455.

Kearney, J.F., A. Radbruch, B. Liesegang and K. Rajewsky. 1978. A new mouse myeloma cell line that has lost immunoglobulin expression, but permits the construction of antibody-secreting hybrid cell lines. J. Immunol. 123: 1548-1550.

Kimura, L.H., M. A. Abad, and Y. Hokama. 1982a. Evaluation of the radioimmunoassay (RIA) for detection of ciguatoxin (CTX) in fish tissues. J. Fish. Biol. 21: 671-680.

Kimura, L.H., Y. Hokama, M. A. Abad, M. Oyama and J.T. Miyahara. 1982b. Comparison of three different assays for the assessment of ciguatoxin in fish diseases: Radioimmunoassay, mouse bioassay and *in vitro* guinea pig atrium assay. Toxicon 20: 907-912.

Köhler, G. and C. Milstein. 1975. Continuous cultures of fused cells secreting antibody of predefined specificity. Nature 256: 494-497.

Nukina, M., L.M. Koyangi and P.J. Scheuer. 1984. Two interchangeable forms of ciguatoxin. Toxicon 22: 169-176.

Scheuer, P.J., W. Takahashi, J. Tsutsumi, and T. Yoshida. 1967. Ciguatoxin: Isolation and chemical nature. Science 155: 1267-1268.

Schreier, M., G. Köhler, H. Hengartner, C. Berek, M. Trucco and L. Forni. 1980. Hybridoma Techniques. Cold Harbor Laboratory, Cold Spring, NY.

Tachibana, K., M. Nukina, Y.G. Joh and P.J. Scheuer. 1986. Recent developments in the molecular structure of ciguatoxin. Biol. Bull., in press.

Tachibana, K., P.J. Scheuer, Y. Tsukitani, H. Kikuchi, D. Van Engen, J. Cardy, Y. Gopichand and F. Schmitz. 1981. Okadaic acid, A cytotoxic polyether from two marine sponges of the genus *Halichondria*. J. Am. Chem. Soc. 103: 2469-2471.

Voller, A., A. Bartlett and D.E. Bidwell. 1978. Enzyme immunoassay with special references to ELISA techniques. J. Clin. Path. 31: 507-520.

Westley, J. W. 1975. The polyether antibiotics: Monocarboxcylic acid ionophores. Ann Rep. Med. Chem. 10: 246-256.

IMMUNOFLUORESCENT REAGENTS: PREPARATION AND ANALYSIS

John F. Daley
Division of Tumor Immunology
Dana-Farber Cancer Institute
44 Binney Street
Boston, MA 02115

INTRODUCTION

Immunofluorescent staining techniques for detecting cell surface antigens have evolved over the past forty-five years (Coons *et al.*, 1941) to the point where it is now possible to detect up to five cell surface antigens simultaneously (Horan *et al.*, 1986). To a great degree, this precision has been dependent upon advances in three areas. First, instruments such as flow cytometers were created that could analyze cellular immunofluorescence with much greater speed and precision than analysis utilizing fluorescence microscopes. Second, the introduction of monoclonal antibodies as probes to detect cellular antigens provided remarkable specificity and uniformity. Finally, chemical modifications of fluorochrome molecules facilitated simple conjugation and purification procedures as well as produced fluorochromes which exhibited distinct spectral emission characteristics.

This discussion will focus upon methods that are employed to create useful immunofluorescent conjugates for a variety of applications. Also to be discussed will be procedures that are routinely employed to evaluate conjugate products.

FLUOROCHROMES

Fluorescein is the most commonly used fluorochrome for immunofluorescent labelling. This is due in part to the high efficiency of the fluorescein molecule in generating strong fluorescence emission. Also, when fluorescein is modified to contain an isothiocyanate group (FITC), conjugation is a simple procedure and generally results in a specific, sensitive and stable conjugate. Traditionally, rhodamine derivatives have been chosen when a second label was desired to simultaneously detect the presence of two independent antigens. Rhodamine derivatives were observed to be substandard when compared to the desirable features exhibited by FITC. Rhodamine conjugates were difficult to purify and displayed considerable non-specific fluorescence when employed as second labels. Rhodamine conjugates were also much less efficient in generating a fluorescent signal that was as intense as the FITC counterparts. The most successful rhodamine derivative developed was a sulfonyl chloride derivative of Rhodamine 101, commonly known as "Texas Red" (Molecular probes, Junction City, Oregon) (Titus *et al.*, 1982). Texas Red has several advantages over other rhodamine derivatives. Most important, is the ability to remove excess Texas Red from conjugated protein when conducting conjugation purification. Addition of alkyl substituents to the Rhodamine 101 molecule resulted in increased quantum efficiency and shifted the excitation and emission maxima to 596 nm and 620 nm, respectively. The excitation and emission properties of Texas Red made it a suitable choice when dual labelling with FITC-conjugated antibodies were desired.

In 1982, Glaser, Stryer, and Oi (1982) reported the use of phycobiliprotein molecules to create immunoconjugates (Oi *et al.*, 1982). Phycoerythrin (PE), Phycocyanin (PC), and Allophycocyanin (APC) were conjugated to a variety of antibodies and also onto avidin. The phycobiliprotein conjugates were found to be extremely sensitive and specific, producing a stokes shift three times greater than FITC conjugates. In addition, flow cytometers had little difficulty in distinguishing the fluorescence emission of phycoerythrin from fluorescein. Unlike Texas Red, phycobiliproteins could be directly conjugated onto antibodies by employing heterobifunctional linking agents. It was noted that direct PE-conjugates displayed a greater fluorescent signal to noise ratio than direct FITC conjugates and exhibited lower non-specific background fluorescence (Glazer and Stryer, 1984).

Another phycobiliprotein that is useful in immunofluorescence is Allophycocyanin (APC). The fact that APC exhibits fluorescence excitation and emission maxima in the deep red region of the visible spectrum (EX_m = 650 nm, EM_m = 670 nm), makes this phycobiliprotein an ideal candidate when three or four color immunofluorescence analysis is desired. Table 1 describes properties of the four major fluorochromes used currently in flow cytometry. The excitation and emission spectra of fluorescein, PE and Texas Red is shown in Figure 1.

Table 1. Major fluorochromes used by flow cytometry for immunofluorescence.

		mol. wt.	max. abs.	max. emsn.
1)	Fluorescein Isothiocyanate	389.1	489nm	514nm
2)	Sulforhodamine 101 (Texas Red)	625.0	596nm	615nm
3)	Phycoerythrin	240,000	488nm 545nm	576nm
4)	Allophcocyanin	104,000	650nm	670nm

METHODS OF STAINING

The detection of cell surface antigens by immunofluorescence is accomplished usually by one of two methods: Direct and Indirect. In the direct method, a fluorochrome conjugated antibody is directly incubated with a single cell suspension or tissue section. After an incubation period typically lasting for 10 to 30 min on ice, excess unbound antibody is removed by extensive washing with media supplemented with 2.5% newborn calf serum. Background fluorescence is determined by incubating a similar cell preparation with a fluorochrome conjugated antibody known to be non-reactive with the cell sample being tested. The sample being incubated with non-specific antibody is commonly called the "negative control" and should be of the same subclass and protein concentration as the test antibody. The ratio of fluorochrome to antibody should be comparable between the negative control and test antibody.

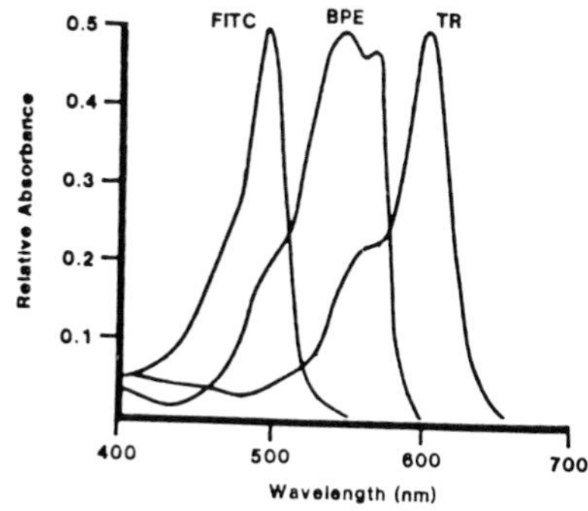

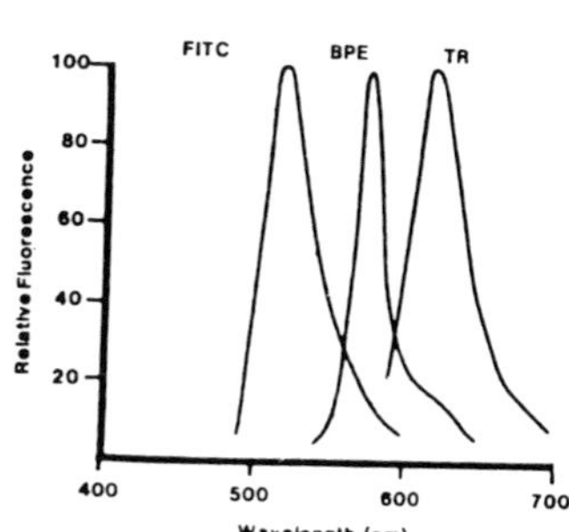

Figure 1. Excitation and emission spectra of FITC, PE and Texas Red (reprinted with permission from Clin. Chem. 29/9: 1583).

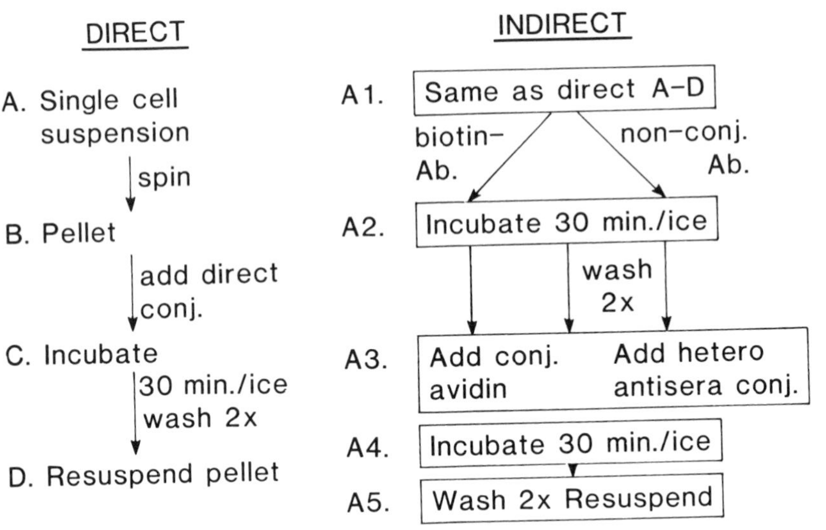

Figure 2. Methods of staining cell samples for immunofluorescence. Steps A-D represent the direct method. Steps A1-A5 represent the indirect method using either fluorochrome conjugated species specific heteroantisera or fluorochrome conjugated avidin.

The indirect method employs a two-step procedure to detect the presence of antigen. First, test antibody is incubated in a manner as described above and then a fluorochrome conjugated reagent is added. The second step reagent is directed against determinants on the primary antibody and should be minimally reactive with antigens on the cell surface. Species specific antisera which is fluorochrome conjugated is frequently used as second step reagents (e.g. FITC-goat anti-mouse to detect mouse monoclonal antibody used in first step reaction).

The biotin/avidin system is another type of indirect labelling (Bayer and Wilcheck, 1976). Biotin conjugated antibody is used as primary antibody. Fluorochrome labelled avidin is then added to bind to any biotin which is covalently attached to the primary antibody. Figure 2 details methods of staining routinely employed in our laboratory.

FLUOROCHROME CONJUGATION OF MONOCLONAL ANTIBODIES

Prior to any conjugation procedures, mouse ascites containing antibody is first tested to insure that antibody activity is present. Flow cytometric analysis provides a convenient method to assay for antibody activity using the species specific indirect method. Typical dilutions for assaying immunoreactivity of monoclonal antibody primary reagent fall between the range of 1/50 to 1/10,000. Once the ascites has been determined to be acceptable with respect to antibody activity and specificity, purification of the immunoglobulin fraction is performed.

In our laboratory, ammonium sulfate purification is routinely employed when FITC or biotin conjugates are desired (Weir *et al.*, 1978). Briefly, an equal volume of saturated ammonium sulfate is added to an aliquot of mouse ascites which contains a monoclonal antibody. The precipitate that is formed is then pelletized by centrifugation and the pellet is resuspended and washed two times with a 50% ammonium sulfate solution. After the second centrifugation wash, the pellet is resuspended with a minimal volume of phosphate buffered saline (PBS) pH = 7.0 and exhaustively dialyzed against the PBS buffer. Ammonium sulfate purified antibody is then ultracentrifuged at 100,000 G for 20 min to remove aggregates and tested again for antibody reactivity and specificity. Purified antibody is then dialyzed against 0.2 M carbonate/bicarbonate buffer containing 1 x PBS (pH 9.0) for four hours at 4°C. The antibody solution is removed and the protein concentration is adjusted to either 10 mg/ml for FITC conjugation or 4.3 mg/ml for biotin conjugation.

For FITC conjugation, 0.5 mg to 1.0 mg of FITC isomer I powder is added to a 1.0 ml antibody solution containing 10 mg of protein. The mixture is then mixed gently in the dark for one hour at room temperature. After one hour, the reaction mixture is then added to a Sephadex G-25 (Pharmacia) column to remove excess unbound dye from conjugated antibody. Column sizes are calculated by multiplying the reaction mixture volume by a factor of 10 (e.g., 1 ml reaction mixture should be passed through a 10 ml column). Free FITC dye, being of low molecular weight is retained at the top of the column. FITC conjugated antibody passes through the column in the void volume and is easily identified. The FITC conjugate is then ultra-centrifuged at 100,000 G for 20 min to remove aggregates.

Calculation of F/P Ratio

FITC: $E_{495} = 0.175$

" : $E_{280} = 0.054$

1) measure $O.D._{280}$ and $O.D._{495}$

2) $$\frac{O.D._{495}}{0.175} = \text{ug } F^{*}/\text{ml}$$

3) $$O.D._{280} - (\text{ug}F^{*}/\text{ml} \times 0.054) = O.D._{280\,\text{protein corrected}}$$

4) $$\frac{O.D._{280\text{ corrected}}}{1.4} = \text{mg/ml Protein}$$

5) $$\text{Molar Ratio} = \frac{\text{ug}F^{*}/\text{ml}}{389} \div \frac{\text{mg/ml protein} \times 10^{3}\ (\text{gram protein})}{1.5 \times 10^{5}\ {}^{**}} = \text{F/P ratio}$$

m.w. FITC = 389

m.w. IgG = 1.5×10^{5}

** if IgM use 9.0×10^{5}

Figure 3. Formula for calculating fluorescein to protein ratio (f/p). Protein concentration can also be determined using this formula.

Absorbance measurements of the FITC conjugate at 280 nm and 495 nm allow determination of the fluorochrome to protein ratio (F/P). Figure 3 shows the formula for calculating the F/P ratio.

Biotin conjugation of antibodies is the simpliest of all conjugation procedures. Antibody which has been dialyzed against the 0.2 M carbonate/bicarbonate buffer described above is adjusted to a concentration of 4.3 mg/ml. 200 microliters of a 1.7 mg/ml solution of N-hydroxysuccinimide ester of biotin (Calbiochem) is added per 1.0 ml of antibody solution. Since the solvent used to dissolve the NHS-biotin is N,N-dimethylformamide, the use of the plastic test tubes and pipettes should be avoided during the conjugation procedure.

Texas Red conjugation is performed in a manner similar to FITC conjugation. Conjugation of Texas Red directly onto antibodies has been unsuccessful in our laboratory and in others (Titus *et al.* 1982). Instead, Texas Red is conjugated onto avidin, utilizing a

Table 2. Major parameters involved for FITC, Texas Red and biotin conjugations.

	FITC	TEXAS RED	BIOTIN
A. AB purity	SAS	SAS	SAS
B. AB conc.	10mg/ml	10 mg/ml	4 mg/ml
C. Buffer	Bicarb/pbs (pH 9.0)	Bicarb/pbs (pH 9.0)	Bicarb/pbs (pH 9.0)
D. Dye:Protein (weight)	0.5:10	1.0:10	1:15
E. Inc. Time	1.0 hr	1.0 hr	4.0 hrs
F. Conj. Pur.	G-25/pbs (pH 7.0)	G-25/pbs (pH 7.0)	Dialysis/pbs (pH 7.0)

onjugation procedure identical to FITC conjugation, except that exas Red is first dissolved in a minimal volume of acetonitrile (50 l to 250 µl). We have found that Texas Red is most efficiently onjugated onto avidin when the avidin concentration is 10 mg/ml and .0 mg to 2.0 mg of Texas Red is added per 10 mg avidin. Table 2 escribes the major parameters involved for the method. Procedures or FITC and biotin conjugation are shown in Figure 4.

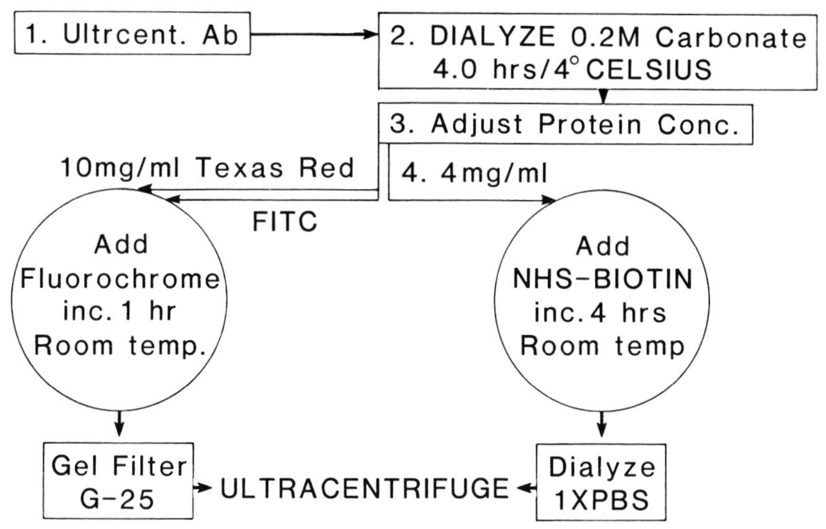

igure 4. Procedures for FITC, Texas Red, and biotin conjugation.

Phycobiliprotein conjugation is a more elaborate procedure than the low molecular weight dyes. First, the protein to be conjugated (antibody or avidin) is dialyzed into a 100 mM sodium phosphate buffer (pH - 7.0). Dialyzed protein is then reacted at pH 7.0 with a heterobifunctional linking reagent (Carlsson *et al.*, 1978) (e.g., SMCC, SMPB, SMB). The reaction converts some of the lysine residues of the protein to thiol reactive maleimides. Next, phycobiliprotein which has been modified with a different heterobifunctional linking reagent (SPDP) is then reduced to create a reactive group that will attach to modified antibody. Both modified phycobiliprotein and antibody are dialyzed exhaustively and then mixed together to yield a stable thioether crosslink. Finally, the reaction mixture is incubated overnight in the dark at room temperature. Purification of phycobiliprotein conjugates is accomplished by molecular size exclusion chromatography using a Sephacryl S-300 column. Resolution between unconjugated phycobiliprotein (104 Kd to 240 Kd) and conjugate (390 Kd to 100 Kd) should be apparent. An alternative to size exclusion chromatography is to dialyze the mixture against 1 nM potassium phosphate buffer and apply to a small (1-5 cc) hydroxyapatite column. At the buffer concentration used, free phycobiliprotein and conjugate will be tightly bound to the upper layers of the hydroxyapatite column. By increasing the phosphate concentration in a linear manner, free phycobiliprotein will be eluted first, followed by phycobiliprotein conjugates. Successful conjugation of phycobiliproteins is commonly done with a phycobiliprotein to antibody ratio of 1:2. Figure 5 illustrates the phycobiliprotein conjugation procedure.

CHOOSING PROPER MOLAR ADDITION RATIOS

Since wide variability may exist between various antibody preparations, or individuals preparing certain critical conjugation steps (e.g., adjusting pH, weighing small amounts accurately, following established procedure...) a range of molar addition ratios should be implemented. It is unlikely that any conjugation reaction is 100% efficient. Therefore, a variation in the amounts of fluorochrome or NHS-biotin added should provide adequate information to ascertain what is the optimal molar excess amount of conjugate that should be added. Figure 6 illustrates the effect of altering the NHS-biotin molar amount when added to monoclonal antibody (α-T8 preparation. Anti-T8 antibody recognizes an antigen that is expressed on the majority of human thymocytes and on a minor subpopulation of peripheral blood lymphocytes. Figure 6 is a

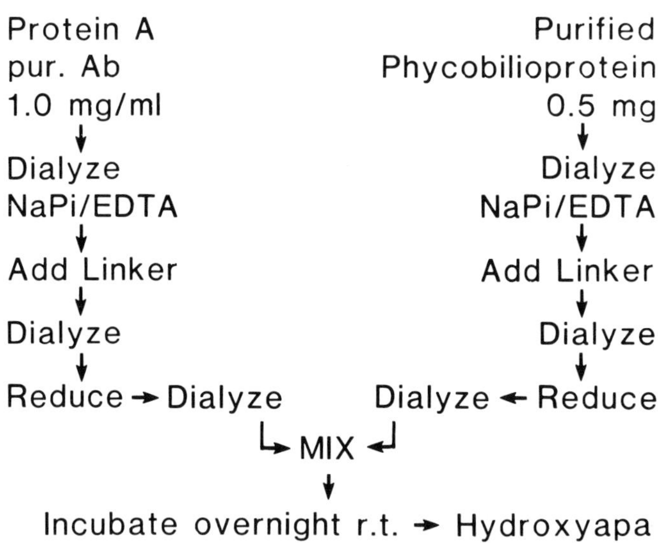

Figure 5. Phycobiliprotein conjugation procedure.

composite of histograms generated by a flow cytometer. Briefly, 1 x 10^6 human thymocytes were incubated with biotinylated T8 antibody and developed with phycoerythrin avidin. 1 x 10^5 scatter gated viable thymocytes were analyzed and graphed. The X axis for each histogram represents relative fluorescence intensity over a 3 decade logarithmic scale. The Y axis pertains to cell number. A 35 fold molar excess of NHS-biotin is the amount commonly used. As seen in Figure 6A, when 10 fold less of the standard amount is added, inadequate reactivity is obtained (18% reactive, peak channel = 26). A molar excess equal to 1/2 the usual amount is successful in resolving the antigen density profile and is comparable in degree of reactivity (76%), but the peak channel is lower than the standard (90 vs 139), indicative of suboptimal biotin attachment. On the other extreme, addition of NHS-biotin amounts for exceeding the amount standardly used creates a situation where the solvent (N-N imethylformamide) concentration denatures the antibody molecule. Such a loss in antibody reactivity is illustrated in Figure 6F. To ircumvent denaturation, water soluble biotin derivatives (Hofmann *et l.*, 1982) can be employed and the buffers used can be adjusted to a ess alkaline pH (pH 9.0 to pH 7.5).

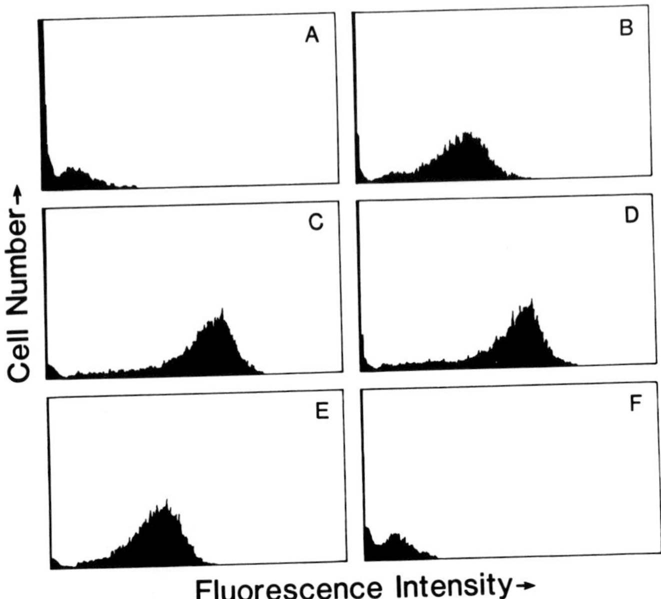

Figure 6. Effects of altering the molar amount of NHS-biotin added to a monoclonal antibody preparation. 6A: 3.5 fold molar excess of biotin to antibody. 6B: 17.5 fold molar excess of biotin to antibody. 6C: 35 fold molar excess of biotin to antibody. 6D: 50 fold molar excess of biotin to antibody. 6E: 100 fold molar excess of biotin to antibody. 6F: 150 fold molar excess of biotin to antibody.

Comparison of the histograms in Figure 6 draws the conclusion that a 35 fold molar excess of NHS-biotin added to T8 antibody yielded the most successful conjugate. This conjugate preparation most closely matched the degree of reactivity (85%) determine by two other independent staining methods and displayed the brightest relative fluorescence intensity which is reflected in the peak channel number of 139.

A similar pilot experiment was conducted to determine what should be the optimal amount of SPDP to conjugate onto phycoerythrin. The SPDP-PE conjugate would eventually be conjugated onto avidin. Figure 7A-D shows the effect of adding a 5, 10, 25 and 40 fold molar excess of SPDP onto phycoerythrin. The 25 fold molar excess preparation provided the brightest staining profile and would be used at this concentration in future conjugations.

Once a conjugate is determined to be useful, it is necessary to calculate what is the proper volume of conjugate to add to each test sample for staining. For FITC and biotin conjugates, a constant volume (100 μl) is commonly added at various dilutions ranging from 1/20 to 1/500. The proper dilution is the one that exhibits maximal fluorescence intensity and minimal background reactivity, as shown in Figure 8. Phycobiliprotein conjugates are usually added undiluted in amounts varying from 1 μl to 20 μl. Figure 9 is a graphic representation of peak channel fluorescence intensity versus volume of PE-Avidin for the pilot experiment illustrated in Figure 8: Graphic analysis reveals that 10 μl of the 25 fold molar excess SPDP conjugate is both the best amount of SPDP to add to PE and is the optimal volume of conjugate to add to a test sample.

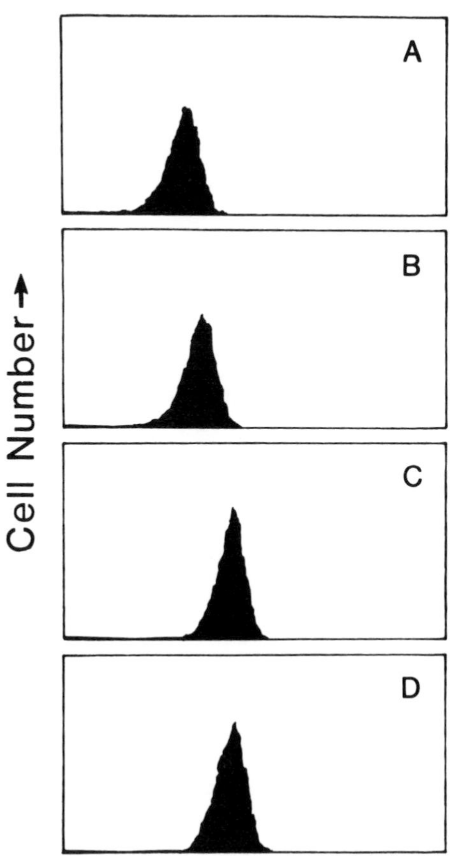

Figure 7. Effect of altering the SPDP molar addition amount added to a phycobiliprotein. 7A: 5 fold molar excess SPDP. 7B: 10 fold molar excess. 7C: 25 fold molar excess and 7D: 40 fold molar excess.

CONJUGATE STABILITY

Critical to the successful use of an immunoconjugate is stability. The conjugate should remain stable while analysis is being conducted and should have a shelf life that will allow repeated applications over an extended period of time. Assuming that the cell surface antigens to be identified are in a metabolically arrested state, then the conjugate should retain specificity and emit a constant fluorescence emission signal when labelled onto cells and excited.

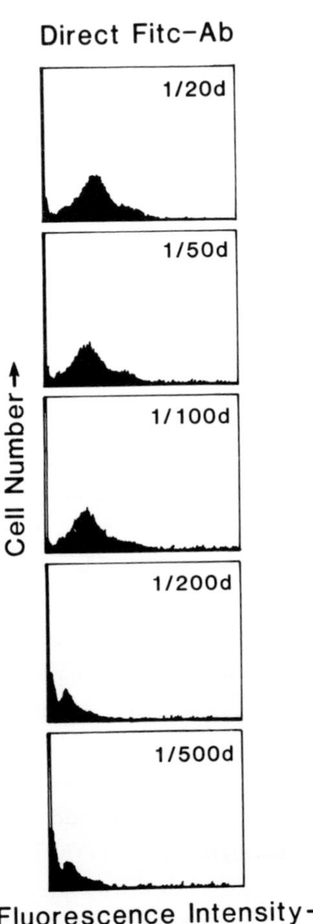

Figure 8. Flow cytometric titration of a FITC-conjugated monoclonal antibody (IF7) with human T cells. Top panel represents 1:20 dilution. Middle panel representing 1:100 dilution is optimal since it displays maximal fluorescence intensity while diluting out background reactivity.

If it is necessary to preserve immunofluorescently labelled cells for extended time periods, then the fluorescence conjugate should not be affected by standard fixation methods (1% formalin, MeOH, EtOH, etc...). There have been instances in our laboratory where formalin fixation of labelled cells resulted in loss of fluorochrome (FITC) attachment onto antibody. A possible explanation for this phenomena could be non-covalent interactions between FITC molecules and hydrophobic regions on antibody molecules. During fixation, protein unfolding occurs resulting in loss of non-covalently bound FITC residing in hydrophobic "pocket regions." To determine if antibody is still present when loss of fluorescence is observed following fixation, species specific fluorochrome conjugated antisera directed against the direct conjugate can be added to see if the

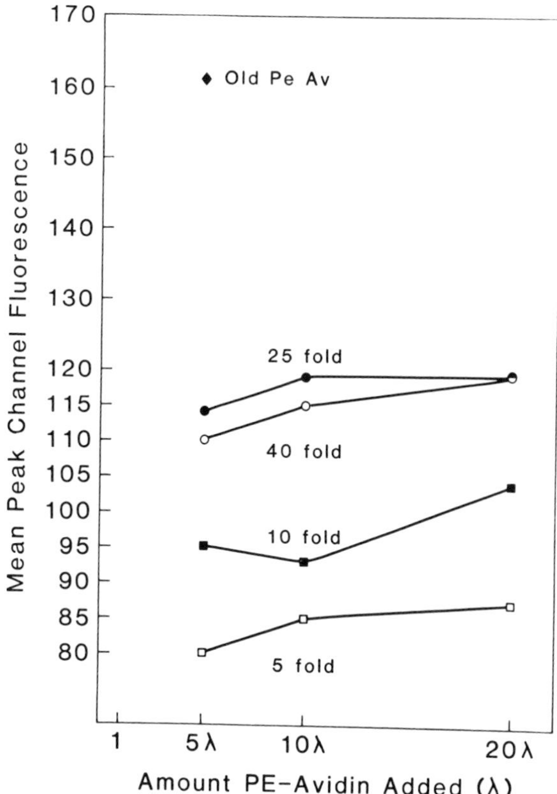

Figure 9. Graph illustrating optimal amount SPDP to add to a phycobiliprotein. Also illustrated is optimal volume of conjugate to add to a test sample, as determined by peak channel fluorescence intensity.

primary antibody is still present. Use of heterofunctional linking agents coupled to FITC may prove effective in alleviating fluorescence loss due to fixation.

STORAGE

As soon as a conjugate has been determined to be suitable for immunofluorescence it should be aliquoted into various volumes (5 µl to 100 µl) and stored frozen at -70°C. If freezing results in less effective immunoconjugates, then storage should consist of at least refrigeration in the dark. 0.01% sodium azide can be added if the reagent is only going to be used for antigen identification and not be subjected to functional studies. The protein conjugate should be at a concentration of at least 1 mg/ml and should be supplemented with bovine serum albumin if the concentration is lower than 1 mg/ml. When using conjugates that have been stored for extended periods of time, ultracentrifugation is recommended to remove aggregates and possible bacterial contaminants. FITC and biotin conjugates have been stored frozen at -70°C in our lab for over five years with minimal loss in reactivity. Phycobiliprotein conjugates have been refrigerated in the dark for over one year with little loss of activity. Conjugate reagents should be tested on a periodic basis to insure reproducible reactivity.

APPLICATION

Direct fluorochrome conjugates are useful when multicolor fluorescent measurements are desired. The fact that direct fluorochrome conjugates do not require a second step for fluorescent development reduces the amount of background staining that could occur. The possibility of cross reactivity of a direct fluorochrome conjugate with a biotinylated antibody or avidin conjugate is very small.

Direct conjugates are useful when competitive inhibition experiments are undertaken. An example of competitive inhibition would be to incubate a non-fluorochrome labelled antibody with a cell sample followed by additional incubation with a fluorochrome labelled antibody which is known to bind to a certain antigen on the cells being tested. If the first antibody is inhibiting the binding of the second fluorochrome antibody, then no detectable fluorescence will be observed.

Another area where direct fluorochrome conjugates are useful is when the cells that are to be labelled for immunofluorescence have been subjected to antibody "panning" techniques (Wysocki and Sato, 1978). The panning technique is a method by which monoclonal antibodies directed against certain cell surface antigens are coated onto petri dishes. Cells are then added to the petri dishes and cells which express the antigen that are identified by the coated antibody adhere to the coated dish. Subsequent immunofluorescence analysis of the enriched populations obtained by panning should be performed only with immunoconjugates which will not react with any residual antibody non-specifically sticking on the cells. Direct fluorochrome or biotin conjugates prove effective in overriding high background fluorescence due to "primary artifacts."

Direct fluorochrome conjugates prove most useful in multicolor immunofluorescence experiments measuring the co-expression of cell surface antigens on human lymphocytes (Blue et al., 1986; Hercend *et al.*, 1985; Lanier and Loken, 1984). Since multicolor immunofluorescence requires multiple fluorochromes, direct conjugates simplify the staining procedure by reducing the number of sample preparation steps.

When high sensitivity is required, the biotin/avidin system should be considered. The affinity of avidin for biotin (10^{-15}M) is much greater than even antibody for antigen (10^{-7}M). Such high affinity enhances the sensitivity of detection by reducing the amount of staining reagent required, thereby reducing the introduction of background fluorescence. Avidin conjugates made with Streptavidin, a form of avidin produced by the bacteria *Streptomyces avidinii*, has been reported to be superior over avidin as an immunoconjugate (Buckland, 1986). The fact that Streptavidin possesses a much lower isoelectric point (pI = 6) than the highly positively charged avidin molecule (pI = 10) minimizes non-antibody specific electrostatic interactions with the slightly negatively charged cell surface.

CONCLUSIONS

In the final analysis, simplicity, specificity, sensitivity and stability must all be ingredients when a fluorochrome conjugate is to be accepted for general use. The development of four distinct fluorochromes (fluorescein, PE, Texas Red, Allophycocyanin) and three staining methods (direct, species specific indirect, biotin/Avidin) at the present time allow many options and possibilities. Careful consideration should be given to the different properties exhibited by each fluorochrome and method when deciding what is the best way to label cells utilizing immunofluorescence. The dyes mentioned above fairly well cover the visible range of the electromagnetic spectrum fairly well. Development of efficient fluorochromes fluorescing in regions outside of the visible spectrum may prove valuable when cells exhibit considerable autofluorescence in the visible region. Such autofluoroescence may be evident when studying marine organisms which contains porphyrins and phycobiliproteins. The stilbene dye SITS (Rothbarth *et al.*, 1978) which is excited maximally at 350 nm and fluoresces at 420 nm may be a useful fluorochrome in alleviating problems due to overlapping autofluorescence.

Flow cytometry provides a valuable means of comparatively evaluating conjugate products. Data derived from flow cytometric analysis can be applied for future flow cytometric experiments or transferred to fluorescence microscopy.

REFERENCES

Bayer, E.A. and M. Wilcheck. 1978. The avidin-biotin complex as a tool in molecular biology. TIBS. 3: N257.

Blue, M.L., J.F. Daley, H. Levine and S.F. Schlossman. 1986. Discrete stages of human thymocyte activation and maturation in vitro: correlation between phenotype and function. Eur. J. Immunol. 16: 771-777.

Buckland, R.M. 1986. Strong signals from streptavidin-biotin. Nature. 320: 557.

Carlsson, J., H. Drevin and R. Axen. 1978. Protein thiolation and reversible protein-protein conjugation: N-succinimydyl 3 (2-pyridyldithio) propionate, a new heterobifunctional reagent. Biochem. J. 173: 723-737.

Coons, H.A., J.H. Creech and N.R. Jones. 1941. Immunological properties of an antibody containing a fluorescent group. Proc. Soc. Exptl. Biol. Med. 47: 200.

Glazer, N.A. and L. Stryer. 1984. Phycofluor probes. TIBS. 9: 483.

Hercend, T., J. Griffin, A. Bensussan, R. Schmidt, M.A. Edson, A. Brennan, C. Murray, J.F. Daley, S.F. Schlossman and J. Ritz. 1985. Generation of monoclonal antibodies to a human natural killer clone. J. Clin. Invest. 75: 932.

Hofmann, K., G. Titus, J.A,. Montibeller and F.A. Finn. 1982. Avidin binding of carboxyl substituted biotin andanalyogues. Biochem. 21: 978-984.

Horan, P.K., S.E. Slezak and G. Poste. 1986. Improved flow cytometric analysis of leukocyte subsets: simultaneous identification of five cell subsets using two color immunofluorescence. Proc. Natl. Acad. Sci. 83: 8361-8365.

Lanier, L.L. and M.R. Loken. 1984. Human lymphocyte subpopulations identified by using three color immunofluorescence and flow cytometry analysis. J. Immunol. 132: 151.

Oi, V., A.N. Glaser and L. Stryer. 1982. Fluorescent phycobiliprotein conjugates for analysis of cells and molecules. J. Cell Biol. 93: 981-986.

Rothbarth, Ph.H., H.J. Tanke, N.A.J. Mul, J.S. Ploem, J.F.C. Uliegenthart and R.E. Ballieux. 1978. Immunofluorescence studies with 4-acetamido-4'-isothiocyanato stilbene-2,2' disulfonic acid (SITS). J. Immunol. Methods. 19: 101.

Titus, A.J., R.P. Haugland, S.O. Sharrow and M.D. Segal. 1982. Texas Red, a hydrophillic, red-emitting fluorophore for use with fluorescein in dual parameter: Flow microfluorometric and fluorescence microscopic studies. J. Immunol. Methods. 50: 193.

Weir, M.D. (editor) 1978. Handbook of Experimental Immunology. 1:Chapter 7.

Wysocki, L.J. and V.L. Sato. 1978. Panning for lymphocytes: a method for cell selection. Proc. Natl. Acad. Sci. 75: 284.

FLOW CYTOMETRY AND IMMUNOFLUORESCENCE IN AQUATIC SCIENCES

Terry L. Cucci
Bigelow Laboratory for Ocean Sciences
McKown Point
West Boothbay Harbor, Maine 04575

David Robins
Institute for Marine Environmental Research
The Hoe
Plymouth PL1 3DH
Devon, England

INTRODUCTION

In aquatic sciences, traditional methods which measure biological parameters of particulates such as biomass, carbon fixation, nitrogen assimilation, etc., have relied on bulk analyses resulting in mean values for heterogeneous populations. Information on the distribution of individual cells or particles about the mean are lacking. Certainly, understanding how individuals or subsets of cells function is essential to the interpretation of the apparent behavior of the population as a whole.

With the advent of the Coulter counter, particle size distributions based on volume were obtainable for all particulate material within a water sample (Sheldon and Parsons, 1967). For biologists, this was a major step towards understanding particle concentration and distribution within the water column. However, the ability to distinguish between the types of particles (living cells, detritus, and/or inorganic material) is crucial for the proper measurement of many biological, chemical and optical properties of aquatic samples.

The application of flow cytometry to aquatic sciences for individual particle analyses has been described (Yentsch *et al.*, 1983; Yentsch *et al.*, 1986). Methods to differentiate cells on the basis of photosynthetic, pigment content, biochemical constituents and optical properties are under development. One group of methods involves immunochemical techniques. The intent of this review paper is to briefly (1) describe the use of flow cytometry in aquatic sciences for rapid analysis and physical separation of individual particles and (2) introduce immunofluorescence methods useful for the differentiation of particle types.

Flow Cytometry

"Flow cytometry" is the rapid measurement of individual particles moving in a fluid stream. Flow cytometers consist of three major component systems; (1) fluidics (sample delivery system), (2) illumination, and (3) detection. A general description of these systems follows, however, for further reading, several books on flow cytometry are available (Melamed *et al.*, 1979; Shapiro, 1985) as well as a series of papers on aquatic flow cytometry (Phinney *et al.*, submitted (a,b)).

Cell suspensions are introduced into the cytometer via the fluidic system whereby the sample is injected into a moving stream of liquid characterized by laminar flow. Thus, the fluid dynamics are such that the sample stream (suspension of particles) is confined within the surrounding sheath fluid. This hydrodynamic focusing causes the cells or particles to flow past the illumination point in single file. One important criterion for optimum instrument performance is that the sample be a single cell suspension of concentration ranging between 5×10^4 to 1.0×10^6 cells per milliliter. High cell concentrations result in more than one cell at a time passing through the excitation beam. Low cell concentrations increase analysis time which may be detrimental to the experimental design due to sample instability. Depending on the size of the flow chamber orifice, cell diameters between 0.5 and 150 microns can be analyzed.

Two major sources of illumination are most common in flow cytometers, lasers and arc lamps. Lasers are highly collimated, monochromatic light sources with high energy output which makes them most desirable for flow cytometry, however, many are water cooled, (in which case either large quantities of water or expensive circulatory cooling units are necessary) and require unusual electrical inputs. Some lasers are also expensive to purchase and operate. Arc lamps require normal electrical inputs and are air cooled, but provide a narrow range of excitation wavelengths (i.e., 488 nm $\pm$ 10 nm) as opposed to monochromatic light and require additional beam shaping optics. whether using lasers or arc lamps, a number of excitation wavelengths are available (Shapiro, 1985).

As particles pass through the excitation beam, scattered light of the of the excitation wavelength and longer wavelength fluorescence, (if the particle fluoresces), are emitted at all angles. Collection lenses focus the scattered and fluoresced light onto photomultiplier tubes (PMT's) or photodiodes. Optical filters are placed in front of fluorescence detectors to exclude scattered light and permit the detection of a single fluorescent signals.

The fluidic, illumination, and detection systems all converge at at the interrogation point, where the particle enters the illumination beam and its optical properties measured. Pulses from the photodetectors are converted to voltage signals and stored in a computer. Shapiro (1985) gives review of signal processing. Commonly measured parameters include forward angle (1.5°-19°) and 90° light scatter, cell volume (impedance volume), and two or three colors of fluorescence, depending on the number of PMT's available and type of flow cytometer.

In addition to analyzing individual particles, some flow cytometers have the ability to sort a desired subpopulation of cells or particles into a collection vessel. This becomes extremely valuable when one wants to validate signals and/or obtain a subpopulation of cells for further study. The flow stream, consisting of cells surrounded by a saline sheath fluid, is broken up into individual droplets by means of a vibrating piezoelectric crystal. After establishing sort criteria on the basis of one to several cellular parameters, droplets containing a cell of interest are positively or negatively charged, and are deflected to the left (-) or right (+) out of the main stream by high voltage deflection plates. The cells may be collected in tubes for further experimentation or on a slide for microscopic identification (Figure 1).

Fluorescence

When certain molecules absorb light at some wavelength electrons are raised to an excited state followed by a return t ground state, at which time energy is released as emitted light or fluorescence. Fluoresced light is always of a longer wavelength than the absorbed light. Fluorescence measurements may be obtaine in biological systems by several methods (1) autofluorescence, (2 staining with fluorescent dyes, (3) binding fluorescent dyes to

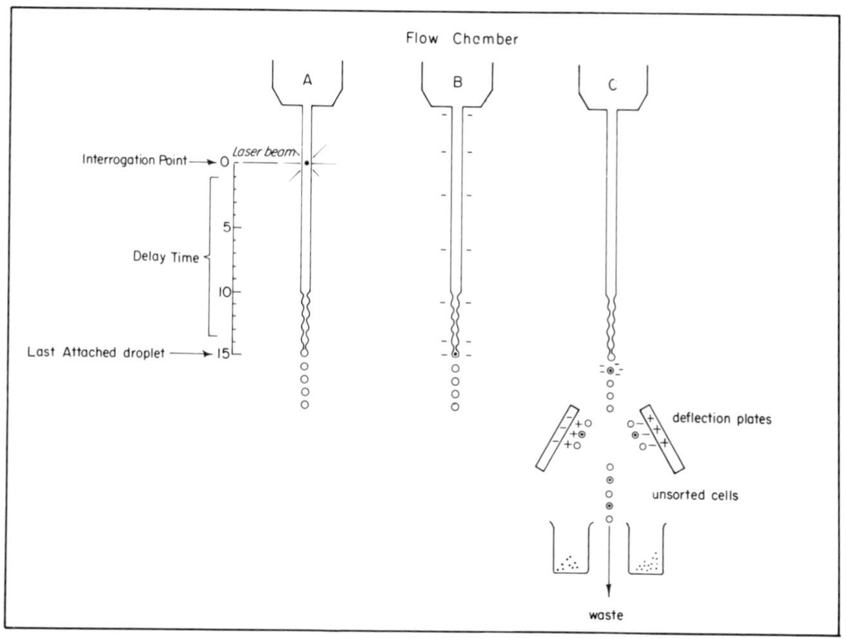

Figure 1. Diagram of flow cell and stream demonstrating the principles of cell sorting. A. Particle enters interrogation point where the light scattering and fluorescent signals are detected. As the particle moves downstream, these signals are analyzed (delay time). B. If the signals meet pre-determined sort criteria, the entire stream is charged as the particle occupies the last attached droplet. C. Droplet containing the particle detaches from the stream carrying the charge. As the droplet passes through deflection plates, it is diverted either to the left or right (depending on the applied charge) into collection vessels. Cells not sorted continue straight to the waste containers.

antibodies, so called immunofluorescence, (4) transforming a non-fluorescent molecule to a fluorescent one by treatment with a reagent (fluorphor), and (5) fluorescence quenching due to the presence of another molecule (Phinney *et al.*, submitted (b)).

Marine and freshwater phytoplankton possess the ability to fluoresce naturally; termed "autofluorescence." Chlorophyll *a*, the major photosynthetic pigment of all autotrophs, absorbs blue light maximally (430nm *in vivo*) and fluoresces in the red (685 nm). Phycoerythrin, an accessory photosynthetic pigment found in bluegreen algae, absorbs blue-green light and fluoresces in the orange spectral region (560 - 575 nm). Because of the fluorescent characteristics of these two pigments, identification and enumeration of phytoplankton in ecological studies have been facilitated utilizing epifluorescence microscopy.

Using flow cytometry, aquatic particles (living cells, detritus, etc.) have been differentiated using the measured parameters of impedance volume, low forward angle light scatter (relative index of size), and the autofluorescent properties of the photosynthetic pigments. Besides the speed of analysis (> 1000 cells per second), a main advantage of flow cytometry is the ability to simultaneously measure these parameters for each particle. The data is visualized by graphic presentation in the form of two parameter analyses such as cell size vs. a fluorescence parameter (allometric analysis) or by two fluorescence parameters (two color or ataxonomic analysis). This multiparameter data analysis helps accentuate discrete subpopulations that may otherwise be indiscriminately lost using a one parameter analysis. Based on the autofluorescing characteristics of phytoplankton and their predominantly single cell morphology, flow cytometry has been applied to the study of phytoplankton ecology (Olson, 1985; Wood *et al.*, 1985), characterization (Paau *et al.*, 1978; Trask *et al.*, 1982), physiology (Yentsch *et al.*, 1983; 1985; Rivkin *et al.*, 1986), and invertebrate grazing of phytoplankton (Cucci *et al.*, 1985; Shumway *et al.*, 1985; Stoecker *et al.*, 1986).

The use of fluorescent stains provides many techniques to differentiate cellular material by fluorescence. A stain may be used to detect differences in cell membrane permeability, cellular enzyme concentrations, nucleic acids, total proteins, and cellular viability. Most fluorescent stains bind with their substrates in fixed cells, thus precluding any further experimentation on the live organism. However, viable stains do exist for measuring certain cellular constituents (Olson *et al.*, 1983).

Many of the fluorescent stains available have been developed for use in mammalian cells, these staining protocols may not be effective on aquatic organisms. Therefore, new protocols may need to be developed. When using fluorescent stains, many aspects must be considered: (1) specificity; what substrate the stain binds to, i.e., protein, RNA, DNA, enzymes, etc., (2) light absorbing and fluorescence characteristics, (3) solubility, (4) stoichiometry, (5 stability of the stain with time, (6) enhanced fluorescence (increased quantum efficiency) of the stain when bound vs. unbound, and (7) necessity to remove the cells from unbound stain in the media prior to analysis by washing. Probably the most important consideration when using fluorescent stains is the specificity of a stain on a molecular or organelle level as opposed to the organismal level. For example, a stain specific for protein would indeed stain all proteinaceous material in a heterogeneous sample, regardless of whether the cell is autotrophic or heterotrophic. If differentiation of these cell types is required, another cellular constituent which is specific to one cell type or the other would be preferrable.

Immunofluorescence techniques provide yet another means for discriminating subpopulations of cells by fluorescence using custom reagents. Over the last decade, the development of flow cytometry and immunofluorescence has been simultaneous in mammalian cell biology (Muirhead *et al.*, 1985). The techniques involve the binding of a fluorochrome to an antibody which, during the antibody-antigen reaction, will be specific for one (monoclonal) or more (polyclonal) reactive sites. Only recently have these techniques been applied to oceanography (Ward and Perry, 1980; Dahle and Laake, 1981; Campbell *et al.*, 1983; Ward, 1984; Ward and Carlucci, 1985; Orellana and Perry, 1985).

Monoclonal antibodies are perhaps gaining the most interest because of their specificity. After the initial production of the antibodies, a screening process follows whereby specificity is determined. When the antibody specificity is suitable for the desired material (antigen) it is purified and rendered fluorescent by binding a fluorochrome to the antibody. Using epifluorescent microscopy or flow cytometry, the antibody-antigen complex can then be visualized.

The simplest techniques involve surface markers, which can be easily used for cell discrimination. Surface markers react and attach only to the material on the cell surface. In some cases, the cell membrane must be perforated, allowing the antibody to react with intracellular material.

An understanding of autofluorescence in phytoplankton is extremely important when selecting an appropriate fluorochrome. The excitation/emission spectra of the photosynthetic pigments and fluorochrome should be known to avoid interference between the phytoplankton autofluorescence and the induced fluorescence from the fluorochrome. Energy transfer from the fluorochrome to the photosynthetic reaction and, vice versa, can alter the fluorescence intensity and/or spectral characteristics (Olson *et al.*, 1983). Choosing an appropriate fluorochrome can usually avoid these problems; however, if no other alternative exists, it may become necessary to extract the photosynthetic pigments out of the cells before attempting the immunochemical assay (Olson *et al.*, 1983).

Standards and controls

As with any instrumentation and experimental design, proper standards and controls must be utilized to insure valid results when using flow cytometry, fluorescent stains, and/or immunofluorescent techniques. Using reagents from a variety of sources can effectively cause variations in results even when the procedures are the same. Without proper standards and control samples, intercomparison of data from one experiment to another becomes impossible. In aquatic flow cytometry, the need for instrument standards, experimental controls,and internal stain standard are becoming apparent. Instrumental variability must be maintained at a minimum if small changes in cell populations are to be measured. Reference particles, highly uniform in size and fluorescence intensity, are used to align the instrument and monitor instrument performance. Experimental (analytical) controls are used as reference points in relative measurements. An important aspect of experimental controls is that they be biologically similar material and of similar fluorescence intensity as the sample material. Internal stain standards are used to evaluate the efficiency of a staining protocol and stability of a sample during the course of an experiment and instrument analysis. For further reading on standards and controls in aquatic flow cytometry, the reader is directed to Phinney *et al.* (submitted (b)).

Reference standards must be used in immunochemical analysis as well. Immunological reagents cannot be effectively characterized by chemical or physical monitoring (Holborow *et al.*, 1982). As in any biochemical standard, they should be similar to the test material, stable for long periods of time, and show linearity in response to changes in dosages. Quantitative and qualitative problems of immunofluorescence standards has been reviewed by Holborow *et al.* (1982).

CONCLUSION

In marine and freshwater ecosystems, research involving the interaction of micro-organisms and their environment has intensified the need to enumerate, classify and isolate individual particles for further analysis and experimentation. Coulter counters have provided an ability to quantify total particle concentration within an aqueous solution including the samples from aquatic systems. However, rapid discrimination between particle or cell types, has not been possible using routine analytical instruments. Utilizing fluorescence as a discriminating parameter, as autofluorescence (phytoplankton) or induced fluorescence (stains, immunofluorescence), individual particles can be differenctiated. Because of its ability to simultaneously measure up to three fluorescence signals in addition to sizing parameters (impedance volume, forward angle light scatter) on individual particles, flow cytometry offers a rapid analytical approach to differentiate aquatic particles. Some instruments are capable of physically sorting subpopulations which may be desirable for experiments requiring further analysis of specific particles.

The problem of cell discrimination utilizing autofluorescence, cell volume, or light scattering characteristics is complicated by the vast array of particle types and sizes encountered in the aquatic environment. Particles having similar size or similar fluorescence intensities cannot be separated. Immunofluorescence techniques offer the advantage of discriminating subpopulations having similar characteristics within complex mixtures of cells in natural populations. No further manipulation of the sample would be required. Although labor intensive to develop, monoclonal antibodies bound to a fluorochrome, designed to be species or particle specific, will provide aquatic scientists with a new and powerful tool for a wide range of research applications.

ACKNOWLEDGEMENTS

The authors are grateful to Clarice M. Yentsch and David A. Phinney for their ideas in developing this paper. We wish to thank Jim Rollins for the graphics and Peg Colby for typing the manuscript. This work was supported by NSF grant No. OCE-8603830 and ONR grant No. N00081C-0043.

REFERENCES

Campbell, L., E.J. Carpenter and V.J. Iacono. 1983. Identification and enumeration of marine chroococcoid cyanobacteria by immunofluorescence. Appl Environ. Microbiol. 46: 553-559.

Cucci, T.L., S.E. Shumway, R.C. Newell, R. Selvin, R.R.L. Guillard and C.M. Yentsch. 1985. Flow cytometry: a new method for characterization of differential ingestion, digestion and egestion by suspension feeders. Mar. Ecol. Prog. Ser. 24: 201-204.

Dahle, A.B. and M. Laake. 1981. Diversity dynamics of marine bacteria: immunofluorescence stain on membrane filters. J. appl. Microbiol. 43: 169-176.

Holborow, E.J., G.D. Johnson and S. Chantler. 1982. Use of international reference preparations for immunofluorescence. IN: Immunofluorescence Technology, Selected Theoretical and Clinical Aspects, G. Wick, K.N. Traill and K. Schauenstein, (eds.). Elsevier Biomedical Press, New York. pp. 1-10.

Melamed, M.R., P.F. Mullaney and M.L. Mendelsohn (eds.). 1979. Flow Cytometry and Sorting, John Wiley & Sons, NY, 716 pp.

Muirhead, K.A., P.K. Horan and G. Poste. 1985. Flow cytometry: present and future. Bio/technology. 3: 337-356.

Olson, R.J., S. Frankel, S.W. Chisolm and H.H. Shapiro. 1983. An inexpensive flow cytometer for analysis of fluorescence signals in phytoplankton: chlorophyll and DNA distributions. J. Exp. Mar. Biol. Ecol., 68: 129-144.

Olson, R.J., D. Vaulot and S.W. Chisholm. 1985. Marine phytoplankton distributions measured using flow cytometry. Deep-Sea Res. 32: 1273-1280.

Orellana, M.V. and M.J. Perry. 1985. Development of an immunofluorescent assay for primary production. EOS 66: 1305 (abstract)

Paau, A.S. and J.R. Cowles. 1978. Applications of flow cytometry to the study of algal cells and isolated chloroplasts. J. Exp. Bot. 29: 1011-1020.

Phinney, D.A., T.L. Cucci and C.M. Yentsch. (submitted, a). Perspectives on aquatic flow cytometry. I: Instrumentation and analysis.

Phinney, D.A., T.L. Cucci and C.M. Yentsch. (submitted, b). Perspectives on aquatic flow cytometry. II: Importance of standards and controls.

Rivkin, R.B., D.A. Phinney and C.M. Yentsch. in press. Effects of flow cytometry on photosynthetic carbon uptake by phytoplankton in culture and from natural populations. Appl. Environm. Microbiol.

Shapiro, H.M. 1985. Practical Flow Cytometry, Alan R. Liss, Inc., NY, 295 pp.

Sheldon, R.W. and T.R. Parsons. 1967. A continuous size spectrum for particulate matter in the sea. J. Fish Res. Bd. Can. 24: 9-9-915.

Shumway, S.E., T.L. Cucci, R.C. Newell and C.M. Yentsch. 1985. Particle selection, ingestion, and absorption in filter-feeding bivalves. J. exp. Mar. Biol. Ecol. 91: 77-92.

Stoecker, D.K., T.L. Cucci, E.M. Hulburt and C.M. Yentsch. 1986. Selective feeding by *Balanion* sp. (Citiata:Balonionidae) on phytoplankton that best support its growth. J. exp. Mar. Biol. Ecol. 95: 113-130.

Trask, B.J., G.J. van den Engh and J.H.B.W. Elgershuizen. 1982. Analysis of phytoplankton by flow cytometry. Cytometry 2: 258-264.

Ward, B.B. 1984. Autotrophic activity of ammonium-oxidizing bacteria: combined autoradiography and immunofluorescence for estimation of single cell activity in the primary nitrate maximum off the coast of Washington. Limnol. Oceanogr. 29: 402-410.

Ward, B.B. and A.F. Carlucci. 1985. Marine ammonium- and nitrite-oxidizing bacteria: serological diversity determined by immunofluorescence in culture and in the environment. Appl. Environ. Microbiol. 50: 194-201.

Ward, B.B. and M.J. Perry. 1980. Immunofluorescent assay for the marine ammonium-oxidizing bacterium *Nitrosococcus oceanus*. Appl. Environ. Microbiol. 39: 913-918.

Wood, A.M., P.K. Horan, K. Muirhead, D.A. Phinney, C.M. Yentsch and J.B. Waterbury. 1985. Discrimination between types of pigments in marine *Synechococcus* spp. by scanning spectroscopy, epifluorescence microscopy, and flow cytometry. Limnol. Oceanogr. 30: 1303-1315.

Yentsch, C.M., T.L. Cucci, D.A. Phinney and J.A. Topinka. 1986. Real-time characterization of individual marine particles at sea: flow cytometry, IN: Tidal Mixing and Plankton Dynamics, Bowman, M., Yentsch, C.M. and Peterson, W.T. (eds.), Lecture Notes on Coastal and Estuarine Studies. Springer Verlag, NY, 17: 414-448.

Yentsch, C.M., T.L. Cucci, D.A. Phinney, R. Selvin and H.E. Glover. 1985. Adaptation to low photon flux densities in *Protogonyaulax tamarensis* var. *excavata*, with reference to chloroplast photomorphogensis. Mar. Biol. 89: 9-20.

Yentsch, C.M., P.K. Horan, K. Muirhead, Q. Dortch, E. Haugen, L. Legendre, L.S. Murphy, M.J. Perry, D.A. Phinney, S.A. Pomponi, R.W. Spinrad, M. Wood, C.S. Yentsch and B.J. Zahuranec. 1983. Flow cytometry and cell sorting: a technique for analysis and sorting of aquatic particles. Limnol. Oceanogr. 28: 1275-1280.

DETECTION SCHEMES FOR ANTIGEN-ANTIBODY REACTIONS

Diether J. Recktenwald, Chia-Huei Chen Chen
Laura Chiu and Morgan Conrad

Becton Dickinson Immunocytometry Systems
2375 Garcia Ave.
Mountain View, CA 94043

INTRODUCTION

Immunochemical assays are based on the fact that the mammalian immune system forms protein molecules, which bind with high affinity to an antigen molecule. All kinds of molecules can be antigen molecules, and the specificity of these proteins, called antibodies, is rather high. When an animal is injected (immunized) with an antigen, the resulting immune response leads to the generation of a variety of antibody molecules against the antigen, a serum. The composition of the serum depends on the history of the animal. Therefore it is not always easy to reproduce the exact antibody mixture of a serum in different immunizations. These antibody mixtures are very specific, because they recognize a number of different sites or epitopes on the antigen molecule. Each of these antibody molecules is made by one B-cell type. These B-cells can be isolated and grown. A colony which is generated from a single B-cell creates one kind of antibody molecule, a monoclonal antibody. Monoclonal antibodies bind or recognize one epitope on an antigen molecule. As a consequence, monoclonal antibodies bind to any molecule which shows this epitope or molecular geometry. Once a B-cell culture is established, monoclonal antibodies can be reproduced in large quantities. If better specificity is needed, it can be achieved by mixing different monoclonal antibodies against the same molecule.

Antibodies and their complexes cannot be detected easily, therefore many detection schemes have been developed to help the use of antibodies in bioanalytical chemistry. These schemes will be discussed in the following section.

METHODS

The detection of antigen-antibody complexes can be performed in two general formats, based on whether a separation step is required for the detection or not. These are the homogeneous and inhomogeneous immunoassays.

In a homogeneous assay the complex formation leads to a change of an easily measurable physical property of either the antibody or the antigen partner of the reaction. One simple example of such a test is the determination of the concentration of an antifluorescein antibody. When the antibody binds to fluorescein, the fluorescein fluorescence is quenched. In an assay with a known amount of initial fluorescence the decrease of fluorescence is proportional to the amount of antibody added. A more applicable assay is the EMIT type assay in which the formation of the complex changes the activity of an enzyme. The resulting enzyme activity indicates the level of an analyte. Other homogeneous immunoassays use fluorescence energy transfer.

Some microparticle-based immunoassays are also regarded as homogeneous because the discrimination between particle and solute is achieved by a special measuring technique, i.e., the measurement of fluorescence polarization, instead of an actual separation step. We will discuss these methods under inhomogeneous assays.

The majority of immunoassays are performed as inhomogeneous, i.e., with a separation between bound and free analyte. One of the most frequently used schemes for this type of assay is a competitive protocol, where a known concentration of labelled antigen competes with the antigen in the analyte for binding sites on the antibody which is immobilized on a solid carrier. The calibration curve for these assays is determined by the affinity of the antibody for antigen and by the concentration of labelled antigen. Another protocol absorbs the antigen with a lot of other molecules on a hydrophobic surface. The amount of bound antigen is determined with a labelled antibody. A variation of this method is the use of two antibodies against one antigen, in which one antibody is used to bind the antigen, and the second antibody is used to detect it. The separation of antigen-antibody complexes from antibody or antigen, which is required for inhomogeneous assays is made possible by the slow dissociation of antigen antibody complexes. Therefore even though the equilibrium of the reaction is perturbed by washing steps, the kinetics of the dissociation reaction does not allow the system to relax to its new equilibrium state within the time of the measurement.

The most obvious and non-disturbing method to quantitate the reaction in an inhomogeneous assay would be a protein determination. This approach, however, is too tedious. Of the many different ways in which the antigen-antibody complex is detected in practice, most use labels. This allows an easy translation of the amount of complex into an electrical signal. For a long time the most popular method has been the use of isotope-labelled antigens or antibodies. This approach allows good sensitivity. Also when small antigen molecules have to be labelled, the use of ^{3}H or ^{14}C does not influence the properties of the labelled molecules. But antigens labelled with radioisotopes have a limited lifetime, their use requires special precautions, and the sensitivity with these labels is limited, as will be discussed below. Other labels have therefore been developed for immunoassays. Most of these labels are attached covalently to protein antigens or to the antibodies. Because of the high molecular weight of proteins, relatively large label molecules can be used successfully. Indeed we have used labels which are several times larger than the antibody molecule, namely fluorecent phycobiliprotein molecules.

We just mentioned fluorescent molecules as adequate labels. Other labels are so called spin labels, which can be detected by electron spin resonance or paramagnetic probes detected by nuclear magnetic resonance. Labels changing ion concentrations in the medium can be detected by electrochemical methods, enzymes can be conjugated to proteins to cause absorbance, fluorescence, or electrochemical changes. A special subset of fluorescence detection is the use of phosphorescence. Factors affecting the sensitivity will be discussed in more detail in the following section.

Sensitivity of Methods

Generally isotope based detection methods are considered to be extremely sensitive. After discussing the different factors which determine the sensitivity of optical detection methods, it will be shown that fluorescence and phosphorescence based assays can be more sensitive than isotope methods, and eventually these methods can be improved to a single molecule sensitivity. We first have to consider one basic concept which influences sensitivity. Some methods use a large signal as baseline, i.e., absorbance measures the change in a bright light level, other methods use a signal baseline close to zero, i.e., the detection of radioactive counts, of fluorescence, or of phosphorescence. In the first case, extreme accuracy in the measurement is required to achieve high sensitivity and a good dynamic range. As an example consider a 1 unit detection in a system with a 10000 unit dynamic range. The accuracy of the method has to be better than 0.01% to achieve these specifications. In a system which measures against zero, high sensitivity can be achieved despite poor accuracy by increasing the gain of the detection system. Another consideration for sensitivity is the size of the signal. Optical methods and isotope methods produce counts. The number of counts per molecule determines the potential sensitivity of the system. Consider an isotope, i.e., ^{125}I, and a fluorescent molecule, i.e., fluorescein, with the same number of isotope and fluorescein molecules per antigen or antibody. The half life of ^{125}I is about 10 days. You will have to wait 10 days on average to get a signal from one ^{125}I. The half life for an excited fluorescein molecule is less than 1 µs. Furthermore, fluorescein can be excited and emit a signal many times. Therefore in 1 s, one can get a million counts from a fluorescein molecule. The probability of getting just a single count from ^{125}I is less than 1 in five million. The situation is even worse for ^{3}H and ^{14}C with much longer half lives. Therefore ^{125}I is used very frequently. Consequently fluorescence methods are potentially more sensitive than isotope methods. Actually fluorescence could be regarded as extremely short lived, low-energy radioactivity generated by light, however, because of the low-energy of the radiation, the detection probability is substantially lower for fluorescence emission than for radioactive decay. Also because of the energies involved, and because many naturally occurring chemicals are quite fluorescent, there is a substantial fluorescent background, which makes it difficult to reach a "vacuum background level" for fluorescence.

As a compromise, some recent systems use time-gated phosphorescence for better sensitivity. The excited state of a phosphorescent molecule can have a lifetime of more than a millisecond. By using pulsed light sources and waiting for a few microseconds before the signal is observed, fluorescent background can be eliminated.

In addition to the background, which is caused by a signal from molecules other than the label molecule, detection of an antigen-antibody reaction is limited by a background generated from improperly deposited label molecules. This is because all molecules have an affinity based on hydrophilic or ionic interactions. These phenomena are normally called non-specific binding. In many cases non-specific binding can be reduced by adding a high concentration of serum albumin or non-relevant immunoglobulin to the assay mixture. Additionally the affinity constant for the antigen-antibody reaction determines the sensitivity of the assay.

Sensitivity can be increased further by using chemical amplification. A very commonly used method is the ELISA technique, in which the antibody is covalently coupled to an enzyme. After the antigen-antibody complex is separated from free antibody, a substrate for the enzyme is added. Typical enzymes convert several thousand substrate molecules into the reaction product. The disappearance of the substrate or the accumulation of product is detected. As discussed above, for statistical reasons it is better to measure the formation of the product. In most cases a colored or a fluorescent product is selected. The amount of antigen-antibody complex is proportional to the rate of the reaction. Therefore ELISA methods require a kinetic measurement.

Chemical and physical signal amplification can improve the detection of labelled antibody, which in an ideal case is quantitatively bound to antigen, but it cannot improve the sensitivity of specific detection if there is nonspecific antibody background (Figure 1). In most cases this background combined with physical background from the sample, i.e., autofluorescence is the limiting factor for the sensitivity of an immunoassay.

For the rest of the discussion, we will describe some state-of-the-art methods for cell surface immunoassays, which are of considerable interest to most biologists. In their procedural part the techniques used for cell surface immunofluorescence are indistinguishable from some particle based assays. The antigen is immobilized on the cell or particle surface; the antibody reacts with the antigen. Two major schemes can be distinguished; direct labelling and indirect labelling.

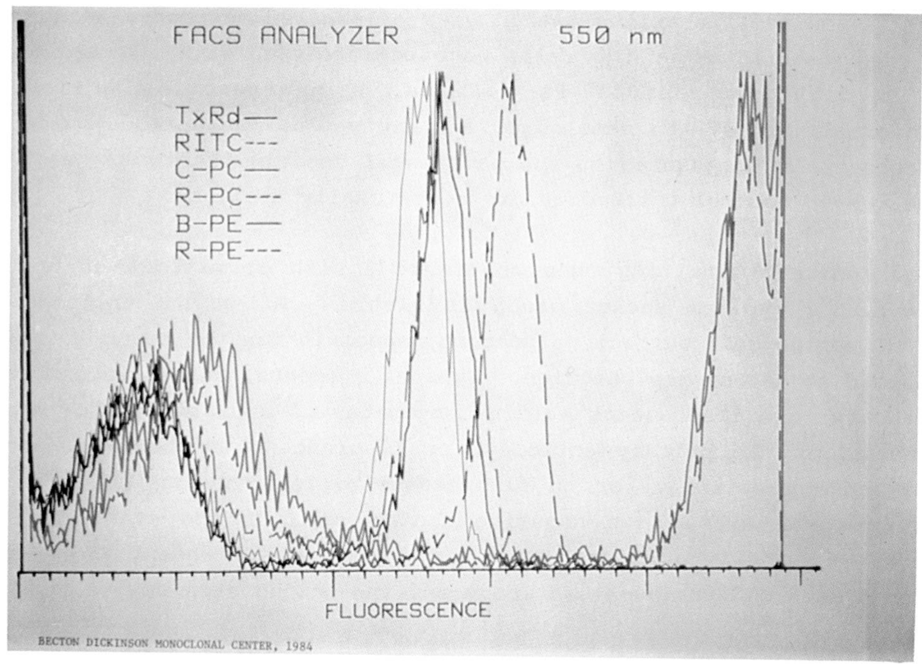

Figure 1. Measurement of lymphocyte cell surface antigens with an anti-leu-3a monoclonal antibody conjugated to different fluorescent dyes.

(TxRd = Texas red, RITC = Rhodamine,
C-,R-PC = c-,R- Phycocyanin,
B-,R-PE = B-,R- Phycoerythrin)

About 50% of the cells have the leu-3a marker. The instrument measures the fluorescence intensity of each cell and displays the measurements obtained from 10,000 cells as number of cells at a given intensity versus the fluorescence intensity (three-decade logarithmic scale for the fluorescence axis, linear scale for the number). The stained cells (population to the right) carry about 50,000 leu-3a antigens (Ledbetter *et al.*, 1981). As shown, the best sensitivity is achieved with phycoerythrins (red lined histograms). Note the shift of the part of the red histogram showing cells without leu-3a, due to non-specific binding.

The direct labelling method uses a labelled antibody. A signal can be measured after the cells have been reacted with the antibody. If a negative control is desired, a non-reacting antibody, preferentially of the same class and isotype as the antibody used in the test, is conjugated to the same label used in the test. Many of the common control antibodies are commercially available.

Indirect labelling uses an antibody with or without an anchor group. A typical anchor group is biotin. After the antibody is bound to the cell surface, a reagent, specific for the bound antibody is used to detect the reaction. Typical reagents, called second step reagents, are fluorescent avidin conjugates if a biotin anchor group is used on the primary antibody, or fluorescent antibodies against the primary antibody, or on fluorescent proteins detecting antibody antigen complexes. Two negative control methods are available for indirect labelling. One uses the second step reagent only, the second uses an irrelevant antibody and the second step reagent.

A number of labels are available to detect cell surface immunofluorescence. These can be grouped into monomeric fluorescent molecules like the fluoresceins, the rhodamines, and the coumarins and polymeric molecules like polylysine derivatives (Block Engineering Inc.) of the monomeric labels and natural fluorescent polymers like the phycoerythrins, the phycocyanins, and the allophycocyanins (Chen, 1985; Loken *et al.*, 1982; Oi *et al.*, 1982; Recktenwald, 1985; Recktenwald *et al.*, 1984; Warner *et al.*, 1983). Small monomeric fluorescent molecules are attached to antibodies by reacting the protein with maleimide-, isothiocyanate-, or sulfonylchloride- derivatives of the fluorescent dye (Samuel *et al.*, 1985; Strottmann *et al.*, 1983), thereby attaching the dye covalently to amino- or sulfhydryl- groups of the protein. The separation of labelled antibody from free dye is achieved by passing the reaction mixture through a desalination column, typically Sephadex G-25 or G-50.

Attachment of polymeric labels to antibodies requires a more gentle chemistry. Agents which crosslink proteins through their sulfhydryl groups are used routinely; a simple scheme using the high affinity of biotin to avidin has also be described (Oi *et al.*, 1982).

Another group of labels are colloidal metals. It has been found that gold and silver colloids can be attached to antibodies and used for cell staining (Alexander *et al.*, 1985; Cramer *et al.*, 1985; DeMey *et al.*, 1982; DeWaele *et al.*, 1986a; DeWaele *et al.*, 1986b; DeWaele *et al.*, 1983; DeWaele *et al.*, 1982; Rosenberg *et al.*, 1984; vandenPol, 1984; Wybran *et al.*, 1985). The staining can be detected by microscopy or by a change of scatter properties of the cells (Bohmer and King, 1984).

With these many labels available, how is the best label for a given application selected? First, the spectral properties of the label are important. If the cell system under investigation has a strong intrinsic fluorescence, labels with emission wavelengths outside of the intrinsic emission range should be used. Lung macrophages, for instance, fluoresce strongly when excited at 488 nm. Use of allophycocyanin with excitation at 633 nm improves the sensitivity for cell surface immunofluorescence measurements significantly (M. Loken, personal communication). Some plant cells show significant fluorescence over the full range of the visible spectrum. The use of colloidal metals should enhance the sensitivity for cell surface immunodetection with these cells. Another option with plant cells would be the pre-bleaching of endogenous fluorescence before the staining with exogeneous label is performed (D. Recktenwald, unpublished).

Another selection criterion is the lack of non-specific reaction of the label with the cell sample under investigation. Most of the monomeric fluorescent labels are rather hydrophobic and tend to "stick" to hydrophobic surfaces. If this interaction becomes a problem, it can sometimes be reduced by adding a high concentration of a protein with hydrophobic sites like serum albumin, or the problem can be avoided by using phycobiliproteins as labels.
These fluorophors offer good non-specific binding properties and optimal sensitivity (Recktenwald, 1985; Recktenwald *et al.*, 1984). With a well-tuned instrument less than one thousand surface molecules can be detected on a single cell (Figure 1). If quantitative information about cell surface antigens is required, and sensitivity is no issue, then monomeric fluorescent labels yield a higher accuracy.

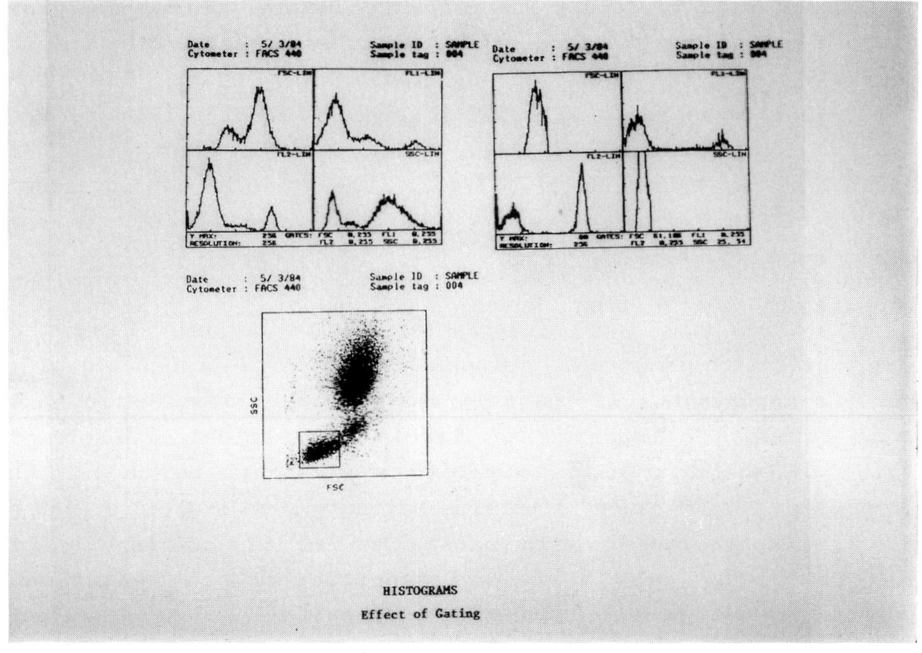

Figure 2. Gating for the analysis of four parameter list-mode data. The dot display in the lower part of the figure shows a gate for a population defined by two angles of light scatter. The histograms in the upper part of the figure show the four parameters before (left) and after (right) the gate was activated.

There is another important application in cell biology which uses a multiplicity of fluorescent labels, the detection of subsets of cell populations, which cannot be identified with a single known antigen, by multicolor immunofluorescence.

The development of multicolor immunofluorescence was driven by the need of immunologists to look at subsets of T-cells, which are defined by a combination of two or more surface markers. It has been found, that in some cases the combination of three or four markers is needed to define subpopulations of lymphocytes (Hardy *et al.*, 1984). Such measurements have to be performed with instruments which can measure many cell parameters, typically 4-8, simultaneously. The information is stored on a cell by cell basis in so-called listmode. This data describes clusters in a multidimensional space. The analysis of the data can be performed by gating. These gates identify a population in a one- or two-dimensional display. The subpopulations of the population in the gate are displayed in one- or two-dimensional displays of other parameters, i.e., two dimensions are observed at a time (Figure 2). Recently we developed a display algorithm, which uses color and movement to allow the researcher to observe four or more dimensions simultaneously (Figure 3 and 4). An interesting application of these multi-color analyses to marine biology should be the simultaneous study of endogenous fluorescence of the photosynthetic systems, combined with the use of markers, which can be identified with antibodies. Multi-parameter immunofluorescence particle analysis might also help to untangle the complexity of food webs.

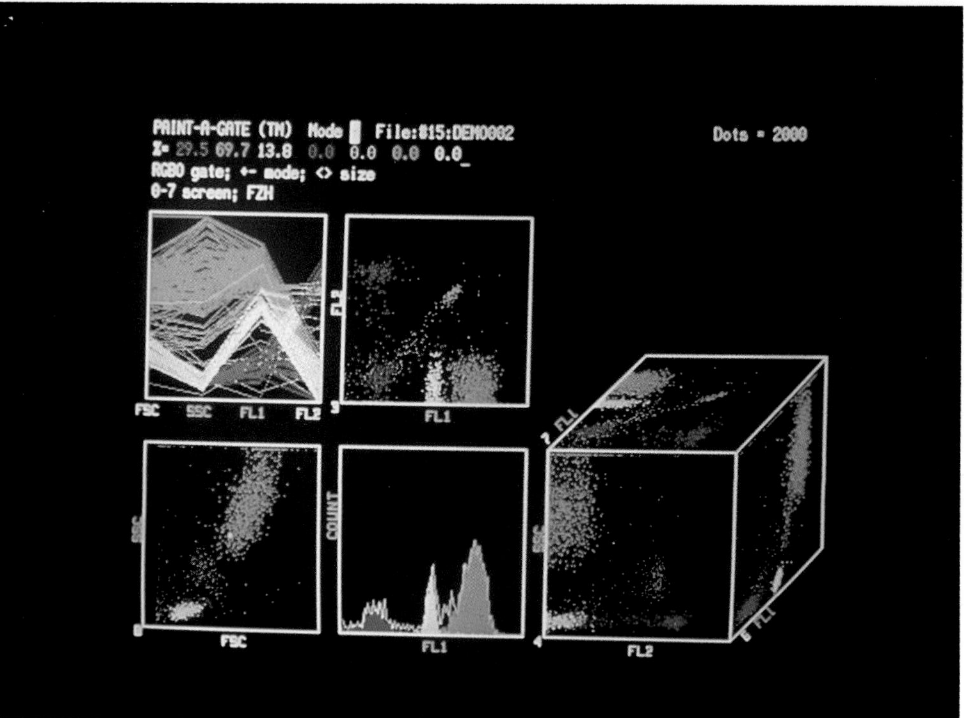

Figure 3. Display of multidimensional data using color and motion on a computer screen. Figure 3 shows projections of four dimensional data displayed in several forms (clusters, correlated dotplots in 2D and on a cube, and histograms). Different clusters are shown in different colors. Blinking can be used to highlight a cluster.

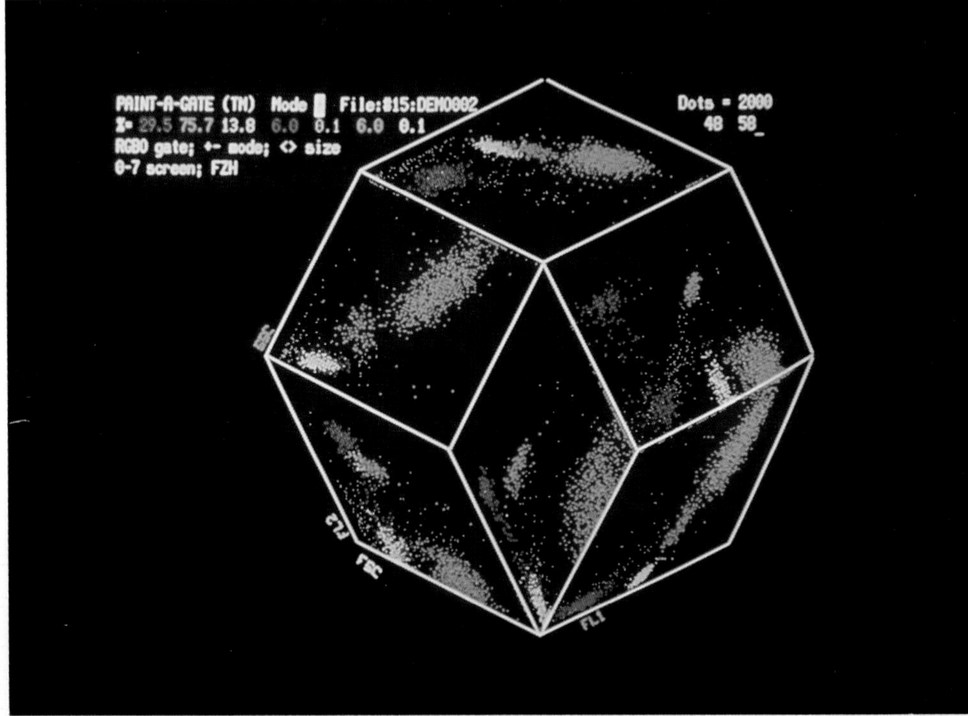

Figure 4. Display of multidimensional data using color and motion on a computer screen. Different clusters are shown in different colors. Blinking can be used to highlight a cluster. This figure shows a display of projections to the 6 visible sides of a dodecaedron.

CONCLUSION

Single molecule detection using fluorescent labels is feasible (Block Engineering Inc.). However, the detection limit for non-fluorescent structures with immunochemical approaches is limited by the affinity of the antibody and by the environment of the structure of interest. At present, it is possible to detect less than 1000 molecules on a cell membrane of a single cell using flow methods or quantitative microscopy.

Multicolor immunofluorescence methods are very new, and have already made significant contributions towards the discovery of new cell subsets in the mammalian immune system. We are sure these methods will make as important contributions in other fields of biology.

REFERENCES

Alexander,R.B., W.B. Isaacs and E.R. Barrack. 1985. Immunogold probes for electron microscopy: evaluation of staining by fluorescence microscopy. J. Histochem. Cytochem. 33(10): 995-1000.

Bohmer, R.M. and N.J. King. 1984. Immuno-gold labeling for flow cytometric analysis. J. Immunol. Methods. 74(1): 49-57.

Block Engineering Inc., Framingham, MA. Dye Tagged Reagent., US. Patent #4166105. Thomas Hirschfeld

Chen, C.H. 1985. Tandem conjugates of phycobiliproteins. Phycobiliprotein Symposium, Seattle WA, September 9th-10th.

Cramer, E., K.B. Pryzwansky, J.L. Villeval, U. Testa and J. Breton-Gorius. 1985. Ultrastructural localization of lactoferrin and myeloperoxidase in human neutrophils by immunogold. Blood. 65(2): 423-32.

DeMey,J., A.M. Lambert, A.S. Bajer, M. Moeremans and M. DeBrabander. 1982. Visualization of microtubules in interphase and mitotic plant cells of *Haemanthus endosperm* with the immuno-gold staining method. Proc. Natl. Acad. Sci. USA. 79(6): 1898-902.

DeWaele, M., J. DeMey, M. Moeremans, L. Broodtaerts, L. Smet and B. VanCamp. 1982. Colloidal gold as a marker for the light microscope detection of leukocyte cell surface antigens with monoclonal antibodies. J. Clin. Immunol. 2(3 Suppl): 24S-31S.

DeWaele. M., J. DeMey, M. Moeremans, M. DeBrabander and B. VanCamp. 1982. Immunogold staining method for the light microscopic detection of leukocyte cell surface antigens with monoclonal antibodies: its application to the enumeration of lymphocyte subpopulations. J. Histochem. Cytochem. 31(3): 376-381.

DeWaele, M., J. DeMey, W. Renmans, C. Labeur, P. Reynaert and B. VanCamp. 1986a. An immunogold-silver staining method for detection of cell-surface antigens in light microscopy. J. Histochem. Cytochem. 34(7): 935-939.

DeWaele, M., J. DeMey, P. Reynaert, M.F. Dehou, W. Gepts and B. VanCamp. 1986b. Detection of cell surface antigens in cryostat sections with immunogold-silver staining. Am. J. Clin. Pathol. 85(5): 573-578.

Hardy, R.R., K. Hayakawa, D.R. Parks, L.A. Herzenberg and L.A. Herzenberg. 1984. Murine B-cell differentiation lineages. J. Exp. Med. 159(4): 1169-1188.

Ledbetter, J.A., A.E. Frankel, L.A. Herzenberg and L.A. Herzenberg. 1981. Human Leu T Cell Differention Antigens: Quantitative Expression on Normal Lymphoid Cells and Cell Lines. IN: Monoclonal Antibodies and T Cell Hybridomas, Perspectives and Technical Notes. G. Haemmerling, U. Haemmerling and J. Kearney. (eds.). Elsevier North Holland, New York, NY.

Loken, M.R., L.L. Lanier, D. Recktenwald and L. Warner. 1982. Flow cytometry monitoring leukocyte sets and subsets. Flow Cytometry and Monoclonal Antibodies for Therapy Monitoring. Montpellier, France. Oct 25-26.

Oi, V.T., A.N. Glazer and L. Stryer. 1982. Fluorescent phycobiliprotein conjugates for analyses of cells and molecules. J. Cell. Biol. 93(3): 981-986.

Recktenwald, D.J. 1985. Phycobiliproteins as multi-color probes for flow-cytometry. Federation of American Societies for Experimental Biology Annual Meeting, Anaheim, CA.

Recktenwald, D.J., H.C. Chen, J. Kimura and N. Warner. 1984. Phycobiliprotein IgG conjugates as tools for multi-color cytometric analysis. International Conference on Analytical Cytology X, Asilomar, CA. June 3-8.

Rosenberg, J.S., E. Weiss and P. Wilding. 1984. Immunogold staining: adaptation of a cell-labeling system for analysis of human leukocyte subsets. Clin. Chem. 30(9): 1462-1466.

Samuel, D., P.L. Amlot and R.A. Abuknesha. 1985. A new method of iodinating ovalbumin, a protein which lacks accessible tyrosine groups, by conjugation to a highly fluorescent coumarin active ester, CASE. J.Immunol. Methods. 81(1): 123-130.

Strottmann, J.M., J.B. Robinson, Jr. and E. Stellwagen. 1983. Advantages of preelectrophoretic conjugation of polypeptides with fluorescent dyes. Anal.Biochem. 132(2): 334-337.

vandenPol, A.N. 1984. Colloidal gold and biotin-avidin conjugates as ultrastructural markers for neural antigens. Q. J. Exp. Physiol. 69(1): 1-33. (REVIEW).

Warner, N.L., J.Y. Kimura and D.J. Recktenwald. 1983. Multiparameter flow cytometry analysis of normal and neoplastic human monocytes with leu monoclonal antibodies. IN: Monoclonal Antibodies. F. Dammaco, G.Doria and A. Pinchera (eds.). Elsevier Biomedical Press. NY.

Wybran, J., J. Rosenberg and F. Romasco. 1985. Immunogold staining: an alternative method for lymphocyte subset enumeration. Comparison with immunofluorescence microscopy and flowcytometry. J. Immunol. Methods. 76(2): 229-238.

Applications

IDENTIFICATION OF MARINE CHROOCOCCOID CYANOBACTERIA BY IMMUNOFLUORESCENCE

Lisa Campbell
Department of Oceanography
University of Hawaii, Manoa
Honolulu, Hawaii 96822

INTRODUCTION

Cyanobacteria are a diverse group of photosynthetic prokaryotes that are widely distributed in terrestrial and aquatic environments. The two major groups of cyanobacteria in the marine plankton are filamentous *Oscillatoria* spp. (formerly *Trichodesmium*) and unicellular strains of the genera *Synechococcus* and *Synechocystis*. Because of their small size (mean diameter, 1 µm), for many years the unicellular forms were overlooked in oceanographic studies. Early work by van Baalen (1962) resulted in the isolation of many species of benthic phycocyanin-containing (PC) forms now assigned to the genus *Synechococcus*. The first oceanic phycoerythrin-containing (PE) *Synechococcus* clone was isolated by Guillard in 1965, but the importance of this group was not realized at that time.

With the introduction of epifluorescence microscopy, the unicellular PE-containing *Synechococcus* spp. were "discovered" in the plankton because of the distinctive autofluorescence of phycoerythrin, their dominant photosynthetic accessory pigment (Johnson and Sieburth, 1979; Waterbury *et al.*, 1979). The PE-containing *Synechococcus* have since been found in most regions of the ocean and at times contribute a major percentage of the photosynthetic biomass (Glover, 1985). The PC-dominant *Synechococcus* spp., however, cannot easily be identified by epifluorescence because the fluorescence emission of phycocyanin overlaps the emission of chlorophyll, so the PC-containing cells cannot easily be distiguished from the other chlorophyll-containing phytoplankton.

Among the important oceanic cyanobacteria, the PE-containing *Synechococcus* spp., distinct species have not been defined. Initial genetic data show that the base pair composition of DNA (expressed as the percentage guanine + cytosine; Rippka *et al.*, 1979) distinguishes the PE-strains from the other clusters of *Synechococcus* strains, but are not sufficient to provide a basis for taxonomy of strains within this group (John Waterbury, in press). Distinguishing morphological and physiological differences have been reported among some of the isolates now in pure culture. The most important morphological variation is motility, which was observed in several clones recently isolated by Waterbury and co-workers (1985). Other than cell size (which ranges from 0.6 to 1.2 µm diameter) and shape (coccoid to rod-shaped), there are no other easily recognizable differences. Variations in cell-wall structures, such as sheaths (Kursar *et al.*, 1981) and the formation of spinae (Perkins *et al.*, 1981; Sarokin and Carpenter, 1981) require obsrvation by transmission electron microscopy. Reports of physiological diversity among the PE-containing strains include: different saturation light intensities for photosynthesis and growth (Glover, 1985); different photoadaptive strategies (Barlow and Alberte, 1985); different temperature optima for growth (Waterbury *et al.*, in press); and various types of phycoerythrins (Wood, 1982; Alberte *et al.*, 1984; Ong *et al.*, 1984; Kilpatrick, 1985). None of these studies are sufficiently comprehensive, however, to permit designation of physiological types or species.

Because none of the observed differences among clones can be recognized in natural populations by standard epifluorescence microscopy, a method for identification of different strains was sought. One approach that has been used to identify specific bacterial groups is immunofluorescence (IF). The IF technique is based on the specificity of the antigen-antibody reaction and permits visual recognition and enumeration of individual strains (serotypes) or species of bacteria. It can be a very powerful tool for autecological studies (Bohlool and Schmidt, 1980) and for identification of taxonomic relationships (e.g. deMacario *et al.*, 1981; Smith, 1982). IF was used successfully to identify freshwater cyanobacteria (Fliermans and Schmidt, 1977), so it appeared to be a potentially useful tool for investigating marine cyanobacteria.

Previous studies were undertaken to determine if IF could be used to identify and to enumerate these small cyanobacteria in the marine phytoplankton (Campbell *et al.*, 1983; Campbell, 1985). Secondly, because the taxonomy of this group is not well defined, the possibility of IF for recognizing distinct clusters of closely related strains (=serogroups) was also investigated by examining the serological relationship among the *Synechococcus* clones in culture (Campbell, 1985). This paper will review the development of an IF assay used to identify marine *Synechocystis* and *Synechococcus* spp. (Campbell *et al.*, 1983), and will discuss the application of this assay to investigate the abundance and distribution of different types of cyanobacteria (Glover *et al.*, 1986; Campbell and Carpenter, 1986).

MATERIALS AND METHODS

Cultures

Axenic cultures of marine phycocyanin (PC)-dominant strains (*Synechocystis* sp. PCC7339; *Synechococcus* coastal isolates WH5701 and WH8101) and phycoerythrin (PE)-containing *Synechococcus* isolates (WH7803, formerly DC-2; WH8016, a Woods Hole Harbor isolate; and WH8113, a motile strain) were provided by J. Waterbury (Woods Hole Oceanographic Institution). Isolation locations for each strain used in this study are given in Tables 1 and 2. Each isolate was grown in f/2 medium without added Silica (Guillard and Ryther, 1962), with additional vitamine B^{12} (20 µg liter^{-1}). The strains used in cross reactivity tests were obtained from J. Waterbury (Woods Hole Culture Collection [WH] and the American Type Culture Collection [ATCC]; see Campbell *et al.*, 1983). An aliquot of each strain was fixed in buffered formalin (5%) and stored refrigerated until used.

Preparation of antisera

Formaldehyde-killed isolates (PCC7339, WH5701, WH8101, WH7803, WH8016 and WH8113) were prepared for whole-cell vaccines according to Taubman and Smith (1974). To summarize, stationary phase cultures were pelleted by centrifugation (6000 xg, 20 min, 10°C), washed six times with sterile saline (0.15 M NaCl), suspended in 200 ml phosphate buffered saline (PBS; 0.02 M phosphate, 0.15 M NaCl, pH 7.5) containing 0.6% formaldehyde, and incubated overnight at room temperature with gentle stirring. The fixed cells were collected by centrifugation (6000 xg, 20 min, 4°C), washed twice with sterile saline, and finally in PBS containing 0.2% formaldehyde. The fixed cells were stored at 4°C in PBS containing 0.2% formaldehyde. Immediately prior to injection of rabbits, an aliquot of vaccine was washed twice with sterile PBS and the cell concentration was adjusted to 10^9 cells ml^{-1}.

For each strain, two female New Zealand white rabbits were used to produce antisera (except for WH5701 and WH8101, for which only 1 each was used). Vaccine was injected into the marginal ear vein of each rabbit according to the following immunization protocol: day 1, 0.5 ml vaccine; day 4, 0.75 ml; day 6, 1.0 ml; and booster injections of 1 ml given every fourteen days. On day 50 the rabbits were exsanguinatedby cardiac puncture. For WH8113, two additional intramuscular injections (1 ml of vaccine with 1 ml Freund's incomplete adjuvant [Difco]) were given on days 87 and 103 to each rabbit to increase the antibody titer, and on day 112 both rabbits were exsanguinated. Sera were prepared (Campbell *et al.*, 1964) and stored frozen (-20° C). Pre-immunization rabbit serum was obtained from each rabbit for control tests.

Serological tests

To determine the presence of precipitating antibodies in the rabbit antisera, double immunodiffusion in agar was performed by a modification of the Ouchterlony technique (Ouchterlony and Nilsson, 1978), as described previously (Campbell *et al.*, 1983). Antigen extracts for immunodiffusion were prepared by autoclaving (Rantz and Randall, 1955) and by acid extracting (Wetherell and Bleiweis, 1975) lyophilized whole cells. The extracts were dialyzed exhaustively against distilled water and reconstituted to 1 mg ml^{-1} in PBS.

The antibody titers of the rabbit antisera to the antigens of each strain were determined by microagglutination using the Cook system (Dynatech, Inc.), as described by Campbell *et al.* (1983). Titers were expressed as the reciprocal of the greatest dilution showing visible agglutination. Results of control tests with preimmunization sera indicated the dilutions that gave false positive results.

Immunofluorescence Assay

To determine the specificity of each antiserum, the Woods Hole Oceanographic Institution *Synechococcus* culture collection (WH) and representatives of other algal and bacterial groups were screened for cross reactions using an indirect immuofluorescence assay. To screen all the PE-*Synechococcus*, a protocol similar to the one used by deMacario *et al.* (1981) for methanogens was followed. One drop of each test strain was heat fixed to glass slides, which were precleaned with 70% ethanol. After washing slides for 5-10 min in

PBS, cells were covered with one drop of antiserum (diluted 1:50), incubated 30 min, and then rinsed with PBS, and washed in PBS for 10 min. Next, the cells were incubated with fluorescein isothiocyanated goat anti rabbit immunoglobulin antiserum for 30 min, rinsed with PBS, and, finally, a coverslip was affixed with a glycerol: carbonate (9:1) mounting medium buffered at an alkaline pH to enhance fluorescence of the FITC. For cross reactivity tests with all other cultures and for labelling field samples, cells were concentrated on 0.4 Nuclepore filters, and the labelling protocol was as described previously (Campbell *et al.*, 1983). Controls consisted of tests using pre-immunization serum followed by the FITC-conjugated antibody.

All samples were examined using a Zeiss Universal epifluorescent microscope with a 50 W mercury lamp. A blue excitation filter (BP450-490), and emission cutoff at 520 nm (Zeiss filter set #48-77-09) was used to view FITC fluorescence, and a green excitation filter (G546) with a 590 nm cutoff filter (Zeiss filter set #48-77-12) was used when counting the PE-containing cells by autofluorescence of their accessory pigment.

Sample collection

Coastal samples were collected monthly from April to December 1982 from 3 stations in the Carmans River estuary of Great South Bay, New York (Figure 1). Oceanic samples were collected from two coastal-to-offshore transects: 8 stations in the eastern tropical North Pacific in November/December 1983 (25° 19.3'N 112° 21.6'W to 24° 21.4'N 115° 5.2'W); and 3 stations in the northwest Atlantic in July/August 1984 (Figure 2). All samples were preserved with 5% buffered formalin and stored refrigerated and protected from light.

RESULTS AND DISCUSSION

Tests of antisera

Analysis by immunodiffusion showed that each rabbit antiserum was specific for the strain against which it had been produced. The antisera directed against PCC7339 (anti-7339) had the highest titers (>20,000), followed by anti-7803 (10,240), and anti-5701, anti-8101 and one replicate anti-8016 (5,120). The antisera directed against the motile strain 8113 and one replicate 8016 serum had the lowest titers (1,280). The results of the agglutination tests showed that rabbit serum dilutions of 1:50 or 1:100 were required to avoid false

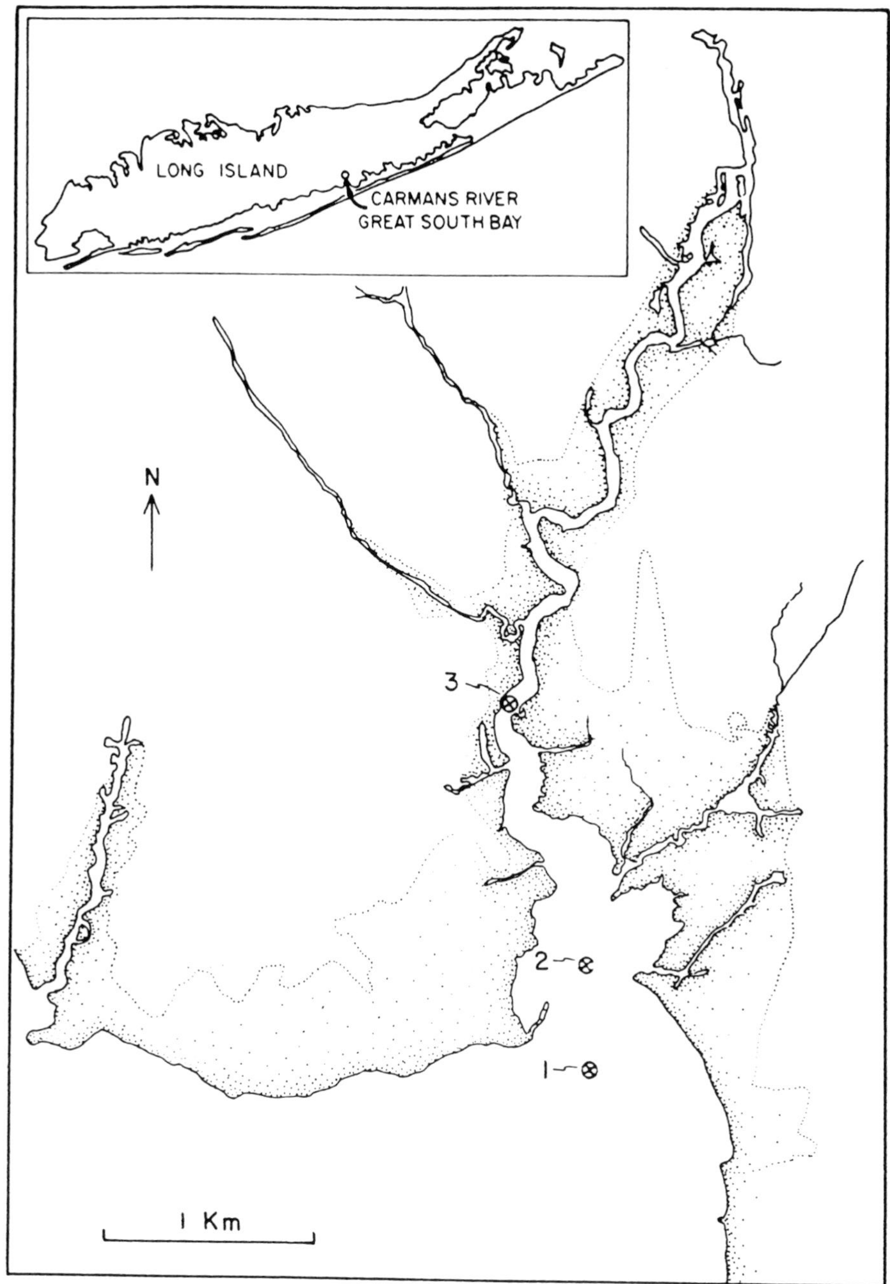

Figure 1. Carmans River estuary sampling sites in Great South Bay, New York. (Reproduced from Campbell *et al.*, 1983.)

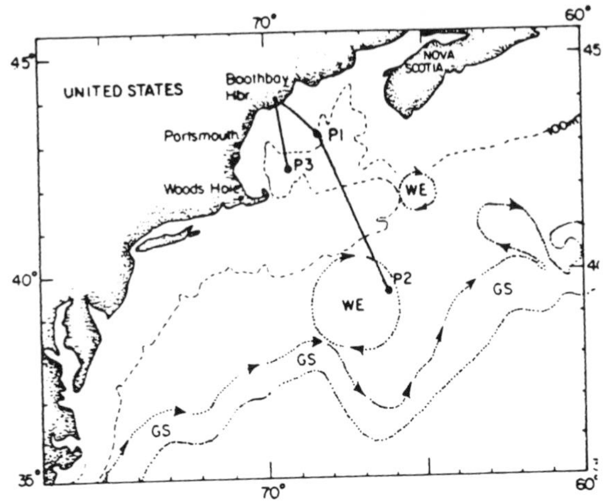

Figure 2. Three stations in the northwest Atlantic Ocean: P1, a neritic frontal region; P2, a warm core eddy; P3, Wilkinson's Basin. (Reproduced from Glover *et al.*, 1986.)

positive results. The required dilution of the FITC conjugate was determined experimentally for each preparation. (Usually, a 1:1000 or 1:2000 dilution was require to avoid non-specific staining). An incubation time of 50 min with the specific antiserum was necessary to obtain 100% labelling of the cells on the filter. No FITC-fluorescence was associated with the cells if incubated with corresponding dilutions of pre-immunization serum or with PBS followed by the FITC-goat-anti-rabbit immunoglobulins, or if incubated with only FITC-goat-anti-rabbit-immunoglobulins.

Cross-reactivity tests

The results of cross-reactivity tests were classified by rating the intensity of fluorescence on a scale of "-" (no reaction) to "++++" (strong positive reaction, very bright fluorescence). A "+" reaction was characterized by very faint fluorescence and only a small percentage (<10%) of the total number of cells labelled. In most cases, the cross-reactivity results of each replicate serum were identical. Where results did vary ("+" or "-"), it was usually in the case of strains with very minor cross-reactions (e.g. WH8201; Table 2).

Each antiserum showed strong IF reactions with its corresponding immunogen and with heterologous strains isolated from similar oceanic environments. Antisera directed against *Synechocystis* PCC7339 (anti-7339) reacted strongly only with the coastal marine isolates (Table 1). Similarly, the antisera directed against the PC-dominant *Synechococcus* spp. (anti-5701 and anti-8101) reacted only with marine PC-*Synechococcus* spp. (Table 1). As defined by the guanine and cytosine base pair composition of the DNA (%G+C; Rippka *et al.*, 1979), these PC-marine strains all belong to one subgroup of *Synechococcus* strains; thus, the cross-reactivity results are consistent with this genetic grouping (Campbell *et al.*, 1983).

Among the PE-containing *Synechococcus* strains, different serological affinities could be recognized based on reactions with each of the antisera directed against PE-containing strains (Table 2). In general, the major cross-reactions were consistent with previous results: with few exceptions, cross-reactions were restricted to those strains isolated from a similar oceanic region.

Of the 37 *Synechococcus* strains in the Woods Hole collection, 19 strains reacted with anti-7803 and 11 of these were specific reactions (Table 2). Half of the isolates (18) were not labelled by this antiserum. The anti-8016 serum reacted strongly (+++ or ++++) with only 5 strains, and all but one of these are coastal isolates (Table 2). All of the 5 strains also showed minor cross-reactions with anti-8113. The majority of isolates in the collection (25) were not labelled by anti-8016, but this is not surprising since most isolations were made from the open ocean.

The antisera directed against the motile *Synechococcus* clone, WH8113, strongly labelled each of the four known motile strains (WH8103, WH8112, and WH8011), and also three other strains (Table 2). In addition, both antisera directed against this motile strain had a large number (12) of minor cross reactions with non-motile strains as well. Possibly these antisera are not directed against the cell proteins or other component associated with motility. WH8113, therefore, must have common antigenic determinants with both WH7803 and WH8016 serogroups. Conversely, only one strain (WH8012) cross-reacted with both anti-7803 and anti-8016, which indicates that, in general, each of these antisera is directed against

Table 1. *Synechocystis* spp. and phycocyanin-dominant *Synechococcus* spp. strain histories and cross-reactivity tests[a].

STRAIN[b]	LOCATION OF ISOLATION	REACTION WITH ANTISERA[c]				
		PCC7339	WH5701	WH8101	WH7803	WH8016
Synechocystis spp. strain						
PCC 7339	Gulf of Mexico	++++	-	-	-	-
ATCC 29108	Brackish water, OR	++++	-	-	-	-
ATCC 27266	Salton Sea, CA	-	-	-	-	-
ATCC 27153	Lake Mendota, WI	-	-	-	-	-
Synechococcus spp. strain (Phycocyanin-dominant)						
WH 5701	Long Island Sound, CT	-	++++	++	-	-
WH 8101	Woods Hole Harbor, MA	-	+++	++++	-	-
WH 8202	29° 15.6' N 85° 54.2'W	-	+++	-	-	-
WH 8007	19° 45'N 92° 25'W	-	++	+	-	-
PCC 7003	Greenwich, CT	+	-	-	-	-
PCC 7002	Magueyes Island, Puerto Rico	+	-	-	-	-
PCC 7418	Solar Lake, Israel	-	-	-	-	-
24.2	Gulf of Mexico	+	-	-	-	-
Anacystis nidulans	freshwater	-	-	-	-	-
C-3	Great South Bay, NY	-	+++	++	-	-
Filamentous cyanobacteria						
Microcoleus sp.	Sippewissett, MA	-	-	-	-	-
Spirulina sp.	Sippewissett, MA	-	-	-	-	-

Table 1. continued

STRAIN[b]	LOCATION OF ISOLATION	REACTION WITH ANTISERA[c]				
		PCC7339	WH5701	WH8101	WH7803	WH8016
Bacillariophyceae						
Thallassiosira sp. (TN2)	Eel Pond, MA	-	-	-	-	-
Cyclotella cryptica (O3A)	Oyster Pond, MA	-	-	-	-	-
Chlorophyceae						
Nannochloris sp. strain GSBn	Great South Bay, NY	-	-	-	-	-
GSB isolate	Great South Bay, NY	-	-	-	-	-
Nitrifying bacteria						
Nitrosomonas marina		-	-	-	-	-
Nitrosococcus oceanus		-	-	-	-	-

[a] Reproduced from Campbell *et al.*, 1983 (Table 2).
[b] PCC, Pasteur Culture Collection; ATCC, American Type Culture Collection; WH, Woods Hole Culture Collection; GSB, Great South Bay
[c] +, relative fluourescence; -, no reaction

Table 2. The *Synechococcus* strains in the Woods Hole Culture Collection: histories, characteristics, and immunofluorescence results.

STRAIN[a] DESIGNATION	ISOLATION LATITUDE	ISOLATION LONGITUDE	SAMPLE DATE	PIGMENT[b] GROUP	%[c] G+C	REACTION WITH ANTISERA: ANTI-7803 1	ANTI-7803 2	ANTI-8016 1	ANTI-8016 2	ANTI-8113 1	ANTI-8113 2
WH 6501	8° 44'N	50° 50'W	1965	II	62.3	++	++	-	-	-	-
WH 7801	33° 44.9'N	67° 29.8'W	7/1/78	II	59.7	+++	+++	-	-	+	+
WH 7802	33° 44.9'N	67° 29.8'W	7/1/78	II	57.9	+	+	-	-	-	-
WH 7803	33° 44.9'N	67° 29.8'W	7/1/78	II	61.3	++++	++++	-	-	-	-
WH 7804	33° 44.8'N	67° 30'W	6/30/78	II		-	-	+++	+++	+	++
WH 7805	33° 44.8'N	67° 30'W	6/30/78	I	59.7	++++	++++	-	-	-	-
WH 7806	38° 19.5'N	69° 34'W	6/29/78	II		-	-	-	-	+++	+++
WH 8001	19° 45'N	92° 25'W	4/80	II	56.4	+	+	-	-	-	-
WH 8002	19° 45'N	92° 25'N	4/80	II	60.0	++	++	-	-	++	++
WH 8003	19° 45'N	92° 25'W	4/80	II	58.5	+++	+++	-	-	-	-
WH 8005	19° 45'N	92° 25'W	4/80	II	58.6	-	-	-	-	-	-
WH 8006	19° 45'N	92° 25'W	4/80	I	57.6	+++	+++	-	-	++	++
WH 8008	19° 45'N	92° 25'W	4/80	I	55.8	++++	++++	-	-	-	-
WH 8009	38° 40'N	69° 19'W	6/26/80	I	57.0	+	+	-	-	+	+
WH 8010	38° 40'N	69° 19'W	6/26/80	I	58.6	+++	+++	-	-	-	-
WH 8011*	34° N	65° W	7/7/80	IV	59.3	-	-	-	-	++++	++++
WH 8012	34° N	65° W	7/7/80	II	62.4	+++	+++	++	++	-	-
WH 8013	34° N	65° W	7/7/80	V		-	-	-	-	-	-
WH 8014	34° N	65° W	7/7/80	II		++++	+++	-	-	-	-
WH 8015	WOODS HOLE HARBOR		6/80	II	55.3	-	-	++++	++++	+	+
WH 8016	(41° 31'N	71° 40'W)	6/80	II	55.5	-	-	++++	++++	+	+
WH 8017	"		6/80	II	54.6	-	-	++++	++++	++	++
WH 8018	"		7/18/80	I	57.6	++++	+++	-	-	-	-
WH 8019	39° 29'N	70° 28'W	6/81	I		-	-	-	-	-	-
WH 8102	22° 29.7'N	65° 36'W	3/15/81	V	60.4	-	-	++	++	+++	+++
WH 8103*	28° 30'N	67° 23.5'W	3/17/81	V	58.9	-	-	+	-	+++	+++

Table 2. continued

STRAIN[a] DESIGNATION	ISOLATION LATITUDE	ISOLATION LONGITUDE	SAMPLE DATE	PIGMENT[b] GROUP	%[c] G+C	REACTION WITH ANTISERA: ANTI-7803 1	ANTI-7803 2	ANTI-8016 1	ANTI-8016 2	ANTI-8113 1	ANTI-8113 2
WH 8104	31° 59'N	68° 18.9'W	3/18/81	II		-	-	-	-	-	-
WH 8105	38° 20.8'N	69° 38.9'W	3/19/81	II	54.9	-	-	++++	++++	++	++
WH 8106	40° 06.8'N	70° 26.3'W	3/20/81	II		+	+	-	+	+	+
WH 8107	39° 28.6'N	70° 27.7'W	6/81	V		-	-	-	-	-	-
WH 8108	39° 28.6'N	70° 27.7'W	6/81	III		-	-	-	-	-	-
WH 8109	39° 28.6'N	70° 27.7'W	6/81	IV		-	-	-	-	-	-
WH 8110	SYDNEY HARBOR, AUST.		8/27/81	I	56.6	-	-	-	-	+++	+++
WH 8111	36° N	66° W	3/82	II		+	+	-	-	-	-
WH 8112*	36° N	66° W	3/82	VI	59.8	+++	+++	-	-	++	++
WH 8113*	36° N	66° W	3/82	VI	60.5	-	-	+	+	++++	++++
WH 8201	20° 44.5'N	109° 4.9'W	5/82	IV		-	-	+	+	++++	++++
						-	+	-	+	-	+

+, relative fluorescence; -, no reaction.

[a] WH, Woods Hole Culture Collection; * motile strains (Waterbury *et al.*, 1985).

[b] pigment data from Waterbury *et al.*, 1986.

[c] % G + C data from Waterbury *et al.*, 1986.

different antigens. No strains were labelled strongly by all three antisera. Six strains were not labelled by any of the antisera, which suggests that at least one additional serogroup exists.

PE-isolates from lower latitudes (tropical and subtropical regions) are predominately of the 7803-serogroup (Table 2), whereas the Woods Hole and N. Atlantic isolates are generally of the 8016-serogroup. Anti-8113 cross-reacts with strains from the Sargasso as well as north Atlantic shelf stations, so a definite serogroup cannot be defined. Because of the number of observed common cross reactions, motility may not be the distinguishing feature of a serogroup.

As mentioned above, from the analysis of the guanine-plus-cytosine (G+C) composition of DNA, three clusters of *Synechococcus* strains with similar percentages have been identified (Rippka *et al.*, 1979). The G+C content of the PE-containing stains ranges from 57-63%, so these isolates probably represent a fourth cluster (Waterbury *et al.*, in press). Although this appears to be a tight cluster, there may be phenotypic differences not yet defined, but recognizable by immunofluorescence. Differences in light-saturation for growth have been observed between cultures of WH7803 and WH8107, a strain not of the 703-serogroup (Campbell, 1985). Other such information, as it becomes available, will permit assessment of differences among serogroups.

To determine whether the intensity of labelling was a function of growth stage or physiological condition of the cell, samples from WH7803 cultures grown at various light levels and under nitrogen (NO_3) limiting or sufficient conditions were labelled (Table 3). In both exponential phase (1.0 division day^{-1}) cultures and in nitrogen-limited, declining cultures, 99-100% of the cells were labelled. Only in the dark incubated cultures did the percentage of stained cells decrease slightly, but still most of the cells were positively labelled (Table 3).

Table 3. Labelling control experiments with *Synechococcus* strain WH 7803 grown under different conditions.

Light Conditions	Initial		Final	
	$[NO_3]$ (μg at l^{-1})	% Labelled	$[NO_3]$ (μg at l^{-1})	% Labelled
Dark	964	100	1033	92
Dark	110	100	133	91
20 $\mu E\ m^{-2}\ s^{-1}$	865	100	723	100
20 $\mu E\ m^{-2}\ s^{-1}$	110	100	1.58	100
55 $\mu E\ m^{-2}\ s^{-1}$	865	100	714	100
55 $\mu E\ m^{-2}\ s^{-1}$	110	100	0.72	99

Field results

Laboratory results indicated the antisera were relatively specific, so samples from several regions were examined for the presence of strains that would be labelled by each antiserum. At the Carmans River site, the *Synechocystis* cells recognized by the anti-7339 serum occurred in spring through late summer at densities of 3 to 4 x 10^4 cells ml^{-1} (Figure 3; Campbell *et al.*, 1983). They were absent during winter months and in June (most likely because of excessive freshwater runoff caused by unusually heavy rains). PC-dominant *Synechococcus* spp. identified by anti-5701 and anti-8101 sera were never abundant ($<10^2$ cells ml^{-1}). The PE-containing *Synechococcus* population was more abundant than the other two groups (up to 10^5 cells ml^{-1}), and they were most abundant at the end of summer and autumn, when *Synechocystis* spp. numbers decreased (Figure 3). Populations of PE-*Synechococcus* labelled by anti-7803 and anti-8016 were also observed (Campbell, 1985; Campbell and Carpenter, 1986).

Overall, unicellular cyanobacteria did not make a large contribution to the photosynthetic biomass at Carmans River. Assuming 20 fg chlorophyll *a* per *Synechocystis* cell and 0.5 fg Chlorophyll *a* per *Synechococcus* cell (Campbell, unpubl. data), the estimated contribution of these cyanobacteria to the total chlorophyll a biomass was less than 5% for much of the year (Figure 4). The largest contribution to the chlorophyll *a* was in the April

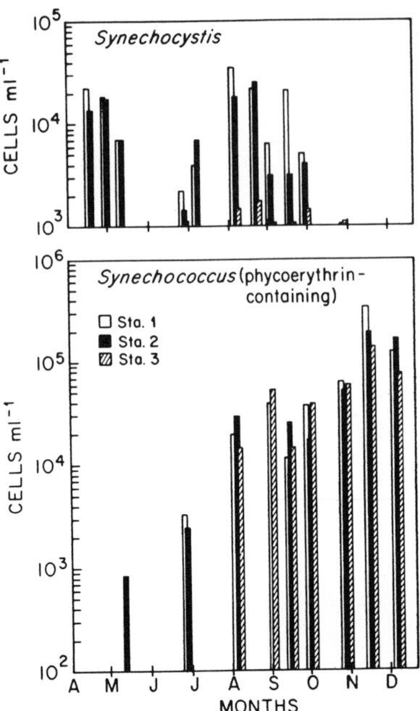

Figure 3. Upper: *Synechocystis* spp. abundance in Carmans River estuary determined by indirect immunofluorescence assay using antisera directed against PCC7339.
Lower: Phycoerythrin-containing *Synechococcus* spp. abundance counted by epifluorescence microscopy. (Reproduced from Campbell *et al.*, 1983).

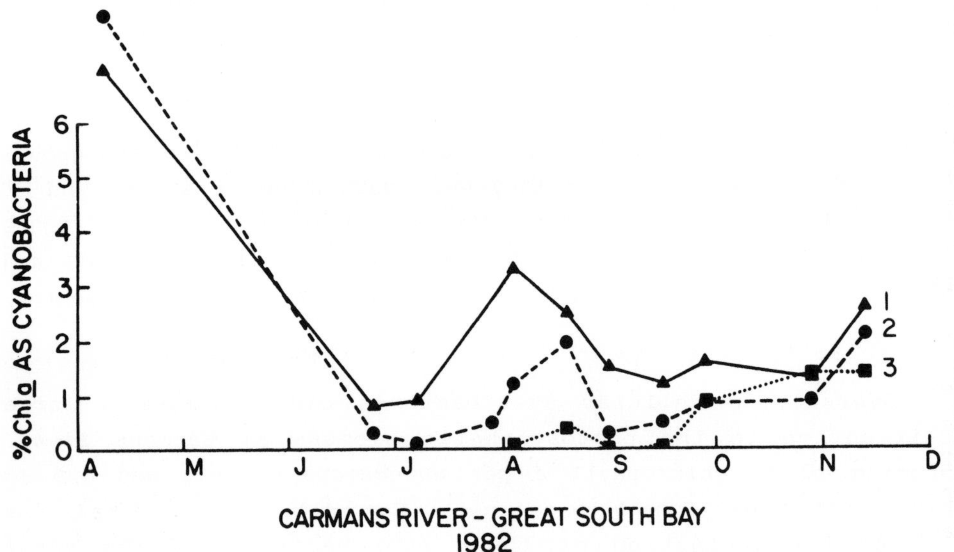

Figure 4. Percentage of the chlorophyll a biomass contributed by the cyanobacterial populations in Carmans River, NY.

sample from station 1 (20%) at the beginning of the spring bloom in Great South Bay. The contribution by marine cyanobacteria to total chlorophyll *a* was much lower at station 3 (Figure 4).

In oceanic samples, PE-*Synechococcus* made a larger contribution to the chlorophyll *a* biomass. Using the same assumption for chlorophyll *a* per cell, cyanobacteria contributed up to 30% of the total chlorophyll a in the samples from the eastern tropical North Pacific in November/December (Campbell and Carpenter, 1986). Based on size-fractionated samples collected in July/August from the northwest Atlantic Ocean, the PE-*Synechococcus* were 25% of the chlorophyll *a* biomass at the warm core eddy station and 46% at the Wilkinson's Basin station (Glover *et al.*, 1986).

Samples from the eastern tropical North Pacific transect were examined for strains labelled by anti-7803 and for strains labelled by anti-5701 or anti-8101 sera. At the offshore stations, the total cell abundance was 10^6 cells l^{-1} (Figure 5a), and often a subsurface maximum occurred, which is typical for oceanic *Synechococcus* populations in stratified water columns (Glover, 1985). The percentage of the population labelled was 10% to 30% (Figure 5c). No definite pattern was observed in the proportion of the population labelled with respect to depth. At station 2 the proportion of cells labelled by anti-7803 was higher at the surface than at depth; but at station 11, the percentage at the surface was the same as at 80 m (Figure 5c). At the coastal stations, PE-Synechococcus were more abundant than at offshore stations (up to 10^7 cells l^{-1}), and the maximum cell number most often occurred at the surface (Figure 5b). In general, the proportion of cells labelled by anti-7803 was constant throughout the euphotic zone (Figure 5d). At two stations this percentage was low (<10%), but at the third station approximately 35% was labelled (Figure 5c).

In the northwest Atlantic, PE-containing *Synechococcus* spp. were more abundant at Wilkinson's Basin, P3, (10^8 cells l^{-1}) than at the neritic frontal station, P1, (10^7 cells l^{-1}) and the warm core eddy, P2, (5×10^7 cells l^{-1})(Glover *et al.*, 1986). The distribution of PE-containing *Synechococcus* with depth in the euphotic zone was the same as at the Pacific stations: cell maximum at the surface at the coastal station, P1, and subsurface maxima at offshore stations, P2

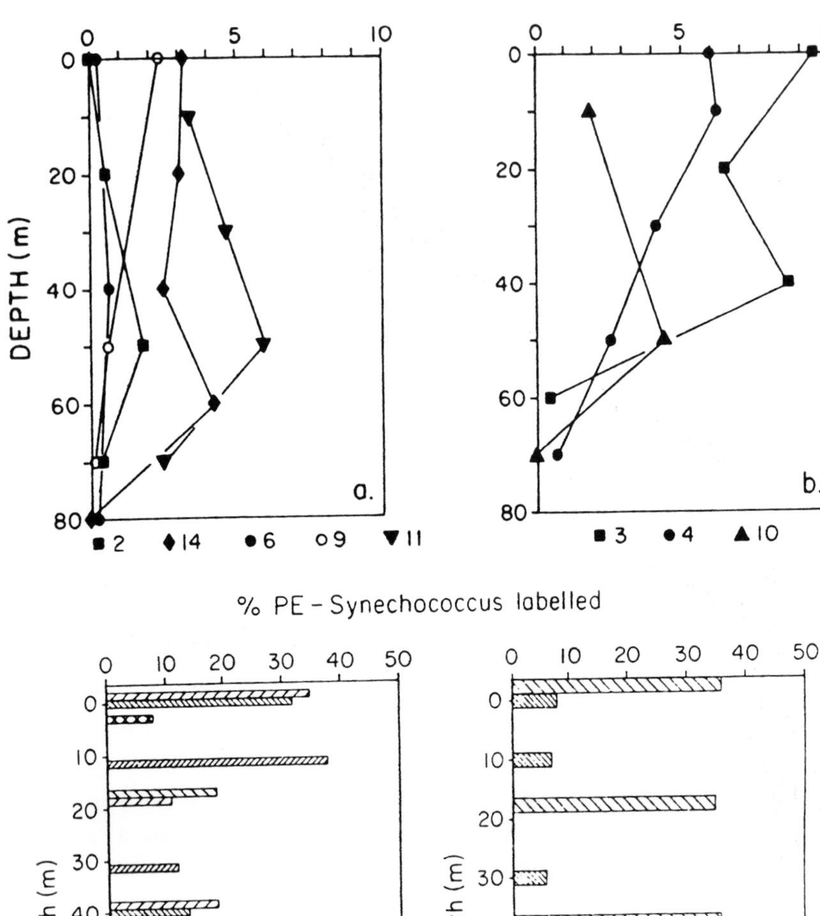

Figure 5. Total PE-*Synechococcus* spp. abundance determined by epifluorescence microscopy in the eastern tropical north Pacific Ocean. a: offshore stations. b: PE-*Synechococcus* spp. abundance at coastal stations. c: Percentage of the populations labelled with the antisera directed against WH7803 at the offshore stations. d: Percentage of the population labelled with anti-7803 at the coastal stations. (Reproduced from Campbell and Carpenter, 1986.)

and P3. Again, the proportion of the population labelled did not vary significantly with depth. Unlike the results at the Pacific tropical stations, the distribution of cells labelled by anti-7803 did vary among stations. At the warm core eddy station, a large percentage, 25% to 30%, was labelled, but at stations P1 and P3, where water temperature was 5° to 10° lower than at P2, in most cases none were labelled (Figure 6; Glover *et al.*, 1986). This difference is consistent with the results obtained by the cross-reactivity tests: anti-7803 predominantly labelled strains isolated from the Sargasso Sea region. The highest percentages of labelling have been counted in samples collected from the Bermuda region (68% to 85%; Campbell and Carpenter, 1986). In general, PE-*Synechococcus* populations were rarely more than 50% labelled by anti-7803. This suggests that other strains (or serogroups) are also important in oceanic waters.

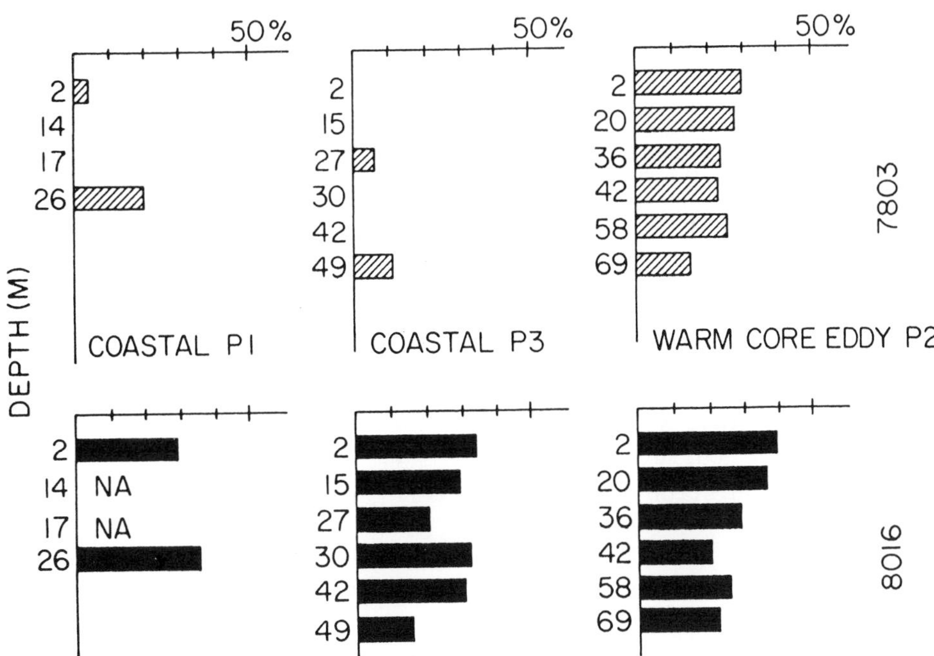

Figure 6. *Synechococcus* spp. labelled by antisera directed against WH7803 and by antisera directed against WH8016 from depth profiles at three stations in the northwest Atlantic Ocean. (Redrawn from Glover *et al.*, 1986.)

Labelling with anti-8016 gave quite different results: anti-8016 labelled approximately 30% of the population at all 3 stations. The presence of strains labelled by anti-8016 at the warm core eddy station was unexpected, especially since no major cross reactions were observed between anti-8016 and Sargasso Sea isolates. However, because of the few minor cross reactions observed, there may have been cells labelled by both antisera. Because of the possibility of common reactivities, the antisera directed against WH7803 and WH8016 will be adsorbed (e.g. Ward, 1982) to ensure recognition of only one serogroup in future work.

Lastly, PC-dominant *Synechococcus* spp. were not found in any of the oceanic samples from these two transects. The absence of these estuarine-type organisms agrees with the field observations of Waterbury *et al.*, (in press) and Murphy and Haugen (1985).

CONCLUSIONS

Based upon the indirect immuofluorescence test results yielding strong positive reactions, antisera to each of the PC-containing and PE-containing strains of cyanobacteria defined clusters of strains, or serogroups. The serogroups appear consistent with the oceanic regions from which the strains were isolated (Tables 1 and 2; Campbell *et al.*, 1983). From the results reported here, it seems that more than two serogroups exist among the PE-containing *Synechococcus*.

The antisera directed against the 3 PE-containing *Synechococcus* strains showed that serological differences exist among 37 clones of the Woods Hole Culture Collection. If serogroups are defined by strains labelled strongly (+++ or ++++), then the three groups can be characterized as follows:

Anti-7803: 11/37 strains; all but 1 are tropical and sub-tropical isolates.
Anti-8016: 5/37 strains; all but 1 are coastal isolates.
Anti-8113: 7/37 strains; strong reactions with 4 known motile strains, and 3 non-motile.

Minor cross-reactions were also observed; therefore, to use these antisera for taxonomic purposes, they should be adsorbed with cross-reacting strains.

PC-dominant *Synechocystis* and *Synechococcus* strains are present in Great South Bay, an estuary on the south shore of Long Island, New York, at densities of 10^4 cells ml^{-1} and $<10^2$ cells ml^{-1}, respectively. There may be a seasonal variation in abundance, but these cells are never of major importance to the total photosynthetic biomass; throughout most of the year the percentage was around 2%.

PC-dominant *Synechococcus* cells were not found in oceanic samples.

In field populations, anti-7803 rarely labelled a majority of the population; usually <50%. This supports the laboratory cross-reactivity tests and suggests that other serogroups of PE-containing *Synechococcus* exist.

ACKNOWLEDGEMENTS

I thank V.J. Iacono for his advice in the development of the IF assay, J.B. Waterbury for providing cultures, M.I. Scranton for sample collection, E.J. Carpenter for supporting my doctoral research (NSF grant OCE 8214764), and H.E. Glover for my support during the preparation of this manuscript (NSF grant OCE 8521170).

REFERENCES

Alberte, R.S., A.M. Wood, T.A. Kursar and R.R.L. Guillard. 1984. Novel phycoerythrins in marine *Synechococcus* spp.: characterization and evolutionary and ecological implications. Plant Physiol. 75: 732-739.

Barlow, R.G. and R.S. Alberte. 1985. Photosynthetic characteristics of phycoerythrin-containing marine *Synechococcus* spp. I. Responses to growth photon flux density. Mar. Biol. 86: 63-74.

Bohlool, B.B. and E.L. Schmidt. 1980. The immunofluorescence approach in microbial ecology. IN: Advances in Microbial Ecology. M. Alexander (ed.). 4: 203-241.

Campbell, D.H., J.S. Garvey, N.E. Cremer and D.H. Sussdorf. 1964. Methods in immunology. W.A. Benjamin, Inc. NY. pp. 22-23, 41-42.

Campbell, L. 1985. Investigations of marine, phycoerythrin-containing *Synechococcus* spp. (Cyanobacteria): Distribution of serogroups and growth rate measurements. 186 pp. Ph.D. Dissertation, State University of New York, Stony Brook, NY.

Campbell, L. and E.J. Carpenter. 1986. Characterization of phycoerythrin-containing *Synechococcus* populations by immunofluorescence. submitted.

Campbell, L., E.J. Carpenter and V.J. Iacono. 1983. Identification and enumeration of marine Chroococcoid cyanobacteria by immunofluorescence. Appl. Environ. Microbiol. 46: 553-559.

deMacario, E.C., M.J. Wolin and A.J.L. Macario. 1981. Immunology of Archaebacteria that produce methane gas. Science 214: 74-75.

Fliermans, C.B. and E.L. Schmidt. 1977. Immunofluorescence for autecological study of a unicellular bluegreen alga. J. Phycol. 13: 364-368.

Glover, H.E. 1985. The physiology and ecology of the marine cyanobacterial genus *Synechococcus*. IN: Advances in aquatic microbiology. H.W. Jannasch and P.J. LeB Williams (eds.). Academic Press, NY. 4: 49-107.

Glover, H.E., L. Campbell and B. Prezelin. 1986. Contribution of *Synechococcus* spp. to size-fractioned primary productivity in three water masses in the Northwest Atlantic. Mar. Biol. 91: 193-203.

Guillard, R.R.L. and J.H. Ryther. 1962. Studies of marine planktonic diatoms. I. *Cyclotella nana* Hustedt, and *Detonula confervacea* (Cleve) Gran. Can. J. Microbiol. 8: 229-239.

Johnson, P.W. and J. McN. Sieburth. 1979. Chroococcoid cyanobacteria in the sea: A ubiquitous and diverse phototrophic biomass. Limnol. Oceanogr. 24: 928-935.

Kilpatrick, K.A. 1985. The development of a method to measure marine cyanobacterial phycoerythrin extracted in solvents. M.S. thesis, Texas A & M University. 74 pp.

Kursar, T.A., H. Swift and R.S. Alberte. 1981. Morphology of a novel cyanobacterium and characterization of light-harvesting complexes from it: implications for phycobiliprotein evolution. Proc. Natl. Acad. Sci. USA. 78 (11): 6888-6892.

Murphy, L.S. and E. Haugen. 1985. The distribution and abundance of phototrophic ultraplankton in the North Atlantic. Limnol. Oceanogr. 30: 47-58.

Ong, L.J., A.N. Glazer and J.B. Waterbury. 1984. An unusual phycoerythrin from a marine cyanobacterium. Science. 224: 80-82.

Ouchterlony, O. and L.A. Nilsson. 1978. Immunodiffusion and immunoelectrophoresis. IN: Handbook of Experimental Immunology. D.M. Weir (ed.). Oxford, Blackwell Scientific Publications. 3: 38.

Perkins F.O., L.W. Haas, D.E. Phillips and K.L. Webb. 1981. Ultrastructure of a marine *Synechococcus* possessing spinae. Can. J. Microbiol. 27: 318-329.

Rantz, L.A. and E. Randall. 1955. Use of autoclaved extract of hemolytic *Streptococci* for serological grouping. Stanford Med. Bull. 13: 290-291.

Rippka, R., J. Deruelles, J.B. Waterbury, M. Herdman, R.Y. Stanier. 1979. Generic assignments, strain histories and properties of pure cultures of cyanobacteria. J. Gen. Microbiol. 111: 1-61.

Sarokin, D.J. and E.J. Carpenter. 1981. Cyanobacterial spinae. Bot. Mar. 24: 389-392.

Smith, A.D. 1982. Immunofluorescence of sulphate-reducing bacteria. Arch. Microbiol. 133: 118-121.

Taubman, M. A. and D.J. Smith. 1974. Effects of local immunization with *Streptococcus mutans* on induction of salivary IgA antibody and experimental dental caries in rats. Infect. Immunity 9: 1079-1091.

Ward, B.B. 1982. Oceanic distribution of ammonium-oxidizing bacteria determined by immunofluorescent assay. J. Mar. Res. 40: 1155-1172.

van Baalen, C. 1962. Studies on marine blue-green algae. Bot. Mar. 3: 129-139.

Waterbury, J.B., S.W. Watson, R.R.L. Guillard and L.E. Brand. 1979. Wide-spread occurrence of a unicellular, marine, planktonic cyanobacterium. Nature 277: 293-294.

Waterbury, J.B., S.W. Watson, F.W. Valois and D.G. Franks. 1986. Biological and ecological characterization of the marine unicellular cyanobacterium *Synechococcus*. IN: Physiological Ecology of Picoplankton. T. Platt and W. Li (eds.). C.J.Fish. Aqua. Sci. 214: in press.

Waterbury, J.B., J.M. Willey, D.G. Franks, F.W. Valois and S.W. Watson. 1985. A cyanobacterium capable of swimming motility. Science 230: 74-76.

Wetherell, J.R., Jr. and A.S. Bleiweiss. 1975. Antigens of *Streptococcus mutans*: Characterization of a polysaccharide antigen from walls of strain G5-5. Infect. Immun. 12: 1341-1348.

Wood, A.M. 1982. Occurrence and ecological significance of different pigment types of marine *Synechococcus*. EOS 63: 960.

AN IMMUNOFLUORESCENCE APPROACH TOWARD THE IDENTIFICATION OF ENDOSYMBIOTIC DIATOMS IN SEVERAL SPECIES OF LARGER FORAMINIFERA

John J. Lee and Ying Chan
Department of Biology
City College of CUNY
New York, NY 10031

Annie Lagziel
Steinitz Marine Biological Lab
Elat, Israel

INTRODUCTION

Larger foraminifera are collectively the hosts for a diverse range of endosymbiotic algae. Individual species of forams are hosts for dinoflagellates, chlorophytes rhodophytes or diatoms (reviewed in Lee, 1980; Lee and McEnery, 1983; Leuteneger, 1984). Although host/ symbiont specificity is considered the norm in invertebrate/algal symbioses the diatom-bearing foraminifera seem to be an exception. Lee, and coworkers (1980a, b, work in progress; Lee and Reimer, 1983) have identified approximately 20 different species of diatoms which they have isolated from species of larger formaminifera. Usually specimens of a host species harbor only a single diatom species at one time but specimens of the same host may harbor a different diatom species at another location in the field or when collected again at the same station.

The specific identification of endosymbiotic algae within their hosts is not yet feasible. Endosymbiotic algae tend to loose many of their diagnostic features (e.g., envelope structure, flagella) when they are within their hosts. Fortunately they regain their characteristics when they are liberated from their hosts and cultivated on suitable media (Lee, 1980). In the cases of larger foraminifera the isolation and cultivation of the endosymbiotic algae is a very labor intensive and lengthy process. The hosts have many surface irregularities (e.g., pits, crevises; Lee, 1983; Figures 3, 4, 9, 12; seen here as Figures 1, 2, 3, and 4) which provide refuge for externally colonizing microphytes and which impede rapid removal by brushing during their initial stages of preparation for symbiont isolation. After individual hosts have been thoroughly brushed with

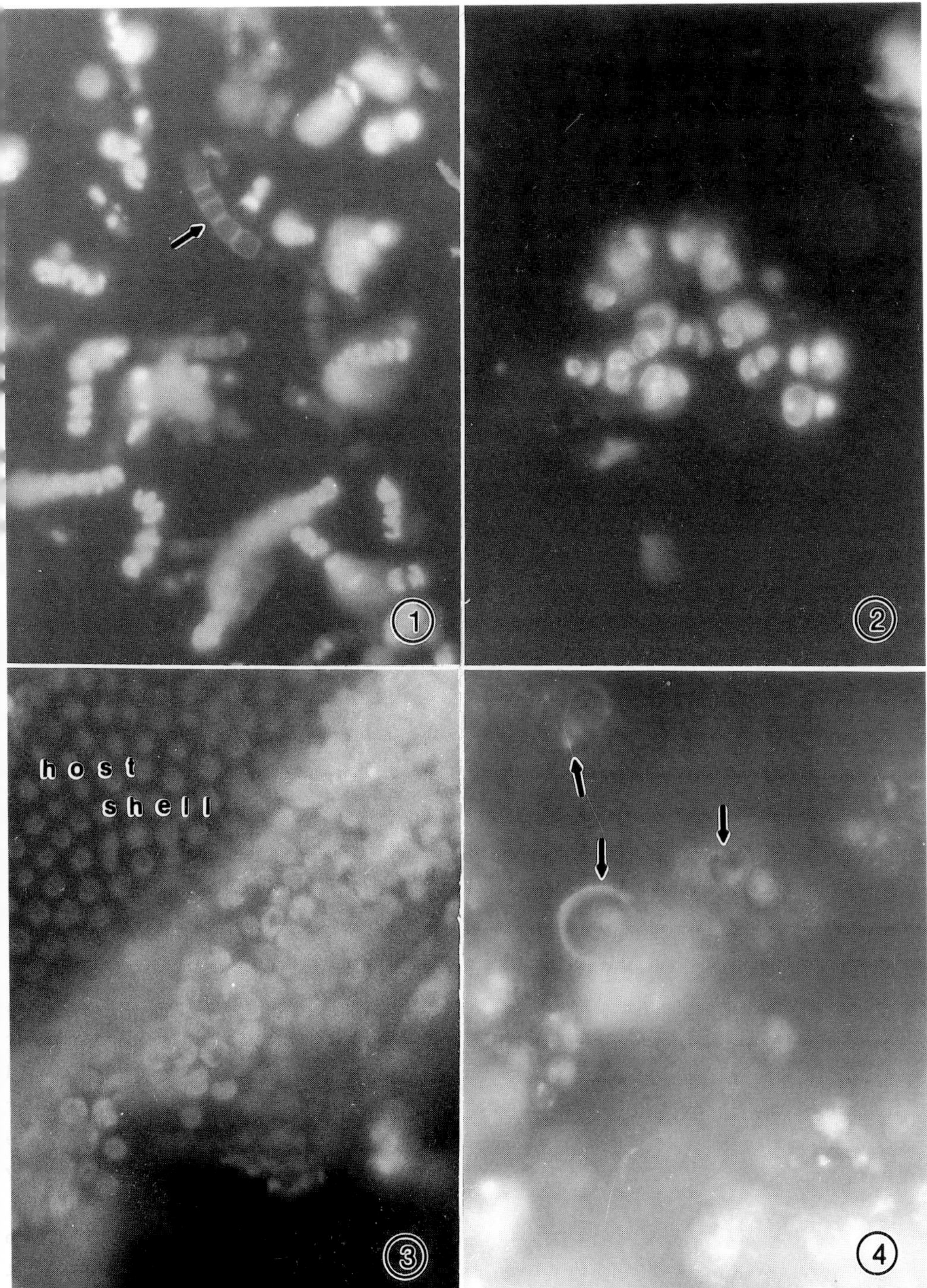
1
2
host
shell
3
4

Figure 1. *Fragilaria shiloi* from a stock culture after incubation with absorbed homologous antiserum X 200.

Figure 2. *Fragilaria shiloi* from a stock culture after incubation with antisera against *Nitzschia panduriformis* X 200.

Figure 3. Negative reaction to antiserum by endosymbionts expressed from a crushed host captured directly from the sea. Fluorescence of the algae is red and due to autofluorescence of their chloroplasts. A fragment of the shell is also visible in the picture X 200.

Figure 4. Positive reaction to antiserum by endosymbionts expressed from a crushed host harvested directly from the sea X 200.

alcohol sterilized 0000 sable brushes and rinsed many times with sterile sea water they are examined at high magnification (60X) in a good quality stereo microscope and a compound microscope with a water immersion lens (Zeiss 40 x 10 = 400 X) to make sure all external algae have been removed. The "cleaned" hosts are then transferred to wells containing sterile sea water in sterile 9-hole spot plates. The foraminifera are grasped with sterilized fine pointed forceps (Dumont #5) and broken and crushed by a similar second pair of forceps. The crushed foraminiferan cytoplasm and endosymbionts are aseptically removed from the spot plate by pipette and transferred to media which will support symbiont growth (Lee *et al.*, 1980a,b; Lee and Reimer, 1983). After 2 or 3 weeks cultures initiated from each host can be examined microscopically. Since the diatoms found as endosymbionts within foraminifera are very small (<10 μm) preparation of frustules for SEM examination is almost routine (Lee and Reimer, 1983). Using this approach takes 4-6 weeks from collection of the hosts to identification of the symbionts. A fine structural approach seems a much less promising alternative (Leutenagger, 1984). Because the preparation of hosts for TEM examination is even more time consuming and labor intensive fewer specimens could be examined (Leutenagger, 1984). Even more disappointing was the fact that although some diversity of characters could be observed, there were not enough differences observable in the TEM to characterize individual species.

The biological bases for the diversity of endosymbiont diatom species in larger foraminideran species are not yet known. Many questions pique ones intellectual curiosity. Is acquisition of a particular symbiont species by zygotes simply a matter of chance? Do physical factors such as depth or season affect the distribution of particular symbiont species in the populations? Are the symbionts harbored species which are abundant in natural communities?

It was with these and similar questions in mind that we started to plan a new series of collections and studies. As we contemplated methodology it seemed reasonable to try immunochemical approaches to the identification of the endosymbiotic algae. If successful they offered promise of greatly simplifying and speeding up the identification process. We report our initial results here.

MATERIALS AND METHODS

Algae

Axenic cultures of *Nitzschia panduriformis*, *Nitzschia frustulum* var *symbiotica*, *Fragilaria shiloi* and *Amphora tennerima*, have been maintained in our lab in initial isolation from their respective hosts (Lee *et al*., 1980a, b; Lee and Reimer, 1983). A new species of *Achnanthes* (MEM) was isolated from a population of *Operculina ammonoides* at 38 M in wadi Taba during the course of this study. Isolation followed the methods outlined in the introduction and published previously (op. cit.).

Preparation of the Antigen

Axenic cultures of the above species were still in log phase (7-10 days) at the time of harvest. They were centrifuged and washed several times in sterile sea water before they were homogenized with the aid of a sterile potter-Elveljem tissue grinder. A cell envelope/frustule fraction of the cells was obtained by centrifuging the homogenate at 500 Xg for 5 min in a refrigerated centrifuge. After 5 serial washings and centrifugation, aliquots of the fraction were examined in a compound light microscope (1,200X) and filtered onto nucleopore membranes for examination in the SEM. We observed only frustrule fragments and membrane debris in the preparations. The protein content of the preparation was assayed by means of the Bradford method (Bio Rad modification) and the volume of the sample was adjusted to yield a protein level of 1 mg/ml.

Preparation and testing of the antisera

A commercial vendor, Pocono Rabbit Farm and Laboratory (Canadensis, PA), prepared the rabbit antisera. They provided initial samples of sera of the rabbits used, sera after initial challenge and after booster shots. When we reported that the antibody titer was high, the rabbits were sacrificed and exsanguinated by the vendor. Three rabbits were used for the preparation of antisera against the envelope (frustule)/membrane fraction of each of the algal species. The antisera were not pooled.

The antisera from each animal were serially diluted (10^{-1}-10^{-5}) and added to 400 µl-polyethylene micro-centrifuge tubes containing aliquots of serially washed algal cells. The antigen and antisera were thoroughly mixed. After incubation for 1 hour at 37° C the tubes were agitated and then centrifuged. The algal cells in the pellet were resuspended in sea water and centrifuged again. The cells freed from unreacted excess antibodies by 5 serial washings, were then mixed with and incubated with 10 ml of diluted (1:10) goat anti-rabbit IgG (whole molecule) FITC conjugate (Sigma F 65005). After incubation for 1 hour at 37° C the cells were centrifuged and washed 5 times in sterile sea water. The cell pellet was resuspended in 100 µl of sea water and pipetted onto a microscope slide. The preparations were examined on a Zeiss photomicroscope II (NYC) or III (Elat) equipped with an epifluorence illuminator/condenser III RS and an appropriate exciter filter/beam splitter/barrier filter set for FITC (Zeiss 450-490 blue glass (42 79 02), FT 510 chromatic splitter (46 6304) and barrier filter LP520 (46 78 73).

Photography

Photographs were taken on either Fuji film (ASA 1,000) or Kodak Kodacolor VR 400 or 1,000 depending upon which was available at the time. The nature of the subject (bright objects in a black background) required the use of the selective spot in the automatic exposure system. Because of the high speeds of the films used it was practical to expose and develop them normally. Black and white prints were made by printing the color negatives on Kodak Panalure II RC paper.

Experimental Organisms and tests

The tests reported here took place in Dec-Jan 84/85 and 85/86 while the first named author was in residence in Elat, Israel. Collections of 5 populations of *Heterostegina depressa* and *Amphistegina lessonii* from 25 M, five populations of *Amphistegina lobifera* from 3, 15, and 20 M, and two *Operculina ammonoides* at 38 M from wadi Taba were tested. The populations tested were from very small carefully collected samples. They consisted of a few (3-5) *Halophila* leaflets (*Amphistegina* Spp samples at 5 - 20 and 25 M) small pieces of coral rubble (*Heterostegina* and *A. lobifera* samples at 38M) or a petri dish-sized scoop of mud (*Operculina* at 38 M). The foraminifera on the samples were observable on the substrates before collection by the divers. Previous experience with isolation of symbionts from populations of foraminifera in similar small samples suggested that usually they share a single endosymbiont species.

Each sample was divided for study. Twenty organisms were selected for immunochemical study. The remaining 10-20 organisms were carefully washed, crushed, and their liberated symbionts were inoculated in culture media for later identification. We found that with appropriate controls we could set up and test immunochemically 20 animals in a day. The initial steps in the parallel culture isolations from the same populations required another 2 days.

We learned very soon in preliminary tests on stock cultures that there was a great deal of cross reactivity of the antisera for different species of endosymbiotic algae. Therefore we first incubated diluted (1:10) antisera with a heterologous endosymbiont for 1 hour at 37°C. After centrifugation to remove the heterologous algal pellet, absorbed serum was mixed with a washed and crushed larger foraminifera in a 400 µl microcentrifuge tube for the reactions described above. Endosymbiotic algae from stock cultures were also tested separately in each experiment to provide both positive and negative controls for the serum.

In 2 experiments we tested antisera against the endosymbionts within *Amphistegina* sp which had been kept in the laboratory culture in light for 1 or 3 weeks. The aim of these experiments was to see if there might be changes in surface antigens of ensymbiotic algae which might be recognized by the antibodies in the antisera as the hosts and their symbionts were placed in stress.

RESULTS

There was quite a variety of endosymbiotic algal species in the small populations of foraminifera we chose for study (Table 1). In both winters *Nitzschia frustulum* var *symbioticum* was the most commonly encountered endosymbiont. *Fragilaria shiloi* and *N. panduriformis* were also abundant. In general the results of the observations of fluorescence in preparations were in consonance with the results obtained by cultural isolation (Table 1). Tests on symbionts expressed from hosts kept in laboratory captivity for 1 or 3 weeks were not significantly different from those on hosts directly from the sea (Tables 1 and 2). However, the autofluorescence of the chloroplasts of the symbionts expressed from captive hosts was considerably weaker, with the least being found in the 3 week populations.

Visual examination of the fluorochrome-tagged endosymbionts directly liberated from hosts gave us some additional details of the symbiosis which would be hard to obtain by any other method. In many preparations part of the algal population reacted with antisera while the rest did not. Careful examination of some of the preparations at high magnification (1,000X) suggested that there were slight volume differences between those which were labeled and those which were not. In some cases only a few specimens (<10) out of hundreds examined were labeled. The labeling of the endosymbionts in one population of host was usual. The fluorescence on the algal surfaces was patchy rather than uniformly distributed completely around them. Although there was slight absorbtion or entrapment of the antiserum and/or FITC conjugate to the crushed shell of the foraminifera to yield a faint fluorescence, it did not interfere with the observation of the symbionts.

DISCUSSION

These preliminary results suggest that the immunochemical approach could be a very useful tool for research on the nature of symbiosis in large foraminifera. It offers promise to greatly accelerate the process of identification of the endosymbionts. The drawback seems to be the complexity of the problem and the dearth of knowledge we have at present on the distribution of the symbionts with in populations of their hosts. In these preliminary tests most of the results were negative. It is clear now that the endosymbionts in the populations we tested were far more varied than we believed at the onset. In our first isolations of symbionts of hosts harvested from the Red Sea at Elat, *Fragilaria shiloi* and *Nitzschia panduriformis* were the most common symbionts (Lee *et al.*, 1980a). It seemed reasonable to make antisera against these 2 species in our first attempts with this approach. In our Dec/Jan 85/86 tests we broadened our antisera battery to 4 species and were more successful. Even a greater variety of antisera should be used in future experiments.

Table 1. Summary of results of immunochemical approach to the identification of endosymbionts within larger forminifera.

		Test		Reactions			Isolations	
Host	pop	Antisera	Number	Total Positive	Partial	Control	# host crushed	Species found
Hetero	38/5	*N. pand.*	9	0	1	OK	25	*NF symb., N. lav.*
Hetero		*F. shil.*	9	0	0	OK		*N. pand.*
Hetero	38/1	*N. pand.*	9	0	2	OK	25	*NF symb.*
Hetero		*F. shil.*	9	0	0	OK		*N. pand.*
A. less		*N. pand.*	5	0	0	OK	15	*NF symb.*
A. lob	25	*N. pand.*	4	0	0	OK	10	*F. shil.*
A. less		*F. shil.*	4	1	0	OK	15	*NF symb., F. shil*
A. lobif		*F. shil.*	5	0	1	OK	10	
A. lobif	8	*N. pand.*	9	0	0	OK	15	*NF symb., A. roet.*
		F. shil.	9	0	0	OK		*N. pand.*
A. lobif	13	*N. pand.*	9	3	0	OK	15	*NF symb., N. lav.*
		F. shil.	9	0	9	OK		*N. pand. F. shil.*
		N. pand.	9	0	2	OK		*N. lav., NF shil.*
Operc.	38	*NF symb.*	9	0	1	OK	30	*N. pand., F. shil.*
		Ach. (MEM)	9	0	0	OK		*Amp. sp.*
		N. pand.	9	3	1	OK		*N. pand.*
A. lobif	20	*NF symb.*	9	1	3	OK	20	*NF symb., N. lav*

Table 1. continued

Host	pop	Test: Antisera	Test: Number	Reactions: Total Positive	Reactions: Partial	Reactions: Control	Isolations: # host crushed	Isolations: Species found
		F. shil.	9	0	2	OK		
		N. pand.	9	2	2	OK		*N. pand.*
A. lobif	20	*NF symb.*	9	2	1	OK	20	*F. shil.*
		F. shil.	9	2	0	OK	20	
	38L	*N. pand.*	9	1	3	OK		*N. pand.*
Hetero		*NF symb.*	8	0	1	OK	20	*NF symb.*
		Ach. (MEM)	8	2	1	OK		*Ach. (MEM)*
		N. pand.	8	0	0	OK	15	*NF symb.*
A. less	38L	*F. shil.*	8	3	0	OK		*F. shil.*
		NF symb.	8	4	0	OK		*N. lav.*

Abbreviations:

Hetero - *Heterostegina depressa*
A. less - *Amphistegina lessonii*
A. lobif - *Amphistegina lobifera*
Operc - *Operculina ammonoids*

Symbionts:

N. pand. - *Nitzchia panduriformis*
F. shil. - *Fragilaria shiloi*
NF symb. - *Nitzschia frustulum var symbiotica*
Ach. (MEM) - *Achnanthes SP (MEM)*
A. roet - *Amphora roettgerii*
Amp. S - *Amphora sp*
Nav. S - *Navicula sp. (Elat)*

Table 2. Immunochemical identification of symbionts within hosts kept in the laboratory for 1 or 2 weeks

Test					Reactions			Isolations	
Host	pop	#	W	Antisera	Total positive	Partial	controls	# host crushed	Species found
A. lobif	T6	4	1	*N. pand.*	0	0	OK	15	*N.F. symb.*
A. lobif	T6	4	1	*F. shiloi.*	0	0	OK		*Navicula Sp.*
A. lessonii	T35	4	1	*N. pand.*	0	0	OK	15	*N.F. symb.*
A. lessonii	T35	4	1	*F. shiloi.*	0	0	OK		*N. pand.*
A. lobif.	T10	4	3	*N. pand.*	0	0	OK	15	*Amp. tenn.*
A. lobif.	T10	4	3	*F. shiloi*	0	0	OK		*F. shiloi*
A. lessonii	T25	4	3	*N. pand.*	0	0	OK	18	*Rhopaloid*
A. lessonii	T25	4	3	*F. shiloi*	0	0	OK		*N.F. symb.*
A. lobif	T15	4	3	*N. pand.*	0	0	OK	20	*N.F. symb.*
A. lobif	T15	4	3	*F. shiloi*	1	0	OK		*F. shiloi*
A. lobif	Z	4	3	*N. pand.*	1	0	OK	20	*N.F. symb.*
A. lobif	Z	4	3	*F. shiloi*	0	1	OK		*Navicula Sp.* *Amphora Sp.* *F. shiloi* *Nitz. pand.*

W - Weeks in captivity
For abbreviations, see Table 1.

Although the fine structure of the endosymbionts of animals cultured in the laboratory appeared normal, Koestler *et al.* (1985) demonstrated that over time the symbiont population was gradually being digested.

The general gradual weakening of the autofluorescence of the chloroplasts of symbionts with hosts maintained in the lab suggests that the laboratory conditions we used were less than favorable for the symbionts.

The large number of negative reactions in our last test leaves several questions about the surface envelopes of the endosymbionts still unresolved. We know that the symbionts under study do not form their frustules when they are in their hosts. Are their silicalemmae (the biological membranes which envelope the silica shell during its appositional expansion) normal? What are the antigens common to the free-living forms isolated in culture and those in their hosts? Appropriate immunochemical studies can tell us much about this aspect of the biology of larger foram/diatom symbiosis.

ACKNOWLEDGEMENTS

This research was supported by grants from NSF (OCE 83-09266), US-Israel Binational Fund (34 18/83) and PSC-CUNY (666 306). Mr. Benno ter Kuille and Dr. James Kendall actively assisted this research by collecting specimens on SCUBA dives. The research conducted at the H. Steinitz Marine Biological Laboratory in Elat was made possible by the facilities and equipment generously made available to us by Drs. Yohnathan Erez and Yehuda Cohen. We gratefully acknowledge a gift from the H.B. Cantor Foundation which provided the Zeiss epifluorescence attachments which were used at City College. Ying Chan was a participant in the City College Research Scholars program. Drs. Eugene Gallagher and Dr. Anthony Garro were very helpful in the initial stages of this research.

REFERENCES

Koestler, R.J., J.J. Lee, J. Reidy, R.P. Sheryll and X. Xenophontos. 1985. Cytological investigation of digestion and re-establishment of symbiosis in the larger benthic foraminifera *Amphistegina lessonii*. Endocyt. C. Res. 2: 21-54.

Lee, J.J. 1980. Nutrition and physiology of the foraminifera 43-46. IN: Biochemistry and Physiology of Protozoa. 2nd Edition. M. Levandowsky and S.H. Hutner (eds.). New York Academic Press.

Lee, J.J. 1983. Perspective on algal endosymbionts in larger foraminifera. Int. Rev. Cytol. Suppl. 14: 49-77.

Lee, J.J. and M.E. McEnery. 1983. Symbiosis in foraminifera 37-68. IN: Algal Symbiosis. L.J. Goff. (ed.). Cambridge University Press.

Lee, J.J., M.E. McEnery, R.F. Rockwell, J. Erez, B. ter Kuile, A. Lagziel and R. Röttger. Identification and distribution of endosymbiotic diatoms larger foraminifera. (in progress)

Lee, J.J., M.E. McEnery, R. Röttger and C.W. Reimer. 1980a. The culture, isolation and identification of endosymbiotic diatoms for *Heterostegina depressa* d'Orgbigny and *Amphistegina lessonii* d'Orbigny (larger foraminifera from Hawaii). Botanica Marina 23: 297-302.

Lee, J.J. and C.W. Reimer. 1983. Isolation and identification of endosymbiotic diatoms from larger foraminifera of the Great Barrier Reef, Australia, Makapuu Tide Pool, Oahu, Hawaii, and the Gulf of Elat, Israel, with the description of new species, *Amphora rottgeri*, *Navicula hanseniana* and *Nitzschia frustrulum* variety *symbiotica*. Proc. VII Inter. symp. Living and fossil diatoms. D.G. Mann. (ed.). Koeltz Scientific Books.

Lee, J.J., C.W. Reimer and M.E. McEnery. 1980b. The taxonomy of diatoms isolated as symbionts from the larger foraminifera from the Red Sea. Botanica Marina. 23: 42-48.

Leutenegger. S. 1984. Symbiosis in benthic foraminifera: specificity and host adaptations. Jour. Foraminifera Res. 14: 16-35.

IMMUNOFLUORESCENT CHARACTERIZATION OF THE EUKARYOTIC ULTRAPLANKTON

Lynda Shapiro
Bigelow Laboratory for Ocean Science
McKown Point
W. Boothbay Harbor, Maine 04575

ABSTRACT

The specificity of the antigen-antibody reaction suggests its use in systematics. Coupling the antibody to a fluorescent tag could provide identification of very small organisms - minute unicells that are difficult to identify in other ways.

More than 50% of marine phytoplankton biomass is organized into cells <3 µm in diameter. Of these, up to 90% belong to the single prokaryotic genus *Synechococcus* and 10% are eukaryotic. The eukaryotic component dominates the phytoplankton biomass in certain environments and may be more important to photosynthesis than its biomass contribution implies. These minute cells are very difficult to identify by other means, but are amenable to immunofluorescence identifications. Antigen targets include the highly specific pigments, cell wall proteins, and possibly proteins of the chloroplast membranes.

PROBES FOR ASSESSING SINGLE-CELL PRIMARY PRODUCTION: ANTIBODIES AGAINST RIBULOSE-1,5-BISPHOSPHATE CARBOXYLASE (RuBPCASE) AND PERIDININ/CHLOROPHYLL *A* PROTEIN (PCP)

M.V. Orellana, M.J. Perry and B.A. Watson

School of Oceanography and
Department of Botany
University of Washington
Seattle, WA 98195

INTRODUCTION

Identification of the factors controlling spatial and temporal variability of phytoplankton production is a major issue in biological oceanography. Although the interest in primary production has a long history, recent technological developments have enabled new approaches to the problem of its measurement. With the availability of new methods, the approach has shifted from gross estimation of primary production to include studying the control processes influencing individual phytoplankton cells. The effect of environmental factors on the metabolic activity and growth at the species level is a key to understanding large-scale patterns of production and distribution of phytoplankton in the ocean.

The ^{14}C method for measuring primary production (Steemann-Nielsen, 1952) and the fluorometric method for determining chlorophyll *a* (Lorenzen, 1966; Holm-Hansen *et al.*, 1965) have played major roles in contributing to our understanding of marine productivity, but they have not provided us with information on phytoplankton dynamics at the level of individual species. Only a few investigators have attempted to associate species identification of individual cells with rates of carbon fixation using the ^{14}C method (Rivkin and Seliger, 1981). The application of such method is limited and tedious.

The application of flow cytometry to problems in oceanography should rapidly change this situation. Flow cytometry permits the simultaneous assessment of several optical parameters of individual particles (Kawamura and Aoyama, 1983; Shapiro, 1985). Large numbers of particles can be processed quickly, generating statistically significant data. The use of this instrumentation for the study of phytoplankton has been described and reviewed in detail (Yentsch *et al.*, 1983; Yentsch *et al.*, 1984; Yentsch and Yentsch, 1984). By coupling flow cytometry with immunofluorescent probes, we can gain information about the metabolic status, potential production for growth and taxonomic classification of phytoplankton cells.

Our goals are to assess variability in primary productivity potential in single cells and to distinguish these cells and their associated productivity according to the major taxonomic groups of phytoplankton. To accomplish this, flow cytometry was coupled with the use of immunofluorescent probes for ribulose,1-5,bisphosphate carboxylase (RuBPCase) and the major light-harvesting pigments in dinoflagellates, peridinin/chlorophyll *a*-binding protein (PCP).

Maximal rates of photosynthesis are regulated by environmental conditions which affect metabolic processes at the cellular level. Two important components of the photosynthetic apparatus in phytoplankton cells are the light-harvesting antennae complex (protein-pigment complex) and the enzymatic component which fixes carbon, primarily RuBPCase. Together, the light harvesting pigment-proteins and RuBPCase, account for more than 50% of the total cellular nitrogen (Thornber and Alberte, 1976; Raven, 1984).

RuBPCase is the only enzyme capable of new, net carbon fixation. The concentration of this enzyme is directly related to the maximal rates of photosynthesis (Bjorkman, 1968; Bowes *et al.*, 1972; Bravdo and Pallas, 1981; Makino, *et al.*, 1983a; von Caemmerer and Farquhar, 1981) and to nitrogen economy (Friedrich and Huffaker, 1980; Peoples *et al.*, 1980, Wittenbach *et al.*, 1980). The enzyme is comprised of an octomeric aggregate of 8 large and 8 small subunits. The large subunit of RuBPCase is highly conservative in an evolutionary context (McFadden, 1980); it is antigenically similar in chlorophyll *a* containing species and it can be assayed immunologically (Collatz *et al.*, 1979; Ku *et al.*, 1979; Perchorowics *et al.*, 1981; Freyssinet *et al.*, 1983).

PCP is ubiquitous to all planktonic, autotrophic dino-flagellates, and thus has potential as a taxonomically distinct feature which may be employed to identify this group in a mixed-assemblage of phytoplankton in the sea. The amount of the light-harvesting pigment binding protein is an indicator of photosynthetic potential (Prezelin and Matlick, 1980; Prezelin and Sweeney, 1978).

In this paper we report preliminary results of our studies of these molecules, and the use of immunofluorescent probes for assaying them. This work demonstrates the significant advances in potential for assessing productivity of single cells and taxonomic groups in the sea.

MATERIALS AND METHODS

Cell cultures

Diatom cultures were grown in modified IMR media (Perry *et al.*, 1981), dinoflagellate cultures were grown in modified F/2 medium (Waaland and Watson, 1980).

The cultures were grown on a 15 h light: 9 h dark cycle and at 300 µE m^{-2} s^{-1} quantum flux density. The temperature was 18° C. Cell densities were determined by hemacytometer counts (improved Neubauer). The culture stocks were diluted daily to maintain maximal growth rates.

Photosynthetic measurements

Chlorophyll *a* determinations were done according to Lorenzen (1969) in a Turner X fluorometer. The maximal photosynthetic rate was determined by ^{14}C incorporation using photosynthesis vs. irradiance (P vs. I) curves according to Talbot *et al.* (1985); the data were fit to anon linear exponential equation to determine the photosynthetic parameter P_{max}, maximal photosynthetic rate (Peterson *et al.*, in press). Nitrate determinations were performed according to Strickland and Parsons, 1972).

Production of antisera

The enzyme RuBPCase was isolated from the diatom *Chaetoceros gracilis* and purified according to Newman and Cattolico (in press). Antisera against RuBPCase was raised in two New Zealand rabbits. The rabbits were injected intradermally according to Vaitukaitis (1981) with 190 µg of antigen, using complete Freund's Adjuvant the first time and incomplete adjuvant the second and third times. The antisera was prepared according to Hurn and Chantler (1980) and stored at -40° C.

PCP was purified from *Glenodinium* sp. to homogeneity by a modification of the method of Prezelin and Haxo (1976). Two New Zealand rabbits were injected intradermally with 500 µg of antigen to raise antiserum against PCP.

AFFINITY PURIFICATION OF ANTISERA

Anti-RuBPCase antisera:

RuBPCase from spinach (Sigma R 8000) was coupled to cyanogen bromide-activated Sepharose (Sigma C 9142) and crosslinked by treatment with 1% gluteraldehyde for 1 h. When 100 µl of anti-RuBPCase antiserum were passed over 2 ml of RuBPCase-Sepharose, more than 80% of the anti-RuBPCase activity of the antiserum was found to be bound to the column. Active anti-RuBPCase antiserum was eluted with 100 mM glycine at pH 3.2 and washed again to a pH 7.2.

Anti-PCP antisera

Purified PCP from *Glenodinium* sp. was coupled to cyanogen bromide activated Sepharose. Crude anti-PCP antiserum was purified by a procedure similar to that for anti-RuBPCase except that elution of the active anti-PCP antibody was with 100 mM glycine at pH 2.7. Recovery of anti-PCP was greater than 50%.

Preparation of cell lysates for SDS-PAGE

Whole cell lysates were prepared by lysis of cell suspensions (0.01 M Tris, 0.02% NaN_3, 0.2 mM phenylmethyl sulfonyl fluoride, 1 mg/l pepstatin, 1mg/l leupeptin, pH 7.5) at 1600 psi in a French pressure cell. The soluble fraction was separated from the membrane fraction by centrifugation at 200,000 xg at 2°C for 1 h. Both fractions were prepared for SDS-PAGE.

ELECTROPHORETIC PROCEDURES

Polyacrylamide gel electrophoresis

High pH discontinuous polyacrylamide gel electrophoresis was performed according to Laemmli (1970). The gels consisted of 5% stacking gel and with either a 12% or 15% resolving gel. Samples (10 µg) were solubilized in 0.1 M Tris HCl pH 6.8, 2% sodium dodecyl sulphate (SDS), 5% mercaptoethanol, 21% glycerol and 0.01% bromophenol blue. Electrophoresis of the gels was done at 30 mA. The gels were stained overnight in 0.25% Coomassie brilliant blue in 45% methanol with 10% acetic acid and destained in 40% methanol with 7% acetic acid.

Western blot analysis

The ability of antisera for RuBPCase and PCP to recognize protein from a variety of algal species was tested by Western blot analysis. Proteins from SDS-PAGE gels were electrophoretically transferred onto nitro- cellulose paper according to the techniques of Towbin *et al.* (1979). The electrophoretic transfer blots were soaked in 3% bovine serum albumen in 0.9% NaCl, 10 mM Tris HCl, pH 7.4 for 1 h, to saturate additional protein-binding sites. Incubation with primary antiserum (1:200 to 1:1000) was carried out for 2 h at 20°C or overnight at 4°C in PBS (8.0 g/L NaCl, 1.15 g/L Na_2HPO_4, pH 7.4) in the presence of 0.1% Tween 20. The antigen-antibody complex was detected using a peroxidase conjugated secondary antibody (goat anti-rabbit IgG at 1:500 in PBS); detection of the ligand protein association was performed with 4-chloro-1-naphthol.

Immunofluorescence staining of RuBPCase in fixed cells

An indirect method was used for testing immunofluorescence. The cells were harvested and fixed in paraformaldehyde in PBS to permeabilize the cell membrane and to cross-link the RuBPCase internally. Following centrifugation at 2000 xg for 5 min the cells were resuspended in 1 ml RuBPCase antiserum (1:300 in PBS) and incubated for 2 h. The cells were then collected by centrifugation and incubated in goat anti-rabbit IgG labelled with fluorescein-isothiocyanate (FITC, Sigma F 6005) at a 1:500 dilution for 1 h, followed by 2 washes in PBS. The cells were mounted for microscopic analysis in an epifluorescence microscope with a 100 W mercury lamp with a 487709 Zeiss blocking filter (excitation at 460-490 nm) and a band pass filter (515-545 nm) or the cells were analyzed by analytical flow cytometry. We analyzed: unfixed cells, fixed cells, fixed cells incubated in the second (e.g., FITC conjugated goat anti-rabbit IgG) antibody for nonspecific cross-reactivity, and fixed cells with both primary and secondary antibodies.

The potential for leakage of RuBPCase during fixation and permeabilization with paraformaldehyde was examined. The supernatant from the post fixation centrifugation was tested for the presence of RuBPCase. Paraformaldehyde was added to a 500 µl suspension of cells in PBS; after centrifugation, the cell pack was 60 µl and the supernatant was 440 µl. No RuBPCase was detected in 100 µl of supernatant by enzyme immunoassay using a dot blot, indicating that leakage of RuBPCase did not occur from fixed cells.

Flow cytometric analysis

Flow cytometric analysis was performed with a FACS Analyzer (Becton Dickinson) equipped with a 110 W mercury lamp and filters adjusted to emit light at 438-500 nm. Two fluorescent wavelengths were simultaneously analyzed. The fluorescence from the FITC antibody-labelled cells was passed through a 510-530 nm band pass filter. For analysis of red autofluorescence of chlorophyll, a 610 nm longpass filter was used. Histograms of the distribution of red autofluorescence/cell, FITC green fluorescence/cell, and the relation between red autofluorescence and FITC fluorescence were obtained.

Assay of RuBPCase nitrogen limited cultures

Relative levels of RuBPCase were determined by immunofluorescence in nitrogen-sufficient and nitrogen-limited cultures. Nitrogen-sufficient cultures were maintained in exponential growth phase at low biomass in IMR medium. Nitrogen-limited cultures were grown in IMR medium lacking added nitrogen for six days; ambient seawater nitrogen content was 45 µM. The culture was sampled daily for cell, chlorophyll *a* and nitrate concentrations. On the sixth day, when the nitrate had been completely depleted, the cells were harvested by gentle aspiration of the medium through a 5 µM Nitex filter. Sub-samples were taken for chlorophyll *a*, cell number, ^{14}C uptake, nitrate, and immunofluorescent antibody binding capacity by RuBPCase.

RESULTS

Taxonomic specificity of the anti-RuBPCase probe

Lane A, in Figure 1 shows the results of analysis of partially purified RuBPCase isolated from *Thalassiosira weissflogii* by SDS-PAGE. Both the large and small subunits of RuBPCase can be seen. The cross-reactivity against *Chaetoceros gracilis* RuBPCase-antiserum was checked by Western blot analysis. The antiserum reacted with the large subunit of *T. weissflogii* (Figure 1, lane B). Western blot analysis for other species indicated that the antibody raised against RuBPCase from *Chaetoceros gracilis* cross-reacts with the large subunit of all the species tested, e.g., *Chaetoceros gracilis*,

Olisthodiscus luteus, *Griffithsia pacifica*, *Ochromonas* sp, *Gonyaulax tamerensis*, *Platymonas* sp., and spinach (Sigma R 8000) (Newman *et al.*, in preparation).

Taxonomic specificity of the anti-PCP probe

Reactivity of affinity purified anti-PCP with proteins present in the soluble and membrane fractions of whole cell lysates from a variety of algal taxa was assessed by Western blot analysis (Figure 2). Lysates from genera of dinoflagellates tested exhibited strong crossreactivity with the anti-PCP antiserum. Electrophoretic analysis of purified PCP from several of these genera shows that the reactive peptides at 14-17 Kd and also at 35 Kd are PCP peptide molecular weight variants (Watson, unpublished results). Minor bands of other molecular weights seem to be attributable to cross-reactivity with non-PCP proteins. The diatoms and chrysophyte lysate tested showed only weakly staining bands. However, the green alga *Platymonas suecica* (lane 19) and the membrane fraction of *Cryptomonas* sp. (lane 22) both exhibited protein bands that react strongly with the PCP antiserum.

Figure 1.

Gel electrophoresis of partially purified RuBPCase from *Thalassiosira weissflogii*.
Lane A. Large (LS) and small (SS) subunit of RuBPCase.
Lane B. Western Blot analysis. Large (L) subunit of *Thalassiosira weissflogii*.

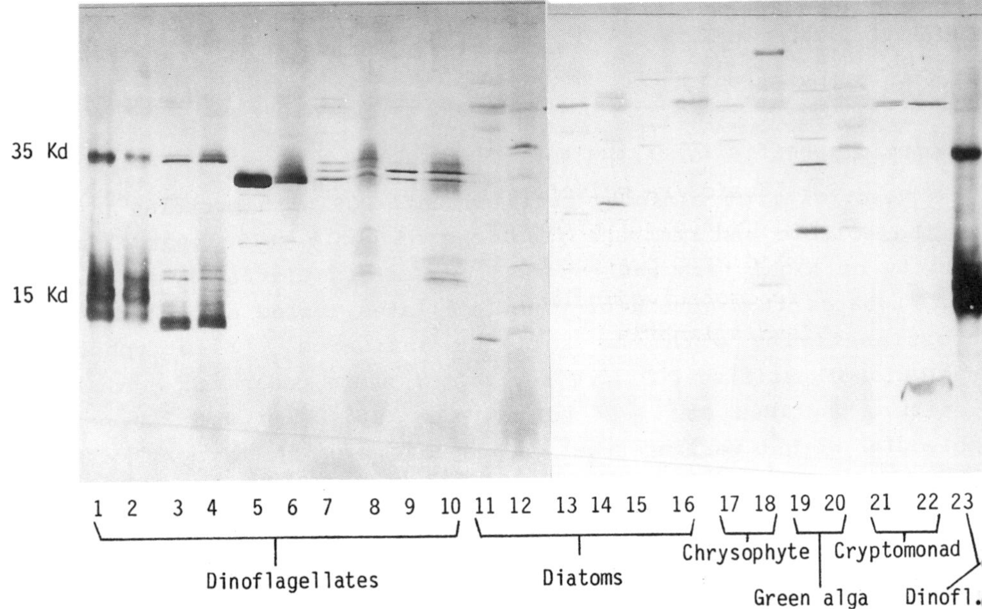

Figure 2. Western immunoblot analysis of cell lysates. Approximately 2-6 µg of soluble protein or 10-30 µg of membrane protein were leaded in each lane. All odd numbered lanes contain soluble fraction proteins; even numbered lanes are membrane fraction proteins. Lane 1 & 2, *Glenodinium* sp.; lanes 3 & 4, *Heterocapsa neie*; lanes 5 & 6, *Gonyaulax tamerensis*; lanes 7-10, *Amphidinium carterae* (two independent protein preparations); lanes 11 & 12, *Nitzschia closteria*; lanes 13 & 14, *Thalassiosira weissflogii*; lanes 15 & 16, *Phaeodactylum tricornutum*; lanes 17 & 18, *Isochrysis galbana*; lanes 19 & 20, *Platymonas suecica*; lanes 21 & 22, *Crytomonas* sp.; lane 23, *Glenodinium* sp. (50 µg).

Immunofluorescence

Microscopic analysis of phytoplankton cells indicated that the anti-RuBPCase serum penetrated the permeabilized cells and showed that green RuBPCase fluorescence was localized inside the chloroplast (Figure 3). In tests of immunostaining with the antiserum, immunofluorescence was detected in all autotrophic cells and no immunofluorescence was detected in heterotrophic cells, e.g., *Giardia lamblia* and *Trichomonas vaginalis* (Table 1).

Table 1. Epifluorescence microscopic detection of RuBPCase immunofluorescence

Species	Immunofluorescence
Chaetoceros gracilis	+
Thalassiosira weissflogii	+
Cylindrotheca fusiformis	+
Amphidinium carterae	+
Trichomonas vaginalis	-
Giardia lamblia	-

Analysis using flow cytometry indicated that cells showed an intrinsic green fluorescence of 2% (of the total relative fluorescence) that increased after fixation to 4%. Controls consisted of cells incubated with primary antiserum alone, which did not fluoresce; control cells incubated with the secondary antibody alone showed some nonspecific binding that fluctuated from 4 to 18% in the different experiments (Table 2).

Table 2. Flow cytometric analysis of RuBPCase intracellular probe in the diatom *Thalassiosira weissflogii*.

Cell Conditions	Green Fluorescence
Fixed + 1st + 2nd AB	100%
Fixed + 2nd AB	2-18%
Fixed	2-4%
Unfixed	2-3%

RuBPCase and Nitrogen Limitation

The immunofluorescence measurements were conducted after 6 d of growth under nitrogen-limited conditions. During this time the culture was monitored for chlorophyll *a*, cell number, and nitrate. By the second day the nitrate was depleted. Chlorophyll *a* per cell decreased from 3.1 pg $(cell)^{-1}$ on the second day to 1.4 pg $(cell)^{-1}$ on the sixth day; the cellular nitrogen decreased from 2.2 pg $(cell)^{-1}$ to 0.43 pg $(cell)^{-1}$ and the maximal photosynthetic rate decreased from 6.3 mg carbon (mg Chlorophyll)$^{-1}$h^{-1} to 1.4 mg carbon (mg chlorophyll)$^{-1}$h^{-1}. Control concentrations were similar to those of day two culture. The flow cytometry analysis showed a decrease in

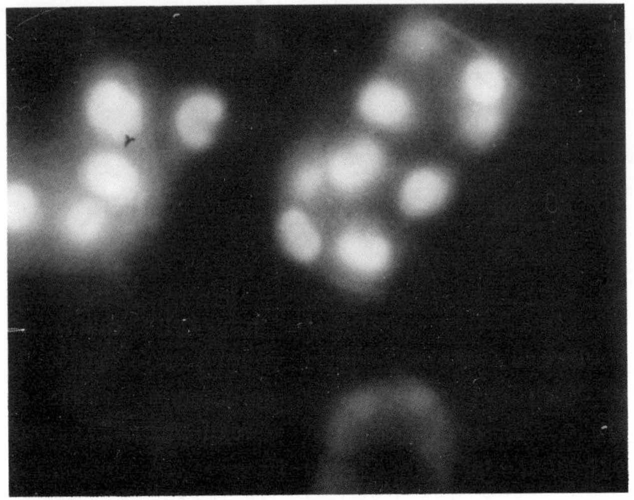

Figure 3. Immunofluorescent microscope detection of RuBPCase in *Thalassiosira weissflogii*.

chlorophyll autofluorescence from 100% (relative fluorescence) in the controls on day two to 28% in the nitrate-limited cultures on day six; RuBPCase immunofluorescence decreased to 20% of the initial value (Table 3; Figures 4a & 4b).

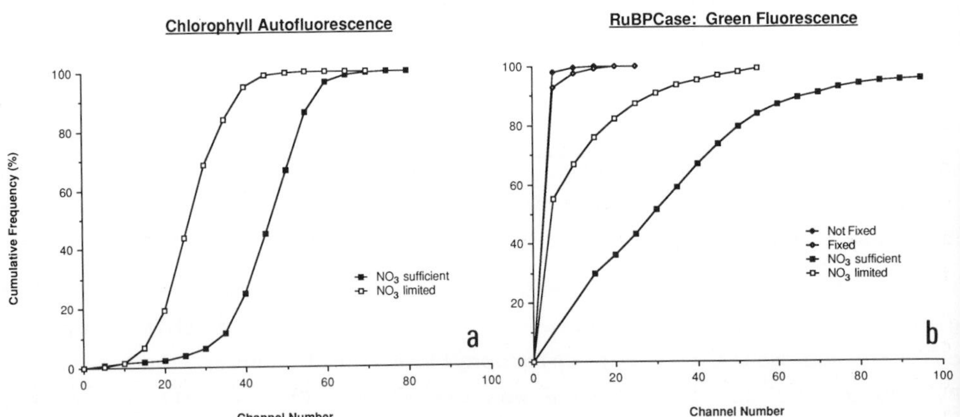

Figure 4 a. Cumulative frequency (CF) of the chlorophyll autofluorescence analysis by flow cytometry.

b. Cumulative frequency (CF) of the RuBPCase immunofluorescence analysis by flow cytometry.

Table 3. Nitrogen effects: Flow cytometric analysis of *Thalassiosira weissflogii*.

Condition	Green Immunofluorescence	Red Autofluorescence
N-sufficient	100%	100%
N-limited	20%	28%

DISCUSSION

The amount of RuBPCase in photosynthetic organisms can be determined by immunological techniques (Collatz *et al.*, 1979; Ku *et al.*, 1979; Perchorowics *et al.*, 1981; Freyssinet *et al.*, 1983; Sugiyama *et al.*, 1984; Fourcroy *et al.*, 1985; Makino *et al.*, 1983a,b). We have used an antibody raised against *Chaetoceros gracilis* RuBPCase to develop a method to quantify the amount of enzyme in individual phytoplankton cells. Application of this probe for quantification of the enzyme in other algal species depends upon equal antigen reactivity in all species. This has been experimentally verified and is a result of the evolutionary conservation (about 85% sequence homology) of the structure of the large subunit among most higher plants and algae (McFadden, 1980; Miziorko and Lorimer, 1983; Akasawa *et al.*, 1984).

In our study, the antiserum to RuBPCase cross-reacted with RuBPCase from all the species of microalgae examined to date (Newman *et al.*, in preparation). These results are consistent with results from other studies, the cross-reactive nature of antibodies made against RuBPCase from one species to the large subunit of other species has been reported over the last decade. For example, the anti-RuBPCase serum against *Euglena* cross-reacts with and the enzyme from lettuce, cucumber and New Zealand spinach (Rabinowitz *et al.*, 1975). Antiserum against the enzyme of *Phaseolus vulgaris* cross-reacts with spinach (Gray and Kekwick, 1974). Other studies also indicate this extensive cross-reactivity (Hatterley *et al.*, 1977; Ku *et al.*, 1979). However, in the study by Plumbley *et al.*, (1986) the large subunit of chromophytes did not cross-react with the large subunit of chlorophytes. The lack of cross-reactivity between these groups is surprising, and may be due to the nature of the antibodies, which were raised against denatured proteins and thus recognized only sequential determinants.

Fixation and permeabilization

Paraformaldehyde permits the diffusion of the antibodies across the cell and the chloroplast membranes. Under the conditions used in this study, it also prevents leakage of RuBPCase out of the cell as shown by preliminary experiments using enzyme immunoassay. This penetration permits the diffusion of the antibodies across the cell and the chloroplast membranes and also enabled specific immunostaining of RuBPCase. This protocol maintained cell size, as determined by forward angle scattering. Furthermore, fixation with paraformaldehyde also results in minimal breakage of cells, as shown by epifluorescent microscopy. Other fixatives previously reported in the literature, such as acetone and ethanol, and permeabilizing agents like Triton X-100, resulted in chlorophyll loss. If autofluorescence of chlorophyll is to be used in combination with anti-RuBPCase in the study of heterogeneous populations, then conditions which minimize chlorophyll loss are a necessity. A study by Navaluna and Perry (submitted) indicates that the chlorophyll autofluorescence in the presence of paraformaldehyde is 50-100% of that in unfixed cells.

Immunofluorescence

There are several factors which may adversely affect the quantification of RuBPCase by immunofluorescence assay. Non-specific binding of either the primary or secondary antibody could yield elevated fluorescence measurements. We observed no binding of anti-RuBPCase to cells which do not contain RuBPCase, thus non-specific interaction of primary antiserum does not seem to be a problem. However, the non-specific binding by the second antibody has been a general problem in immunofluorescent work with plant material (Bauwe, 1984), especially for quantitative evaluation of immunofluorescence. Our experience shows that this background varies from 2 to 18% of the total immunofluorescence using our probe for RuBPCase. This background fluorescence can be monitored by appropriate controls and subtracted from experimental values. Another solution would be direct labelling of the first antibody.

Immunofluorescent methods have been used to localize RuBPCase in leaves of higher plants (Hatterley *et al.*, 1977; Reed and Chollet, 1985) and to localize RuBPCase in guard cells of 41 different plants (Madhavan and Smith, 1982; Zemel and Gepstein, 1985). Recently antibody probes have been used to localize RuBPCase in pyrenoids

(Kiss *et al.*, 1986). Therefore, the utilization of anti-RuBPCase antibody to measure quantitatively RuBPCase in single cells, while at the same time measuring chlorophyll by autofluorescence, is a feasible method for rapid assessment of the photosynthetic potential in individual cells in mixed species populations.

Nitrogen limitation

Nitrogen is a limiting nutrient in the marine environment (Eppley and Peterson, 1979); its availability limits growth and production. Nitrogen-starvation results in a reduction of chlorophyll (Rosen and Lowe, 1984) and also of accessory pigments (Stevens, *et al.*, 1981) per cell. Nitrogen-limitation produces a reduction of plastid stroma in *Cryptomonas*, including deterioration of the thylakoid membranes (Rhiel *et al.*, 1985), and a reduction of phycobiliproteins in cyanobacteria (Daley and Brown, 1973). Loss of chlorophyll due to nitrogen starvation has also been reported in higher plants, including disorganization of chloroplast lamellae (Hurkman, 1979). RuBPCase is an abundant protein, it accounts for 50% of the soluble proteins in plants and algae (Ellis, 1979). A depletion of nitrogen supply should show an impact on the enzyme protein level. Lapointe and Duke (1984) show that nitrate loading affects the amount of protein in *Gracilaria tikvahiae*. Makino *et al.* (1983a,b) showed that in rice plants exposed to nitrate limitation, the RuBPCase activity and the enzyme protein decrease; Kuppers and Weidner (1980) observed a seasonal variation of the enzyme activity which is coupled with environmental nitrate. Smith *et al.*, 1983 observed an increase in RuBPCase activity per unit chlorophyll *a* and per cell during a spring bloom. Previously, Ku *et al.*, 1979 had shown a direct relationship between the enzyme activity and the enzyme concentration. The effect of nitrogen deficiency on the light-harvesting and enzymatic components of the photosynthetic apparatus of algae has been investigated in our laboratories (Cleveland and Perry, in press; Perry *et al.*, a, submitted), showing that nitrate deficiency decreases the concentration of light harvesting pigments, decreases the activity of RuBPCase and PEPCase (phosphoenolpyruvate carboxylase) and results in lower photosynthetic rate. In this study we have compared exponentially growing cells in nitrogen-sufficient conditions to cells grown under nitrogen-deficient conditions with respect to their autofluorescent and immunofluorescent patterns. Our results showed reproducible differences in the bivariate plots of chlorophyll autofluorescence per cell and RuBPCase immunofluorescence per cell. Chlorophyll and

RuBPCase per cell were shown to be much lower for cells grown under nitrogen-limited conditions relative to the controls grown in nitrogen-sufficient media (Table 3). These results are in accordance with the literature.

The decrease in RuPBCase protein level is significant for survival of individual cells. It has been suggested that RuBPCase is a major reserve of amino acids (Wittenbach *et al.*, 1980), which are mobilized during nitrogen limitation (Makino, 1983b). Our experiment showed that, even though the nitrogen supply had been limited for several days, the cells continued to divide. This could be due to the mobilization of amino acids derived from RuBPCase coincident with a decrease in RuBPCase, as suggested by Perry *et al.*, b, (submitted).

We also saw a decrease in photosynthetic rate which was correlated with the decrease in RuBPCase. Nitrogen content and photosynthetic activity have been correlated in higher plants (Evans, 1983) and in phytoplankton (Cleveland and Perry, in press; Prezelin and Matlick, 1983; Prezelin, 1982; Eppley and Renger, 1974; Laws and Bannister, 1980). Perry *et al.*,a, (submitted) showed a very high correlation of photosynthetic rate with RuBPCase activity. Makino *et al.* (1983a), showed a high positive correlation between photosynthetic rate and RuBPCase content in rice and an uncoupling of photosynthetic capacity when RuBPCase decreases. They concluded that the level of RuBPCase can be a limiting factor throughout the life of the leaves.

In summary, the optimum protocol for RuBPCase analysis in single cells depends on several factors. These factors include fixation, permeabilization, the nature of the anti-RuBPcase labelling probe, the duration of the anti-RuBPCase labelling period, and the cell type and its physiology. Although the procedure has worked successfully, it is apparent that the ability to get the antibody inside the cell and to obtain reproducible labelling with anti-RuBPCase varies from one cell type to another. The application of this procedure to a mixed population in the ocean may require modifications to optimize the fixation of different cell types.

Taxonomic specificity and cross-reactivity of PCP

Antisera raised against unique pigment-protein complexes can be used for the taxonomic identification of specific groups in mixed phytoplankton assemblages. For example, the antiserum to PCP, from

Glenodinium sp. could be used to distinguish dinoflagellates and to quantify PCP in individual dinoflagellates. For both these purposes, it is essential that the anti-PCP probe reacts with PCP present in all dinoflagellate genera and does not react to a significant degree with proteins other than PCP, either in dinoflagellates or in other phytoplanktonic organisms. Our preliminary results regarding the specificity of the anti-PCP probe are encouraging in that all dinoflagellate genera tested exhibited positively staining PCP peptides. This intergeneric cross-reactivity of the anti-PCP probe is a strong reaction and indicates that a broad spectrum of dinoflagellates will be recognized by the probe. However, reaction of the probe with non-PCP proteins in other distantly related taxa indicates that the specificity of the probe for PCP will need to be increased. We anticipate that this will be possible by the use of further affinity chromatography purification steps to adsorb the antibody component responsible for the observed non-PCP cross-reactivity. To this end we are currently purifying the cross reactive peptides from the green alga, *Platymonas suecica*. These proteins will be immobilized on agarose and employed to adsorb out the non-specific component of the antibody preparation.

The need for field techniques for studying physiological and biochemical relationships and for quantification of photosynthetic potential in single cells has motivated us to develop immunofluorescent probes. The combined use of the anti-RuBPCase probe with the anti-PCP probe will yield information on the cellular content of both the light-harvesting and the dark or enzymatic components of the photosynthetic apparatus in dinoflagellates. We believe that this new approach will greatly improve our understanding of phytoplankton dynamics in terms of the taxonomic composition in phytoplankton assemblages, their metabolic activity, and their potential for productivity.

ACKNOWLEDGEMENTS

We thank M. C. Talbot for the photosynthetic measurements and Dr. Thomas Paquette of University of Washington Diabetes Research Center for valuable discussions and for assistance in raising antibodies against PCP. This project was supported by NSF Grant OCE-8600310 and OCE-8700110. Contribution 1675, School of Oceanography, University of Washington Seattle, WA 98195.

REFERENCES

Akazawa, T., Takabe and H. Kobayashi. 1984. Molecular evolution of ribulose-1,5-bisphosphate carboxylase/oxygenase (Rubisco). Trends Biochem Sci. 9: 380-383.

Bauwe, H. 1984. Photosynthetic enzyme activities and immunofluorescence studies on the localization of ribulose 1,5-bisphosphate carboxylase/oxygenase in leaves of C3, C4 and C3-C4 intermediate species of *Flaveria* (Asteraceae). Biochem. Physiol. Pflanzen 179: 253-268.

Bjorkman, O. 1968. Carboxydismutase activity in shade-adapted and sun-adapted species of higher plants. Physiol. Plant. 21: 481-485.

Bowes, G., W.L. Ogren and R.H. Hageman. 1972. Light saturation, photosynthesis rate, RuDP carboxylase activity, and specific leaf weight in soybeans grown under different light intensities. Crop Sci. 12: 77-79.

Bravdo, B. and J.E. Pallas, Jr. 1981. Photosynthesis photorespiration and RuBP carboxylase/oxygenase activity in selected peanut genotypes. Photosynthetica 16: 36:42.

Cleveland, J.S. and M.J. Perry. (in press) Quantum yield, relative specific absorption, and fluorescence in nitrogen-limited *Chaetoceros gracilis*. Mar. Biol.

Collatz, G.J., M. Badger, C. Smith and J.A. Berry. 1979. A radioimmune assay for RuP2 carboxylase protein. Carnegie Inst Yearbook. 78: 171-175.

Daley, R.J. and S.R. Brown. 1973. Chlorophyll, nitrogen, and photosynthetic patterns during growth and senescence of two blue green algae. J. Phycol. 9: 395-401.

Ellis, J. 1979. The most abundant protein in the world. Trends Biochem. Sci. 4: 241-244.

Eppley, R.W. and B.J. Peterson. 1979. Particulate organic matter flux and planktonic new production in the deep ocean. Nature. 282: 677-680.

Eppley, R.W. and E.M. Renger. 1974. Nitrogen assimilation of an oceanic diatom in nitrogen-limited continuous culture. J. Phycol. 10: 15-23.

Evans, J.E. 1983. Nitrogen and photosynthesis in the flag leaf of wheat (*Triticum aestivum* L.) Plant. Physiol. 72: 279-302.

Fourcroy, D. Klein-Eude and C. Lambert. 1985. Phytochrome control of gene expression in radish seedlings. II. Far-red light mediated appearance of ribulose-1,5-bisphosphate carboxylase and the mRNA for its small subunit. Pl. Sc. Lett. 37: 235-244.

Freyssinet, G., M. Freyssinet and D.E. Buetow. 1983. Immunodetection of ribulose-1,5-bisphosphate carboxylase in mutants of *Euglena gracilis* impaired in photosynthesis. Plant Sci. Lett. 32: 61-72.

Friedrich, J.W. and R.C. Huffaker. 1980. Photosynthesis, leaf resistances, and ribulose-1,5-biphosphate carboxylase degradation in senescing barley leaves. Plant Physiol. 65: 1103-1107.

Gray, J.C. and R.G.O. Kekwick. 1974. An immunological investigation of the structure and function of ribulose-1,5-bisphosphate carboxylase. Eur. J. Biochem. 44: 481-489.

Hattersley, P.W., L. Watson and C.B. Osmond. 1977. *In situ* immunofluorescent labelling of ribulose-1,5-bisphosphate carboxylase in leaves of C3 and C4 plants. Aust. J. Plant Physiol. 4: 523-539.

Holm-Hansen, O., C.J. Lorenzen, W.R Holmes and J.D.H. Strickland. 1965. Fluorometric determination of chlorophyll. J. Cons. Int. Exp. Mer. 30: 3-15.,

Hurkman, W.J. 1979. Ultrastructural changes of chloroplast in attached and detached, aging in primary wheat leaves. Am. J. Bot. 66: 64-70.

Hurn, B.A.L. and S.M. Chantler. 1980. Production of reagent antibodies. Methods in Enzymology. 70: 105- 142.

Kawamura, A., Jr., and Y. Aoyama. 1983. Immunofluorescence in Medical Science. University of Tokyo Press, Springer-Verlag, New York. 262 pp.

Kiss, J.Z., A.C. Vasconcelos and R.E. Triemer. 1986. Paramylon synthesis and chloroplast structure associated with nutrient levels in *Euglena* (Euglenophyceae). J. Phycol. 22: 327-333.

Ku, M.S.B., M.R. Schmitt and G.E. Edwards. 1979. Quantitative determination of RuBP carboxylase oxygenase protein in leaves of several C3 and C4 plants. J. Exp. Bot. 30: 89-98.

Kuppers, U. and M. Weidner. 1980. Seasonal variation of enzyme activities in *Laminaria hyperborea*. Planta 148: 222-230.

Laemmli, U.K. 1970. Cleavage of structural proteins during the assembly of the head of bacteriophage T4. Nature 227: 680-685.

Lapointe, B.E. and C.S. Duke. 1984. Biochemical strategies for growth of *Gracilaria tikvahiae* (Rhodophyta) in relation to light intensity and nitrogen availability. J. Phycol. 20: 488-495.

Laws, E.A. and T.T. Bannister. 1980. Nutrient and light-limited growth of *Thalassiosira fluviatilis* in continuous culture, with implications for phytoplankton growth in the ocean. Limnol. Oceanogr. 25: 457-473.

Lorenzen, C. J. 1966. A method for the continuous measurement of *in vivo* chlorophyll concentration. Deep Sea Res. 13: 223-227.

Madhavan, S. and B.N. Smith. 1982. Localization of ribulose bisphosphate carboxylase in the guard cells by an immunofluorescence technique. Plant Physiol. 69: 273-277.

Makino, A., T. Mae and K. Ohira. 1983a. Photosynthesis and ribulose-1,5-bisphosphate carboxylase in rice leaves. Plant Physiol. 73: 1002-1007.

Makino, A., T. Mae and K. Ohira. 1983b. Effect of nitrogen, phosphorus or potassium on the photosynthetic rate and ribulose-1,5-bisphosphate carboxylase content in rice leaves during expansion. Soil Sci. Plant Nutr. 30: 63-70.

McFadden, B.A. 1980. A perspective of ribulose bisphosphate carboxylase/oxygenase, the key catalyst in photosynthesis and photorespiration. Acc. Chem. Res. 13: 394-399.

Miziorko, H.M. and G.H. Lorimer. 1983. Ribulose-1,5-bisphosphate carboxylase-oxygenase. Ann. Rev. Biochem. 52: 507-535.

Navaluna, N., and M.J. Perry. (submitted) The effect of chemical fixation on some optical properties of phytoplankton.

Newman, S.M. and R.A. Cattolico. (in press) Is RuBPCase in *Olisthodiscus luteus* an evolutionary divergent enzyme? Plant. Physiol.

Newman, S.M., M.V. Orellana, R.A. Cattolico and M.J.Perry. (in preparation) Structure of ribulose-1,5-bisphosphate carboxylase-oxygenase among plants.

Peoples, M.B., V.C. Beilharz, S.P. Waters, R.J. Simpson and M.J. Dalling. 1980. Nitrogen redistribution during grain growth in wheat (*Triticum aestivum* L.) II. Chloroplast senescence and the degradation of ribulose-1,5-bisphosphate carboxylase. Planta. 149: 241- 251.

Perchorowicz, J.T., D.A. Raynes and R.G. Jensen. 1981. Light limitation of photosynthesis and activation of ribulose bisphosphate carboxylase in wheat seedlings. Proc. Natl. Acad. Sci. USA. 78: 2985- 2989.

Perry, M.J., M.C. Talbot and R.S. Alberte. 1981. Photoadaptation in marine phytoplankton: Response of the photosynthetic unit. Mar. Biol. 62: 91-101.

Perry, M.J., J. Smith, M.C. Talbot and N. Welschmeyer. (submitted a) Nitrogen limitation in marine phytoplankton: response of the photosynthetic apparatus. Mar. Ecol. Prog. Ser.

Perry, M.J., J. Smith, and N. Welschmeyer. (submitted b) Resource exploitation in phytoplankton: the role of photosynthetic nitrogen. Amer. Nat.

Peterson, D.H., M.H. Perry, K.E. Bencala and MC. Talbot. (in press). Phytoplankton productivity in relation to light intensity: a simple equation. Est. Coastal Shelf Sci.

Plumley, F.G., D.L. Kirchman, R.E. Hodson and G.W. Schmidt. 1986. Ribulose biphosphate carboxylase from three chlorophyll *c*-containing algae: physical and immunological characterizations. Plant Physiol. 80: 685-691.

Prezelin, B.B. 1982. Effects on light intensity on aging of the dinoflagellate *Gonyaulax polyedra*. Mar. Biol. 69: 129-135.

Prezelin, B.B. and F.T. Haxo. 1976. Purification and characterization of peridinin-chorophyll a-proteins from the marine dinoflagellates *Glenodinium* sp. and *Gonyalax polyedra*. Planta. 128: 133-141.

Prezelin, B.B. and H.A. Matlick. 1980. Time course photoadaptation in the photosynthesis-irradiance relationship of dinoflagellate exhibiting photosynthetic periodicity. Mar. Biol. 58: 85-96.

Prezelin, B.B. and H.A. Matlick. 1983. Nutrient dependent low-light adaptation in the dinoflagellate *Gonyaulax polyedra*. Mar. Biol. 74: 141-150.

Prezelin, B.B. and B.M. Sweeney. 1979. Photoadaptation of photosynthesis in *Gonyaulax polyedra*. Mar. Biol. 48: 27-35.

Rabinowitz, H., A. Reisfeld, D. Sagher and M. Edelman. 1975. Ribulose diphosphate carboxylase from autotrophic *Euglena gracilis*. Plant Physiol. 56: 345-350.

Raven, J.A. 1984. Energetics and transport in aquatic plants. MBL Lectures in Biology. 4: 587 pp.

Reed, J.E. and R. Chollet. 1985. Immunofluorescent localization of phosphoenolpyruvate carboxylase and ribulose-1,5-bisphosphate carboxylase/oxygenase proteins in leaves of C3, C4 and C3-C4 intermediate *Flaveria* species. Planta. 165: 439-445.

Rhiel, E., E. Morschel and W. Wehrmeyer. 1985. Correlation of pigment deprivation and ultrastructural organization of thylakoid membranes in *Cryptomonas maculata* following nutrient deficiency. Protoplasma 129: 62-73.

Rivkin, R.B. and H.H. Seliger. 1981. Liquid scintillation counting for ^{14}C uptake of single algal cells isolated from natural samples. Limnol. Oceanogr. 26: 780-785.

Rosen, B.H. and R.L. Lowe. 1984. Physiological and ultrastructural responses of *Cyclotella meneghiniana* (Bacillariophyta) to light intensity and nutrient limitation. J. Phycol. 20: 173-183.

Shapiro, H.M., 1985. Practical Flow Cytometry, Alan R. Liss, Inc., NY, 295 pp.

Smith, J.C., T. Platt and W.G. Harrison. 1983. Photoadaptation of carboxylating enzymes and photosynthesis during a spring bloom. Prog. Oceanogr. 12: 425-459.

Steeman-Nielsen, E. 1952. The use of radioactive carbon (^{14}C) for measuring organic production in the sea. J. Cons. Int. Explor. Mer. 18: 117-140.

Stevens, S.E., D.L. Balkwill and D.A.M. Paone. 1981. The effects of nitrogen limitation on the ultrastructure of the cyanobacterium *Agmenellum quadruplicatum*. Arch. Microbiol. 130: 204-212.

Strickland, J.D.H. and T.R. Parsons. 1972. A practical handbook of seawater analysis. 310 pp. Otawa. Fish. Res. Board Can. Bull. 167.

Sugiyama, T., M. Mizuno and M. Hayashi. 1984. Partitioning of nitrogen among ribulose-1,5-bisphosphate carboxylase/ oxygenase, phosphoenolpyruvate carboxylase, and pyruvate orthophosphate dikinase as related to biomass productivity in maize seedlings. Plant. Physiol. 75: 665-669.

Talbot, M.C., D.S. Thoreson and M.J. Perry. 1985. Photosynthesis vs. light intensity measurements: A miniaturized incubator. ONR Technical Report 406: 1-13.

Thornber, J.P and R.S. Alberte. 1976. Chorophyll-proteins: membrane-bound photoreceptor complexes in plants. IN: The Enzymes of Biological Membranes. M. Martonosi (ed.). 3: 163-190.

Towbin, H.,T. Staehelin and J. Gordon. 1979. Electrophoretic transfer of proteins from polyacrylamide gels to nitro-cellulose sheets: procedure and some applications. Proc. Natl. Sci. USA 76: 4350-4354.

Vaitukaitis, J.L. 1981. Production of antisera with small doses of immunogen: Multiple intradermal injections. Methods in Enzymology. 73: 46-75.

von Caemmerer, S. and G.D. Farquhar. 1981. Some relationships between the biochemistry of photosynthesis and the gas exchange of leaves. Planta 153: 376-387.

Waaland, S.D. and B.A. Watson. 1980. Isolation of a cell-fusion hormone from *Griffithsia pacifica* Kylin, a red alga. Planta 149: 493-497.

Wittenbach, V.A., R.C. Ackerson, R.T. Giaquinta and R.R. Hebert. 1980. Changes in photosynthesis, ribulose bisphosphate carboxylase, proteolytic activity and ultrastructure of soybean leaves during senescence. Crop Sci. 20: 225-231.

Yentsch, C.M., P.K. Horan, K. Muirhead, Q. Dortch, E. Haugen, L. Legendre, L.S. Murphy, M.J. Perry, D.A. Phinney, S.A. Pomponi, R.W. Spinrad, M. Wood, C.S. Yentsch and B.J. Zahuranec. 1983. Flow cytometry and cell sorting: a technique for analysis and sorting of aquatic particles. Limnol. Oceanog. 28: 1275-1280.

Yentsch, C.M., T.L. Cucci and D. A. Phinney. 1984. Flow cytometry and cell sorting: problems and promises for biological ocean science research. IN: Marine Phytoplankton and Productivity. Holm-Hansen, O., L. Bolis and R. Gilles, (eds.). Springer-Verlag, Berlin.

Yentsch, C.M. and C.S. Yentsch. 1984. Emergence of optical instrumentation for measuring biological properties. Oceanogr. Mar. Biol. Ann. Rev. 22: 55-98.

Zemel, E. and S. Gepstein. 1985. Immunological evidence for the presence of ribulose bisphosphate carboxylase in guard cells chloroplasts. Plant. Physiol. 78: 586-590

EXAMINING NITRATE REDUCTION BY PHYTOPLANKTON WITH AN IMMUNOASSAY

W.M. Balch
A-018, Scripps Inst. Of Oceanography
La Jolla, CA 92037

C.M. Yentsch
Bigelow Laboratory for Ocean Sciences
McKown Pt., W. Boothbay Harbor, ME 04575

Beatrez Reguera
Instituto Oceanografico
Orillamar 47, Vigo, Spain

Wilbur Campbell
Department of Biological Sciences
Michigan Technological University
Houghton, MI 49931

INTRODUCTION

The uptake of nitrate by phytoplankton is a central issue in biological oceanography, mainly because of its ramifications to primary production and particle fluxes to the sediments. Dugdale and Goering (1967) first conceived of the idea of "new" production as that primary production which was growing on nitrogen sources from outside the euphotic zone (such as nitrate diffusing upwards). They further defined "regenerated production" as the primary production growing on regenerated nitrogen (such as ammonium and urea). These definitions were important for several reasons. For an oceanic ecosystem, apparently at equilibrium with regards to the concentration of particulate carbon, the rate of new production directly relates to the sinking flux of biogenic material (Eppley and Peterson, 1979). This implies that the rate of nitrate utilization should set an upper limit to the rates of supply of carbon to the benthic communities; for deep basins in the Southern California Bight, this prediction is supported by experimental data (Jahnke, in prep).

Estimating new production has not been easy. Direct estimates were provided by measuring the uptake of $^{15}N\text{-}NO_3$. This technique is not optimal because a relatively large spike of $^{15}N\text{-}NO_3$ is needed which neglects the requirement of a trace addition (Dugdale and

Wilkerson, 1986). The result was an overestimate of new production. New production has also been measured through seasonal oxygen changes in the surface and oxygen minimum waters (Jenkins and Goldman, 1984). The advantage of this technique is its implicit seasonal time scale but it is not practical to use in highly advected coastal areas. Sediment traps have offered a direct measure of the sinking flux but sinking flux estimates can be biased by problems in advection, preservation of the settled material and animals swimming into the traps. It remains extremely difficult to reliably estimate new production, hence settling organic flux, in nitrate-depleted surface waters.

A new way to estimate new production is to measure net nitrate depletion using the chemiluminescent technique of Garside (1982). Nitrate can be measured with a precision of 2 nM -- this represents an improvement of about an order of magnitude in sensitivity over the older colorimetric technique (Strickland and Parsons, 1972). Garside (1985) provided data on the nitrate levels in the nutrient-depleted surface waters of a warm-core ring. He measured nitrate concentrations as low as 20 nM at the surface and exponentially increasing concentrations with depth. Balch *et al.* (in press) and Eppley and Renger (in prep.) used the chemiluminescent technique to measure net nitrate depletion in mixed-layer samples of the Gulf of Maine and the Southern California Bight. They measured surface concentrations as low as 11 nM which were rarely depleted below this level. Eppley and Renger (submitted) demonstrated that although the ^{15}N and chemiluminescent techniques give similar results when performed on the same water sample, an addition of 100 nM $^{15}N\text{-}NO_3$ can artificially increase the nitrate uptake rate. The net uptake rate measured with Garside's technique can be very high if 100 nM $^{15}N\text{-}NO_3$ is added and unmeasurable with no nitrate addition. Therefore, new production estimates made with ^{15}N in nutrient depleted surface waters may be artificially high.

With the advent of more precise ways to measure nitrate, it is mandatory to have more precise ways of measuring the biological utilization of nitrate in nutrient-depleted surface waters. As already mentioned, ^{15}N techniques often do not meet the requirement for a trace addition (Dugdale and Wilkerson, 1986). A new method for measuring the biological transport of nitrate has recently become available. Balch (submitted) used ^{36}Cl-chlorate as a radioanalogue of nitrate and was able to study nitrate transport of phytoplankton.

Balch *et al.*, (in press) show that nitrate transport by natural phytoplankton populations in the surface mixed layer is constitutive with no apparent induction period. The analogue techniques are highly sensitive but can provide, at best, only a qualitative estimate of new production.

Another step in the biological utilization of nitrogen is assimilation. This has been studied extensively in marine phytoplankton (see Collos and Slawyk, 1980; Syrett, 1981; Falkowski, 1983; Wheeler, 1983; Dunn-Coleman *et al.*, 1984; Paasche *et al.*, 1984; and Paasche, 1986 for reviews of this subject). An established technique for detecting nitrate assimilation is to measure the activity of the reducing enzyme, nitrate reductase. This enzyme is responsible for reducing nitrate to nitrite; thus its activity in phytoplankton should provide an estimate of nitrate utilization, hence new production. Eppley *et al.* (1969) detected nitrate reductase (NR) activity in natural phytoplankton; this activity was apparently inhibited by ammonium. A combination of ^{15}N and NR measurements revealed that the relation between NR activity and nitrate uptake is not constant but varies with light and nutrient conditions (Blasco *et al.*, 1984). They interpreted NR activity as representing the past nitrate history of the phytoplankton whereas nitrate uptake represents the present incubation conditions.

One of the factors which affects NR activity most is light (Packard *et al.*, 1971; Packard, 1973; Packard and Blasco, 1974). In the latter paper, the NR activity was observed to be light-repressed or light-activated on time scales of about 30 minutes. Moreover, NR activity of darkened cells approached zero in some examples (Figure 1B; lower panel from Packard and Blasco (1974)). This light dependence may be species-specific. Our own data for cultures of the coccolithophorid, *Emiliana huxleyi* demonstrate that there may be two forms of nitrate reductase, a light labile and a dark-stabile form. Figure 1A shows that light-treated cells, exposed to darkness, will lose about 50% of their NR activity in as little as 12 minutes. Cells were also exposed to far-red light which had the same effect as darkness. Following 24 hours in darkness the NR activity remained between 30 and 60% of the activity in the light control (data not shown). Activation of NR in *E. huxleyi* was on a time scale of 5 to 30 minutes (Figure 1B). These data and the above references suggest that if NR activity is measured in natural populations, the past light history of the cells must be known to correctly interpret the data. Moreover, the experimental manipulations necessary for sample collection and assay may produce large artifacts. These have been persistent problems in studies of NR.

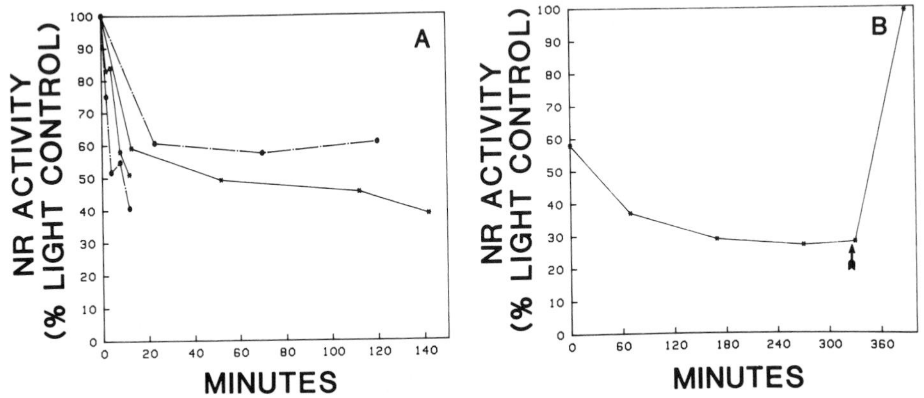

Figure 1. A) Deactivation of nitrate reductase activity in darkness (NR activity given as a percent of white light control). Cultures of *Emiliana huxleyi* were grown in IMR/2 media (Eppley *et al.*, 1967) in a 12 h light: 12 h dark light regime (growth irradiance 130 µEin m^{-2} s^{-1}). Experiments were started well into the light cycle. The experimental light source was a Unitron UH1 microscope lamp equipped with incandescent lamp (∿30 µEin m^{-2} s^{-1}). Cells were illuminated up to time-zero of the experiment afterwhich an aliquot was placed in darkness (●) or far-red light (∿15-50 µEin m^{-2} s^{-1}; 725 nm) (*).

B) Activation of nitrate reductase activity in *E. huxleyi* (grown as described above). The cells were in darkness since the previous night. Activity is given as percent of white light control. The arrow denotes the time at which 30 µEin m^{-2} s^{-1} white light was turned on. Samples were filtered 5 and 30 minutes after the light was turned on.

Another major problem in the interpretation of NR activity data is that the rates of nitrate reduction can account for less than half of the nitrate uptake measured with ^{15}N (Dortch *et al.*, 1979; Dortch and Maske, 1982). This could result from loss of NR activity during the assay or possibly a second, unmeasured nitrate reductase which uses a cofactor other than NADH (Collos and Slawyk, 1977). Moreover

NR activity is frequently undetectable, even in waters containing high nitrate (see Eppley *et al.*, 1969; their Table 2). It is clear that a more sensitive technique is needed to assay for this enzyme; a simple calculation can further illustrate this point.

Lewis *et al.* (1986) has estimated the vertical eddy diffusivity coefficient, Kv, at the pycnocline of the oligotrophic Sargasso Sea to be about 4×10^{-5} m^2 s^{-1} and the corresponding nitrate gradient, dNO_3/dZ, to be 0.04 mM m^{-4}. The resulting flux of nitrate through the pycnocline can be estimated using the equation for Fickian diffusion:

$$\text{Flux} = Kv \times dNO_3/dZ$$

where the flux is given in units of mM m^{-2} s^{-1}. The resultant flux would be about 1.6×10^{-6} mM m^{-2} s^{-1} or about 1.4 mM m^{-2} d^{-1}. If this is taken up uniformly by the phytoplankton in the mixed layer ($\sim$50 mg Chl m^{-2}) then the chlorophyll specific uptake rate (and assimilation rate) would be about 0.12 nmol (µg Chl h^{-1})$^{-1}$. This same calculation can be applied to the data of the Southern California Bight (Eppley *et al.*, 1979; Kv $\sim 3 \times 10^{-5}$ m^2 s^{-1} and dNO_3/dZ=0.56 mM m^{-4}) and the expected flux is 1.5 mM m^{-2} d^{-1} or in terms of chlorophyll, 1.2 nmol (µg chl h^{-1})$^{-1}$. The sensitivity of the NR technique (as given in Eppley *et al.*, 1969 and Eppley, 1978) is about 0.2 to 1 nmol h^{-1} l^{-1} given a sample containing 1 µg of chlorophyll *a*. Thus, in the oligotrophic stations, the NR assay is probably not sensitive enough to detect nitrate reduction, whereas in coastal waters the assay is operating near its limit of sensitivity, especially for nitrate-depleted surface samples.

Another sink for nitrate in oxygen-depleted waters is denitrification where nitrate is used as a terminal electron acceptor and is reduced by nitrate reductase. Denitrification has been studied in the oxygen-deficient waters below the Peruvian upwelling system using an assay for nitrate reductase (Packard *et al.*, 1978; 1983). Typical denitrification rates of approximately 1 nmol NO_3 reduced h^{-1} l^{-1} have been estimated. As previously stated, however, such rates are approaching the limits of detection of the NR assay given a 1 liter sample size. Obviously, larger samples can be filtered but this becomes cumbersome. Current estimates of denitrification may be underestimated due to the lack in sensitivity in the NR activity assay.

A more sensitive technique to detect nitrate assimilation is needed for several reasons. The chemiluminescent nitrate technique can only be used to measure net nitrate depletion or production. Phytoplankton uptake rates may be offset by production of nitrate by nitrifying bacteria and this would not be detected with the chemiluminescent method. A sensitive estimate of total NR protein would provide an upper limit to the nitrate assimilation rate--that is, an estimate of potential new production. A measure of total NR protein would be preferable to that of NR activity; this is because many experimental factors (light, ammonium, etc.) affect NR activity and the activity often does not account for the nitrate reduced. A more sensitive technique might be used to examine the size fractions of algae responsible for the new production, or even the species (given a suitable single-cell method).

One technique attractive for measuring potential nitrate assimilation is an immunoassay. Perhaps it should measure the total nitrate reductase protein present (as opposed to the activity) which can be reasonably be converted into a "potential activity" given standard data for a purified enzyme. A radioimmunoassay (RIA) can detect the binding of 100 pg of antibody while an enzyme linked immunosorbant assay (ELISA) can detect the binding of 1 ng of antibody to the targeted protein (Nisonoff, 1982). This roughly translates to the detection of pico- to nanograms of nitrate reductase protein. This would increase considerably our ability to detect nitrate reductase in phytoplankton.

NITRATE REDUCTASE

Nitrate reductase was first reported by Evans and Nason (1953). Since then many discoveries have been made concerning its structure, regulation, and function. It has been purified from plants using affinity chromatography with blue Sepharose C16B columns (Solomonson, 1975; Redinbaugh and Campbell, 1983) and appears to be a multimeric protein with a total weight of 200-500 kilodaltons (Smarrelli and Campbell, 1981). Nitrate reductase has been associated with membrane and cytoplasm fractions of plant cells (Falkowski, 1983). The structure of NR is well conserved through higher plant lines (Campbell, 1985). Two forms of nitrate reductase are currently known, a dye-dependent NR and a NADH-dependent NR (Campbell, 1985). The activity of NR is assayed in phytoplankton using the protocols of Eppley (1978) (revised by Harrison (1975) and Hocheman *et al.*, (in press)).

Immunological techniques have already been applied to the study of nitrate reductase. Funkhouser and Ramadoss (1980) first purified antibodies to *Chlorella* NR and noted cross-reacting material in cell extracts and a cross-reacting NR precursor protein. Structural comparisons of NR have been made using immunological techniques (Smarrelli and Campbell, 1981). Immunofluorescence has proved useful to examine the intracellular location of NR in fungus (Roldan *et al.*, 1982), soybeans (Vaughn *et al.*, 1984), and green algae (Lopez-Ruiz *et al.*, 1986). The enzyme appears to be associated with the plasmalemma and plastids near the plasmalemma. An immunoassay has already been developed for NR of corn by Campbell and Ripp (1984) and Campbell and Remmler (1986).

Vaughn *et al.* (1984) provided data on the sensitivity of an immunoassay for NR. They suggest that 0.5 ng of NR can be detected with their ELISA technique. This is equivalent to 5 x 10^{-5} units of enzyme which is approximately equal to 0.5 pmol NO_2 formed min^{-1} or about 30 pmol NO_2 formed h^{-1}. A RIA may be 10X more sensitive than an ELISA (Nisonoff, 1982) or 3 pmol NO_2 formed h^{-1}. The sensitivity of the technique used by Eppley *et al.* (1969) with a sample size of 1 µg chlorophyll *a* is from 0.2 to 1 nmol NO_2 formed h^{-1}. Thus, and ELISA and RIA should be about 10-50 and 100-500 times more sensitive respectively than the activity assay. An ELISA or RIA would be more quantitative than an fluorescence immunoassay since the ELISA and RIA are bulk measurements. The fluorescence immunoassay, using single cells, can be quantitative when the cells are permeabilized to allow the antibodies to enter. In a non-quantitative manner the fluorescence immunoassay would be better used to detect which species contained nitrate reductase.

PHYTOPLANKTON STUDIES WITH NITRATE REDUCTASE ANTIBODIES

We have performed some preliminary experiments with fluorescent NR antibodies to visualize NR in single cells. A rabbit was immunized with purified squash nitrate reductase, the antiserum was isolated, freeze dried and kept frozen for 2 years. It was reconstituted in a solution of 0.1 M acetate, 0.05 M PO_4, 0.15 M NaCl and 0.1 mM EDTA in de-ionized water. The titer was checked by examining inhibition of NR activity by the antibodies.

The antibodies were biotinylated by dialyzing in a buffer solution (17.3 g $NaHCO_3$ and 8.6 g Na_2CO_3 per liter phosphate buffer saline (PBS) at pH=9). N-hydroxysuccimimide ester of biotin was prepared by adding 1.7 mg of NHS-Biotin to 1 ml dimethylformanide. Two hundred milliliters of this mixture were added to 3 mg of antibodies and reacted at room temperature for 4 hours. This was then dialyzed with 3 changes of phosphate buffered saline (PBS) and ultracentrifuged at 40 K for 20 minutes. Fluoresceinated avidin, which binds to biotin, was used to visualize the antibodies (diluted 40x and 40 µl were combined with 10^6 cells). Antibody-biotin-avidin conjugates were visible under a Nikon epifluorescence microscope containing a 418 nm excitation filter, 515 nm barrier filter and 500 nm dichroic filter.

Cultured dinoflagellates (*Gymnodinium catenatum* clone 87 v from the Instituto Oceanografico at Vigo, Spain and *Protoperidinium affinia* clone 53) were grown in K media (Keller and Guillard, 1985) with nitrate or ammonium as their nitrogen source. Cells were harvested while in log growth. Natural populations were sampled from 10 m in the Ria Vigo (42° 24' N X 8° 42'W) and the Ria Pontevedra (42° 24'N X 8° 42'W) during May 1986. The Ria Vigo sample was dominated by *Leptocylindrus danicus* and *Detoniella pumila* and the Ria Pontevedra sample was dominated by *Nitzschia seriata*.

The antibody-NR conjugates were prepared using a modified protocol of Ward and Perry (1980). Each step in the procedure was followed by copious rinsing in PBS (pH=8). All manipulations were performed with the cells held on a 12 µm Nuclepore filter in a glass filter tower (thus only large cells were caught). All solutions were added to the filter apparatus with no vacuum applied; to remove the solution, vacuum was applied. Next, the filter was immersed in 5 ml of methanol for 1 hour to extract the chlorophyll. Then 0.5 ml of PBS and 50 µl of NR antibody solution were added to the filter and incubated for 30 minutes before filtering and rinsing in PBS. One half milliliter of PBS plus 50 µl of FITC avidin were added to the filter for 30 min (avidin bound to the biotinylated antibodies during this time) and then filtered and rinsed. Finally the filter was given a rinse in carbonate buffer and transferred to a microscope slide. A few drops of mounting solution (Ward and Perry, 1980) were added before the coverslip was placed over the filter.

The results of these preliminary experiments showed that: 1) NR antibodies bound to the phytoplankton cells (inside the cells as bright spots and on the plasmamembrane as a diffuse glow (Figure 2). This agrees with earlier observations which indicate that a large fraction of the activity is membrane bound in phytoplankton (T. Packard, personal communication). 2) Controls with FITC avidin and no antibody showed no fluorescence. 3) There was some binding of the preimmunization antiserum to the plasmamembrane and inside the cells which suggests that either the rabbit already contained NR antibodies or that there were other antibodies not specific to NR.

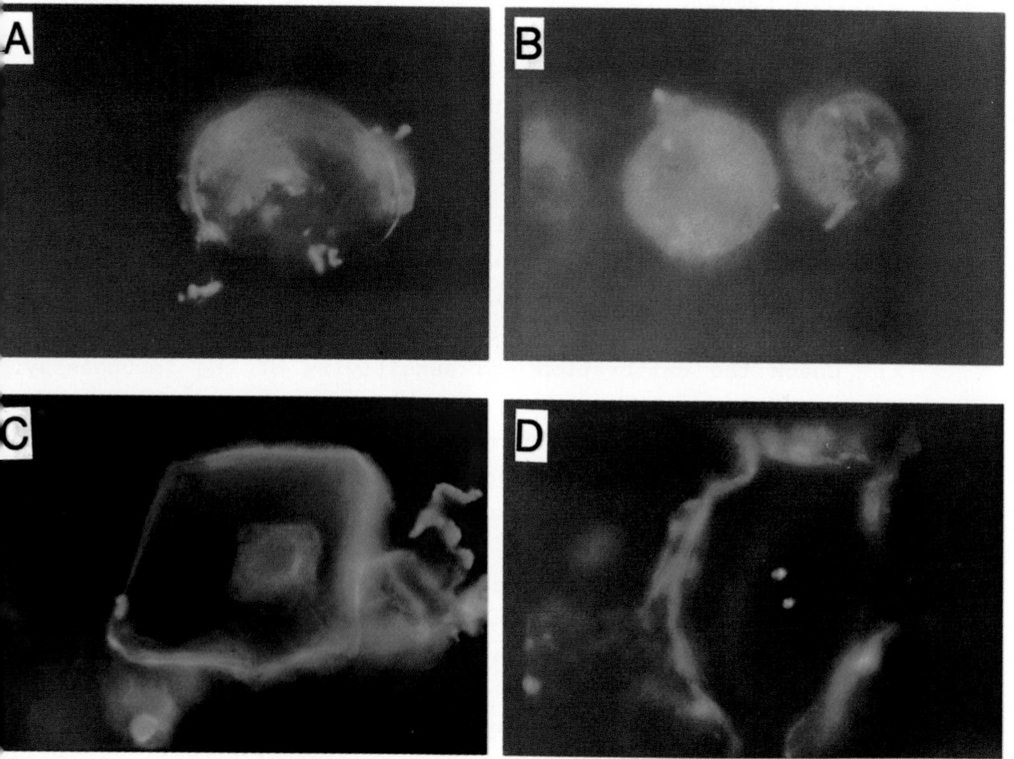

Figure 2. Fluorescence micrographs of dinoflagellates treated with fluorescently labeled NR antibodies (details of preparation given in text). a) *Gymnodinium catenatum* from culture; b) *Protogonyaulax affinis* from culture; c) *Protoperidinium* sp. from Ria Pontevedra, and d)*Dinophysis* sp. from Ria Vigo.

If a cross-reacting precursor protein (Funkhouser and Ramadoss, 1980) was always present in phytoplankton cells, then this could limit the usefulness of the NR immunoassay as used here. More specific antibodies should be used in subsequent work. 4) Many species of diatoms and dinoflagellates in rias of Spain cross-reacted with the NR antibodies although the pattern of staining varied from species to species. *Nitzschia* sp. showed one or two fluorescein spots per cell; *Chaetoceros* sp. showed a bright fluorescein spot on every 3 or 4 cells and the dinoflagellates were typified by a membrane "glow" plus one or two bright spots inside the cells (Figure 2). Thus, the antibody technique for tagging nitrate reductase of phytoplankton gave positive results in these preliminary trials and is worthy of further experimentation.

FUTURE RESEARCH

The three major objectives for future oceanographic research with NR antibodies should be:

1) To develop a quantitative immunoassay for NR protein in phytoplankton. This will involve NR purification, immunization, antibody isolation, verification of NR antibody specificity and checks of cross-reactivity of the antibody.

2) Check the calibration and sensitivity of the immunoassay. Examine the relation of NR protein to the measured activity.

3) Examine the occurence of NR protein in the marine environment. How does NR protein vary with the degree of eutrophication, depth, time of year, light, etc? What species of phytoplankton contain NR? For the latter question, it will be necessary to refine the fluorescence immunoassay described above. Are the phytoplankton in the surface mixed layer utilizing the nanomolar concentrations of nitrate and, if so, at what potential rate? One idea which is gaining increasing popularity is that the phytoplankton in the deep chlorophyll maximum are the "new producers" while those above are the "regenerated producers". This would be very testable with an immunoassay for NR provided that the presence of NR was related to nitrate utilization by the cells. Denitrification rates should be re-assessed using the more sensitive antibody measurements. An upper limit to denitrification rates in the oxygen minimum zones should be established which is critical for constructing global nitrogen budgets. By far the most important goal of future work with NR antibodies should be to define the rates

of potential new production in the surface waters, for this will set the upper limits to the particulate flux of organic carbon to the aphotic regions of the oceans.

ACKNOWLEDGEMENTS

We would like to acknowledge John Daley (Dana Farber Cancer Institute, Boston, MA) performed biotinylation of the antibodies. Santiago Fraga (Instituto Oceanografico, Vigo, Spain) assisted with the laboratory and field sampling. D. Phinney (Bigelow Laboratory, Boothbay Harbor, ME) also assisted with the antibody preparation. P. Matrai read a preliminary draft of the manuscript. Some of the ideas presented here came about through discussions with Bess Ward.

REFERENCES

Balch, W.M. submitted. Studies of nitrate transport by marine phytoplankton using $^{36}Cl\text{-}ClO_3$ as a transport analogue. I. Physiological Findings. J. Phycol.

Balch, W.M., C. Garside and E.H. Renger. in press. Studies of nitrate transport by marine phytoplankton using $^{36}Cl\text{-}ClO_3$ as a transport analogue. II Field Observations. Deep Sea Res.

Blasco, D., J.J. MacIsaac, T.T. Packard and R.C. Dugdale. 1984. Relationship between nitrate reductase and nitrate uptake in phytoplankton in the Peru upwelling region. Limnol. Oceanogr. 29: 275-286.

Campbell, W.H. 1985. The biochemistry of higher plant nitrate reductase. IN: Nitrogen Fixation and CO_2 Metabolism. P.W. Ludden and J.E. Burris. (eds.). Elsevier Science Publishing Co., Inc., NY pp. 143-151.

Campbell, W.H. and J.L. Remmler. 1986. Regulation of corn (*Zea Mays*) leaf nitrate reductase: I. Immunochemical methods for analysis of the enzymes protein component. Plant Physiol. 80: 435-441.

Campbell, W.H. and K.G. Ripp. 1984. An ELISA for higher nitrate reductase. Ann. New York. Acad. Sci. 435: 123-151.

Collos, Y. and G. Slawyk. 1976. Significance of cellular nitrate content in natural populations of marine phytoplankton growing in shipboard culture. Mar. Biol. 34: 27-32.

Collos, Y. and G. Slawyk. 1977. Nitrate reductase activity as a function of *in situ* nitrate uptake and environmental factors of euphotic zone profiles. J. Exp. Mar. Biol. Ecol. 29: 119-130.

Dortch, Q., S.I. Ahmed and T.T. Packard. 1979. Nitrate reductase and glutamate dehydrogenase activities in *Skeletonema costatum* as measures of nitrogen assimilation rates. J. Plank. Res. 1: 169-185.

Dortch, Q. and H. Maske. 1982. Dark uptake of nitrate and nitrate reductase activity of a red-tide population off Peru. Mar. Ecol. Prog. Ser. 9: 299-303.

Dugdale, R.C. and J.J. Goering. 1967. Uptake of new and regenerated forms of nitrogen in primary productivity. Limnol. Oceanogr. 12: 196-206.

Dugdale, R.C. and F.P. Wilkerson. 1986. The use of ^{15}N to measure nitrogen uptake in eutrophic oceans: experimental considerations. Limnol. Oceanogr. 31: 673-689.

Dunn-Coleman, N.S., J. Smarrelli, Jr. and R.H. Garrett. 1984. Nitrate assimilation in eukaryotic cells. IN: Internatl. Rev. Cytol. G.H. Bourne, J.F. Danelli and K.W. Joen (eds.). Academic Press, NY. 92: 1-50.

Eppley, R.W. 1978. Nitrate reductase in marine phytoplankton. IN: Handbook of Phycological Methods. J.A. Hellebust and J.S. Craigii. (eds.). Cambridge University Press. pp. 217-223.

Eppley, R.W., J.L. Coatsworth and L. Solorzano. 1969. Studies of nitrate reductase in marine phytoplankton. Limnol. Oceanogr. 14: 194-205.

Eppley, R.W., R.W. Holmes and J.D.H. Strickland. 1967. Sinking rates of marine phytoplankton measured with a fluorometer. J. Exp. Mar. Biol. Ecol. 1: 191-208.

Eppley, R.W. and B.J. Peterson. 1979. Particulate organic matter flux and planktonic new production in the deep ocean. Nature 28: 677-680.

Eppley, R.W. and E. Renger. submitted. New production above the nitracline: plankton utilize nanomolar concentrations of nitrate in nutrient-depleted surface waters. Oceanographie Tropicale.

Eppley, R.W., E.H. Renger, W.G. Harrison and J.J. Cullen. 1979. Ammonium distribution in southern California coastal waters and its role in the growth of phytoplankton. Limnol. Oceanogr. 24: 495-509.

Evans, H.J. and A. Nason. 1953. Plant Physiol. 28: 233-254.

Falkowski, P. 1983. Enzymology of nitrogen assimilation. IN: Nitrogen in the Marine Environment, E.J. Carpenter and D.G. Capone (eds.). Academic Press, NY. pp. 839-868.

Funkhouser, E.A. and C.S. Ramadoss. 1980. Synthesis of nitrate reductase in *Chlorella*. II. Evidence for synthesis in ammonia-grown cells. Plant Physiol. 65: 944-948.

Garside, Chris. 1982. A chemiluminescent technique for the determination of nanomolar concentrations of nitrate, nitrate and nitrite or nitrite alone in seawater. Mar. Chem. 11: 159-167.

Garside, Chris. 1985. The vertical distribution of nitrate in open ocean surface water. Deep Sea Research. 32: 723-732.

Harrison, W.G. 1975. Assay of nitrate reductase from plasmolyzed marine phytoplankton. Annual Report of the Food Chain Group. Scripps Institute of Oceanography, La Jolla, CA. 473-483.

Hochman, A., A., Nissanyk, D. Wynne, B. Kaplan and T. Berman. in press. Nitrate reductase: an improved assay method for phytoplankton. Limnol. Oceanogr.

Jenkins, W.J. and J.C. Goldman. 1984. Seasonal oxygen cycling and primary production in the Sargasso Sea. J. Mar. Res. 43: 465-491.

Keller, M.D. and R.R.L Guillard. 1985. Factors significant to marine dinoflagellate culture. IN: Toxic Dinoflagellates. D.M. Anderson, A.W. White and D.G. Baden (eds.). Elsevier Science Publishing Co., Inc. pp. 113-116.

Lewis, Marion, W.G. Harrison, N.S. Oakey, D. Herbert and T. Platt. 1986. Vertical nitrate fluxes in the oligotrophic ocean. Science. 234: 870-873.

Lopez-Ruiz, A., J.P. Verbelen, J.M. Roldan and J. Diez. 1986. Nitrate reductase of green algae is located in the pyrenoid. Plant Physiol. 79: 1006-1010.

Nisonoff, A. 1982. Introduction to Molecular Immunology. Sinauer Associates Inc.: Sunderland, MA. 204 pp.

Paasche, E. 1986. Pelagic Primary Production in Nearshore Waters. IN: Nitrogen Cycling in Coastal Marine Environments.

Paasche, E., I. Bryceson and K. Tangen. 1984. Interspecific variation in dark nitrogen uptake by dinoflagellates. J. Phycol. 20: 394-401.

Packard, T.T. 1973. The light dependence of nitrate reductase in marine phytoplankton. Limnol. Oceanogr. 18: 466-469.

Packard, T.T. and D. Blasco. 1974. Nitrate reductase activity in upwelling regions 2. Ammonia and light dependence. Tethys. 6: 269-280.

Packard, T.T., D. Blasco, J.J. MacIsaac and R.C. Dugdale. 1971. Variations of nitrate reductase activity in marine phytoplankton. Inv. Pesq. 35: 209-219.

Packard, T.T., R.C. Dugdale, J.J. Goering and R.T. Barber. 1978. Nitrate reductase activity in the subsurface waters of the Peru Current. J. Mar. Res. 36: 59-76;.

Packard, T.T., P.C. Garfield and L.A. Codispoti. 1983. Oxygen consumption and denitrification below the Peruvian upwelling. IN: Coastal Upwelling, pt. A. E. Suess and J. Thiede (eds.). Plenum Publishing Corporation. pp. 147-173.

Redinbaugh, M.G. and W.H. Campbell. 1983. Purification and squash NADH: nitrate reductase by zinc chelate affinity chromatography. Plant Physiol. 71: 205-207.

Roldan, J.M., V.P. Verbelen, W.L. Butler and K. Tokuyasu. 1982. Intracellular localization of nitrate reductase in *Neurospora crassa*. Plant Physiol. 70: 872-874.

Smarrelli Jr., J. and W.H. Campbell. 1981. Immunological approach to structural comparisons of assimilatory nitrate reductases. Plant Physiol. 68: 1226-1230.

Solomonson, L.P. 1975. Purification of NADH-NR by affinity chromatography. Plant Physiol. 56: 853-855.

Syrett, P. 1981. Nitrogen metabolism of microalgae. IN: Physiological Bases of Phytoplankton Ecology. T. Platt (ed.). Can. Bull. Fish. Aquat. Sci. 210: 182-210.

Strickland, J.D.H. and T. Parsons. 1972. A Practical Handbook of Seawater Analysis, Second edition. Fish. Res. Board Can., Ottawa, Bull. 167: 310 pp.

Vaughn, K.C., S.O. Duke and E.A. Funkhouser. 1984. Immunochemical characterization and localization of nitrate reductase in norflurazon-treated soybean cotyledons. Physiol. Plant. 62: 481-484.

Ward, B.B. and M.J. Perry. 1980. Immunofluorescent assay for the marine ammonium-oxidizing bacterium *Nitrosococcus oceanus*. Applied and Environmental Microbiology. 39: 913-918.

Wheeler, P.A. 1983. Phytoplankton nitrogen metabolism. IN: Nitrogen in the Marine Environment, E.J. Carpenter and D.G. Capone (eds.). Academic Press, NY. pp. 309-346.

A POSSIBLE IMMUNOLOGICAL APPROACH TO THE STUDY OF PLASMALEMMA REDOX ENZYMES IN PHYTOPLANKTON

Gary J. Jones, Francois M.M. Morel
Massachusetts Institute of Technology 48-213
Cambridge, MA 02139

Clarice M. Yentsch
Bigelow Laboratory for Ocean Sciences
West Boothbay Harbor, ME 04575

Edward A. Funkhouser
Department of Biochemistry
Texas A & M University
College Station, TX 77843

ABSTRACT

The phytoplankton plasmalemma contains redox enzymes which will reduce external trace metal complexes, redox dyes and molecular oxygen (Jones and Morel, 1986; Jones *et al.*, 1985; Palenik *et al.*, 1986). The chemical properties of the trace metal reductase are consistent with those of NAD(P)H:cytochrome c (ferricyanide) reductase which is found as a prosthetic group in many cellular redox enzymes e.g., nitrate reductase, sulfite reductase and oleate desaturase. We are now attempting to substantiate these initial observations through the use of anti-serum, prepared against algal nitrate reductase, as a probe for the plasmalemma trace metal reductase. Preliminary experiments indicate that nitrate reductase anti-serum completely inhibits plasmalemma redox activity in the marine diatoms *Thalassiosira weissflogii* and *T. pseudonana*.

REFERENCES

Jones, G.J., B. Palenik and F.M.M. Morel. 1987. Trace metal reduction by phytoplankton: the roll of plasmalemma redox enzymes. J. Phycol. (in press)

Jones, G.J., T.D. Waite and J.D. Smith. 1985. Light-dependent reduction of copper (II) and its effect on cell-mediated, thiolo-dependent superoxide production. Biochem. Biophys. Res. Comm. 128: 1031-1036.

Palenik, B., O.C. Zafiriou and F.M.M. Morel. 1987. Hydrogen peroxide production by a marine phytoplankton. Limnol. Oceanogr. (submitted)

ANTILUCIFERASE IMMUNOGOLD LABELING OF TWO DIFFERENT ORGANELLES IN THE MARINE DINOFLAGELLATE *GONYAULAX POLYEDRA*

M.-T. Nicolas
Department of Cellular
and Developmental Biology
Harvard University
16 Divinity Avenue
Cambridge, MA 02138
and
Laboratoire de bioluminescence
CNRS
Gif-sur-Yvette, France

J.-M. Bassot
Laboratoire de bioluminescence
CNRS
Gif-sur-Yvette, France

Carl Hirschie Johnson and J. Woodland Hastings
Department of Cellular
and Developmental Biology
Harvard University
16 Divinity Avenue
Cambridge, MA 02138

INTRODUCTION

Bioluminescent marine dinoflagellates emit brief (100 msec) and bright (10^{10} photons) flashes upon stimulation (Krasnow *et al.*, 1980). Although it has been known that such flashes originate from subcellular sites, their identity and cellular relationships had not been established (Fogel *et al.*, 1972; Johnson *et al.*, 1985).

In order to visualize the emitting organelles, we employed a polyclonal antibody raised in rabbits against soluble luciferase from *Gonyaulax polyedra* (Dunlap and Hastings, 1981), combined with immunogold labeling after fast freeze fixation and freeze substitution with acetone and osmium tetroxide (Escaig, 1982). EM sections were first treated with the antiluciferase and then with a goat anti-rabbit antibody labeled with gold particles, using the techniques of De Mey (1983).

RESULTS

Two structures are specifically labeled (Figure 1); cytoplasmic dense bodies associated with the vacuolar membrane, which we identify as the luminous organelles, and refer to as scintillons, and mature trichocysts, in the space surrounding the crystalline shaft (Nicolas *et al.*, 1985).

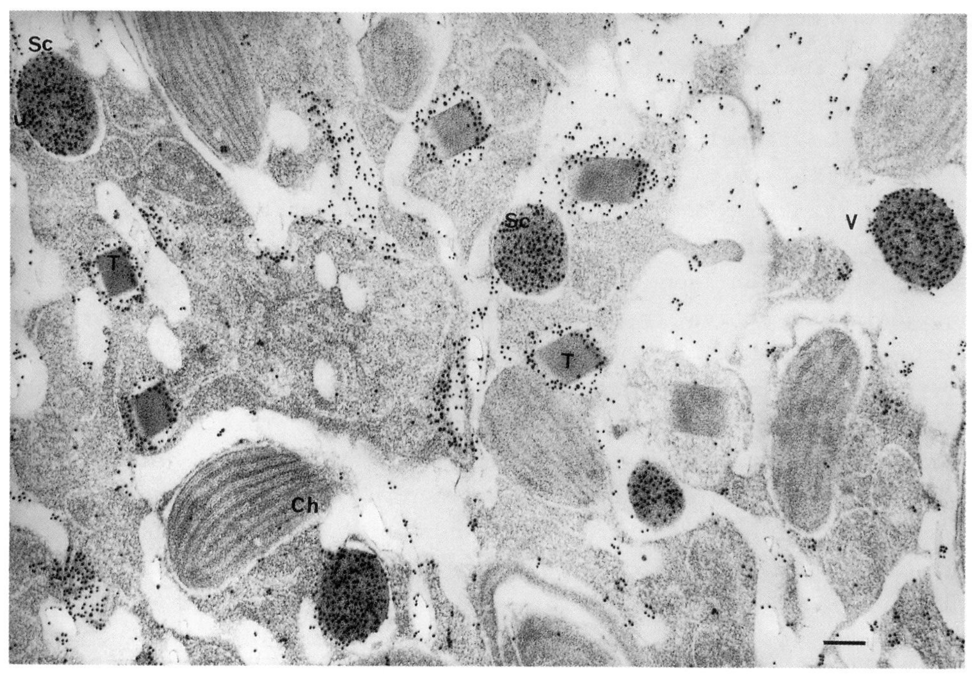

Figure 1. Thin section of *Gonyaulax polyedra*, prepared for electron microscopy by fast freeze-freeze substitution, labeled with affinity purified antiluciferase and goat anti-rabbit antibody with 10 nm gold particles attached. The micrograph shows numerous trichocysts (T) and scintillons (Sc), the latter mostly surrounded by the vacuole (V). Labeling occurs on trichocysts in spaces arount the shaft, and on the dense matrix of scintillons. ch, chloroplast. Bar: 0.1 µM.

In *G. polyedra* mature scintillons are made of a dense heterogenous matrix surrounded by the vacuolar membrane, actually hanging in the vacuole like a drop but maintaining connections via narrow cytoplasmic bridges. In their origins in the Golgi region, "prescintillons" appear as small cytoplasmic condensations which label with antiluciferase but lack a limiting membrane. It is postulated that during development they migrate radially and establish an association with the vacuolar membrane, ultimately protruding into the vacuole.

Mature trichocysts are elongated with a condensed crystalline shaft ready to be ejected (Hausman, 1978). Labeling occurred exclusively on the space between the shaft and the sheath, and in the tip region beyond the shaft.

The labeling of two different organelles by antibody raised against a single antigen is unexplained. As mentioned above, luminescence is believed to be associated with the dense bodies, not trichocysts. But trichocyst labeling occurs also with liquid chemical fixation, and with affinity purified antiluciferase antibody as well. Trichocyst labeling is thus not non specific, but the nature of its antigen is unclear. It could be luciferase or an antigenically active fragment thereof, present in this compartment for whatever reason. But the cross reacting material could be a different protein with a different function, but related both antigenically and evolutionarily to luciferase. This possibility is suggested by the findings that in non-luminous dinoflagellates the dense bodies (scintillons) are absent, but the trichocysts are present and label similarly with antiluciferase, yet there is no luciferase activity detectable in extracts (Nicolas *et al.*, 1986). The possibility that a completely different protein is responsible seems unlikely; as mentioned above, affinity purified antibody labels the trichocysts, and the antibody recognizes only a single band on both 1-D and 2-D (Western) immunoblots (Johnson *et al.*, 1984).

Scintillons clearly originate and remain in the cytoplasmic compartment. But their intimate association with the vacuole, which is acidic in dinoflagellates (Nawata and Siboaka, 1979), as in other plant cells (Marty *et al.*, 1980), suggests a specific model concerning the mechanism of the coupling between the conducted action potential (resulting from the stimulation) and the bioluminescent flash response. The action potential, propagated along the vacuolar membrane (Eckert and Sibaoka, 1968), is postulated to result in the opening of specific ion channels, allowing protons to flow from the vacuole to the scintillon, thus triggering the flash as modeled in and predicted by the pH-dependent properties of the isolated biochemical system (Hastings, 1986). Such a model is in agreement with both biochemical and physiological data.

The vacuolar membrane in the region of the scintillon could be functionally specialized, containing for example a proton translocating ATPase whose function would be to move protons from the scintillons back to the vacuole. The cytoplasmic bridges could allow for recharging by the transfer of cytoplasmic components, such as luciferin needed for the luminescent reaction and ATP required for the ATPase.

The vacuole, in spite of its complex topography, thus constitutes a continuous internal compartment, and all of the scintillons of the cell are thereby related and interconnected. A single stimulus, generating an action potential, would thus result in a concerted albeit slightly asynchronous firing of most or all of the cellular scintillons (Eckert and Reynolds, 1967; Krasnow *et al.*, 1980; Widder and Case, 1982).

ACKNOWLEDGEMENTS

This work was supported by grants to J.W. Hastings from the National Institutes of Health (GM-19536) and the National Science Foundation (PCM 83-09414).

REFERENCES

DeMey, J. 1983. Colloidal Gold Probes. IN: Immunocytochemistry. J.M. Polak and S. Van Noorden (eds.). Wright-PSG, Bristol, London, Boston. pp. 82-112.

Dunlap, J. and J.W. Hastings. 1981. The biological clock in *Gonyaulax* controls luciferase activity by regulating turnover. J. Biol. Chem. 256: 10509-10518.

Eckert, R. and G.T. Reynolds. 1967. The subcellular origin of bioluminescence in *Noctiluca miliaris*. J. Gen. Physiol. 50: 1429-1458.

Eckert, R., and T. Sibaoka. 1968. The flash triggering action potential of the luminescent dinoflagellate *Noctiluca*. J. Gen. Physiol. 52: 258-282.

Escaig, J. 1982. New instruments which facilitate rapid freezing at 83°K and 6°K. J. Microscopy. 126: 221-229.

Fogel, M., R. Schmitter and J.W. Hastings. 1972. On the physical identity of scintillons: Bioluminescent particles in *Gonyaulax polyedra*. J. Cell. Sci. 11: 305-317.

Hastings, J.W. 1986. Bioluminescence in bacteria and dinoflagellates. IN: Light Emission in Plants and Bacteria. J. Amesz Govindjee and D.C. Fork. (eds.). Academic Press, NY. pp. 363-398.

Hausman, K., 1978. Extrusive organelles in Protists. Int. Rev. Cytol. 52: 198-276.

Johnson, C.H., S. Inoue, A. Flint and J.W. Hastings. 1985. Compartmentation of algal bioluminescence: autofluorescence of bioluminescent particles in the dinoflagellate *Gonyaulax* as studied with image intensified video microscopy and flow cytometry. J. Cell Biol. 100: 1435-1446.

Johnson, C.H., J.F. Roeber and J.W. Hastings. 1984. Circadian changes in enzyme concentration account for rhythm of enzyme activity in *Gonyaulax*. Science. 223: 1428-1430.

Krasnow, R., J. Dunlap, W. Taylor and J.W. Hastings. 1980. Circadian spontaneous bioluminsecent glow and flashing of *Gonyaulax polyedra*. J. Comp. Physiol. 138: 19-26.

Marty, F., D. Branton and R.A. Leigh. 1980. Plant Vacuoles. IN: The Biochemistry of Plants. P.K. Stumpf and E.E. Conn. (eds.). Academic Press. 1: 625-658.

Nawata, T and T. Siboaka. 1979. Coupling between action potential and bioluminescence in *Noctiluca*: Effects of inorganic ions and pH in vacuolar sap. J. Comp. Physiol. 134: 137-149.

Nicolas, M.-T., C.H. Johnson, J.-M. Bassot and J.W. Hastings. 1985. Immunogold labeling of organelles in the bioluminescent dinoflagellate *Gonyaulax polyedra* with anti-luciferase antibody. Cell Biology International Reports. 9: 797-802.

Nicolas, M.-T., B.M. Sweeney and J.W. Hastings. 1986. Ultrastructural localization of luciferase in *Phyrocystis* (Pyrrophyta) by immuno-gold labelling. J. Cell Sci. in press.

Widder, E.A., and J.F. Case. 1982. Luminescent microsource activity in bioluminescence of the dinoflagellate *Pyrocystis fusiformis*. J. Comp. Physiol. 145: 517-527.

ENZYME-LINKED IMMUNOSORBENT ASSAY FOR METAL-BINDING PROTEINS OF *MYTILUS EDULIS*

G. Roesijadi*
Department of Biology
Pennsylvania State University
208 Mueller Laboratory
University Park, PA 16802

J.E. Morris
Battelle, Pacific Northwest Laboratory
Biology and Chemistry Department
Richland, WA 99352

INTRODUCTION

Metallothioneins (MTs) are cysteine-rich, low molecular weight, metal-binding proteins whose functions, although not clearly understood, appear to be involved with detoxification, storage, and regulation of copper, zinc, cadmium, and mercury (reviews by Webb, 1979; Cousins, 1985). These proteins have been detected in diverse organisms and are ubiquitously distributed in the animal kingdom (Kojima and Kagi, 1978; Roesijadi, 1981; Klaaverkamp *et al.*, 1984), as well as in some fungi (Hamer *et al.*, 1985) and prokaryotes (Olafson *et al.*, 1980). In numerous studies, marine and freshwater fish and invertebrates have been shown capable of synthesizing increased levels of MT's or similar proteins in response to metal exposure.

Recent studies on the marine mussel, *Mytilus edulis*, have shown that metallothionein-like, metal-binding proteins are present in organs such as gills, digestive gland, and kidneys and probably associated with pathways for metal uptake, storage, and excretion (George and Pirie, 1980; Viarengo *et al.*, 1981; Roesijadi, 1982). To date, specific and sensitive immunoassays for such proteins in non-mammalian species such as the mussel have not been available. Immunochemical assays developed for mammalian metallothioneins (Garvey *et al.*, 1982) are not appropriate for use on similar proteins in fish or invertebrates because of the lack of cross-reactivity of the antibodies (J.S. Garvey, J. Kay, personal communications). We report here the development of antibodies against a low molecular weight metal-binding protein isolated from the marine mussel, *Mytilus edulis*, and antigen detection based on an enzyme-linked immunosorbent assay (ELISA).

As observed previously (Roesijadi and Drum, 1982), the major low molecular weight, mercury-binding protein peak in mussel cytosol elutes at a position of about 20 KD on Sephadex G-75. This M_r is twice that expected for monomeric MT, which normally migrates at 10 KD on Sephadex, and most likely represents a dimer. The 20 KD peak further separates into three protein peaks when subjected to anion exchange chromatography with DEAE-cellulose. One of the peaks, designated $HgBP_{20}II$, was rechromatographed on Sephadex G-75, partially characterized, and used as antigen for development of antibodies and immunochemical detection procedures.

From SDS-PAGE analysis (Laemmli, 1970) the estimated metal-free M_r from SDS-PAGE was 12.7 KD for the dimer. From amino acid analyses (Table 1), it was estimated that this protein had a minimum metal-free M_r of 12.9 KD, which is similar to that obtained by SDS-PAGE. The amino acid composition of $HgBP_{20}II$ clearly resembled that of an MT, with high cysteine (26%) and glycine (17%) contents (Table 1). Lacking were aromatic residues, histidine, and methionine. Mercury analysis of the protein indicated 12 g-atoms Hg/dimer; 6 g-atom/subunit if Hg is assumed to be distributed even between two subunits. The cys:Hg ratio was 2.4.

Injection of this protein into goats yielded antisera of suitable titer three months after initial injection. Ouchterlony analysis of the antiserum against the immunizing antigen resulted in a single precipitation line. Pre-immune goat serum was not reactive. The IgG fraction of antiserum was purified by ammonium sulfate precipitation and anion exchange chromatography on DEAE-cellulose. The purified IgG fraction containing goat anti-($HgBP_{20}II$) formed single precipitation lines when reacted against horse anti-goat IgG and rabbit anti-whole goat serum in Ouchterlony assays and immunoelectrophoresis.

Ouchterlony analysis of the IgG fraction obtained above showed a single precipitation line between it and $HgBP_{20}II$, the primary antigen, and cross-reactivity with $HgBP_{20}I$ and $HgBP_{20}III$, the two other metal-binding proteins obtained during isolation of $HgBP_{20}II$. The antibodies were not cross-reactive with rabbit MT, an observation confirmed by ELISA.

The assay subsequently developed was an indirect, competitive enzyme-linked immunosorbent assay run under non-equilibrium conditions (Voller, 1979; Signorella and Hymer, 1984). The basic steps included the following: (1) Add a known quantity of the reference antigen ($HgBP_{20}II$) for binding to the wells of microtiter

Table 1. Amino acid composition of $HGBP_{20}II$

Amino Acid	Residue %	Nearest Integer[a]
1/2 CYS	26.3	29
ASX	9.2	10
THR	7.1	8
SER	9.4	10
GLX	4.2	5
PRO	4.1	5
GLY	16.7	19
ALA	4.4	5
VAL	3.9	4
MET	0.3	0
ILE	2.8	3
LEU	0.9	1
TYR	0.3	0
PHE	0.3	0
HIS	0.4	0
LYS	7.9	9
ARG	1.8	2
TOTAL		110[b]

[a] One LEU/molecule
[b] Minimum metal-free M_r = 12,900

plates; (2) Concurrently, incubate unknown sample or standard amounts of competing antigen with the primary antibody in a microcentrifuge tube; (3) Add the incubation mixture to each of the wells to allow unreacted antibody to bind with the immobilized antigen in the well; (4) Introduce anti-goat IgG conjugated to horseradish peroxidase; (5) Add substrate and quantify the horseradish peroxidase activity. Appropriate plate washes are included between each of the steps.

The optimized assay conditions included 106 ng antigen added to each well for binding, 0.106-106 ng antigen for the standard curve, 168 ng goat anti-($HgBP_{20}$II) IgG in the incubation mixture, and 1:500 dilution of the conjugate (purchased from Dynatech laboratories). Standard curves were constructed from plots of logit Y versus $\log_{10}$ standard antigen, with logit Y = ln (Y/1 1 - Y), Y = B/BO, B = sample absorbance minus 0% absorbance, and BO = 100% absorbance minus 0% absorbance (Signorella and Hymer, 1984). 0% absorbance represented negative controls in which plate-bound antigen and primary antibody were omitted; 100% absorbance represented controls in which no competing antigen was added.

The results using known amounts of competing antigen as standards are shown in Figure 1. A plot of B/BO versus log antigen dose indicated a sigmoid dose-response relationship (Figure 1a). Logit Y transformation of B/BO (Figure 1b) linearized the intermediate portion of the curve with 1.06 and 21.2 ng approaching the lower and upper limits of the linear range, respectively. These values corresponded to 0.082 and 1.642 pmol $HgBP_{20}$II using the dimer M_r of 12.9 KD for calculations.

Analysis of cytosolic extracts indicated detectable levels of 0.5 µg metal-binding protein/g wet wt in control gills and highly elevated levels in metal-exposed samples, with up to 1780 µg/g in gills of mussels exposed to mercury. Under conditions in which mercury accumulation and induction of metal-binding proteins had occurred as a result of exposure, linearity between metal-binding protein and tissue mercury concentrations was observed (Figure 2a). Constitutive levels of metal-binding proteins associated with the controls did not exist as mercury-binding proteins and were not on the linear portion of this curve. In contrast, cadmium was bound to the metal-binding proteins of controls (data not shown), and the observed linearity (Figure 2b) included the value of control mussels. Gills of mussels exposed to 5 µg/l copper measured 3.7 µg/g metal-binding proteins.

The sensitivity of the ELISA described here (working range of 1.06-21.2 ng; 0.082-1.6 pmol) is in the same general range as that of the radioimmunoassay described by Garvey *et al.* (1982) for mammalian MT and is capable of detection of metal-binding proteins in tissues from control as well as exposed mussels. This sensitivity will

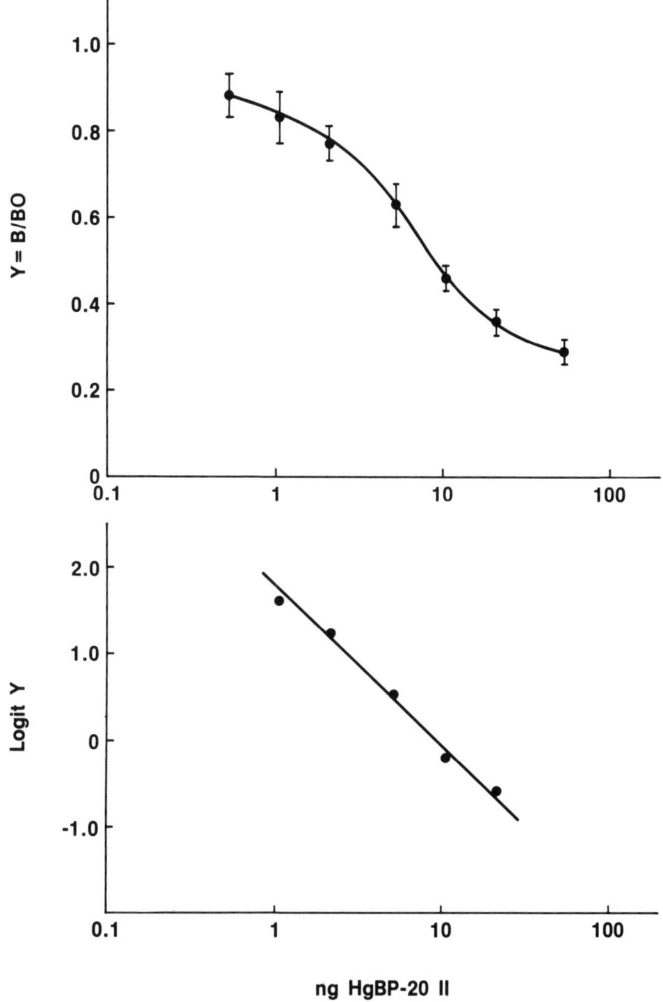

Figure 1. ELISA standard curves with (a) Y = B/BO versus standard antigen concentrations and (b) Logit Y versus standard antigen concentrations. $HgBP_{20}II$ is standard antigen. B is the background-corrected absorbance of standard or sample, and BO is the corrected-absorbance of a positive control.

enable detection of low levels of the proteins and in small samples such as hemolymph and hemocytes, as well as in the cytosol. The values for metal-binding proteins in gills of *Mytilus edulis* were in the same general range of values obtained for livers of metal-exposed

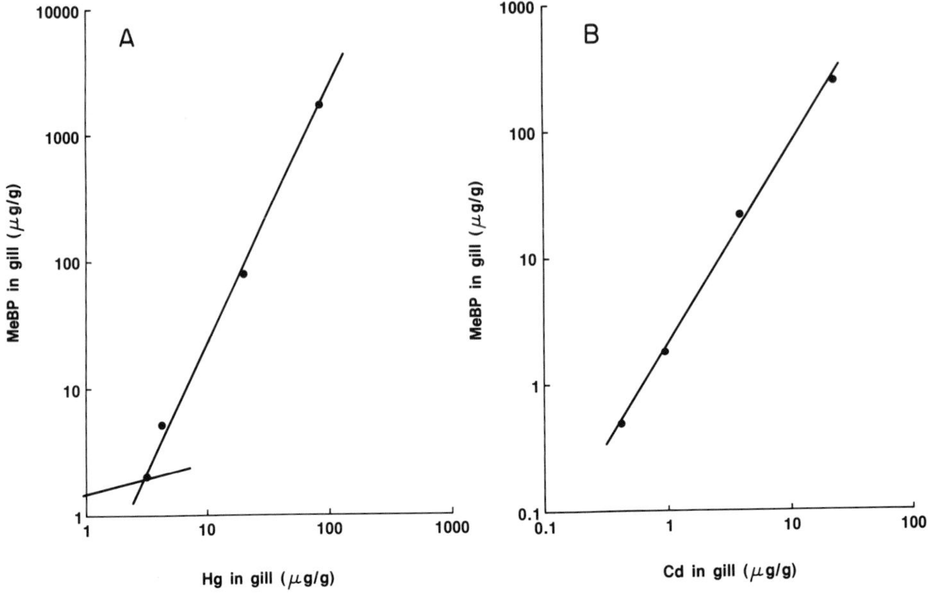

Figure 2. A. Relationship between concentrations of metal-binding proteins and mercury in gills of mercury-exposed mussels. Metal-binding proteins were estimated by ELISA. Mussels were exposed to mercury for 28 days to the concentrations (in µg/l) shown in parentheses on the graph. The value for the control (not shown) lies below the linear portion of the curve.
B. Relationship between concentrations of metal-binding proteins and cadmium in gill or cadmium-exposed mussels. Metal-binding proteins were estimated by ELISA. Mussels were exposed to cadmium for 28 days to the concentrations (µg/l) shown in parentheses on the graph. "C" designates the control.

and unexposed fish and rats by other techniques for detection of MT's (i.e., polarography, Hg and Cd radioassay, radioimmunoassay) (Roch *et al.*, 1982; Waalkes *et al.*, 1985).

Application of this assay in future work will facilitate the detection and quantification of metal-binding proteins in *Mytilus edulis*. The antibodies will also be useful for the cellular localization of these proteins with cytochemical procedures. At the present time, it appears that the antibodies described here will be applicable only to the metal-binding proteins of the mussel due to the high specificity of the antibodies. The mussel has served as a model in numerous studies on molluscan metal-binding proteins and trace metal metabolism and has also been used as an indicator species in marine pollution monitoring programs. Procedures for the specific detection and localization of MT-like, metal-binding proteins in the mussel will contribute to a better understanding of the normal metabolism of metals as well as responses to elevated metal exposure.

ACKNOWLEDGEMENTS

This study was partially supported by the U.S. Department of Energy under Contract No. DE-AC06-76RLO-1830 and Grant No. DE-FG02-86ER60410. *Current address of author is: University of Maryland, Center of Environmental and Estuarine Studies, Chesapeake Biological Laboratory, Solomons, MD. 20688.

REFERENCES

Cousins, R.J. 1985. Absorption, transport, and hepatic metabolism of copper and zinc: special reference to metallothionein and ceruloplasmin. Physiol. Rev. 65: 238-309.

Garvey, J.S., R.J. Vander Mallie and C.C. Chang. 1982. Radio immunoassay of metallothioneins. IN: Methods in Enzymology. J.J. Langone and H. Van Vunakis (eds.). Academic Press, NY. pp. 121-138.

George, S.G. and B.J.S. Pirie. 1980. Metabolism of zinc in the mussel, *Mytilus edulis* (L.): a combined ultrastructural and biochemical study. J.Mar. Biol. Ass. U.K. 60: 575-590.

Hamer, D.H., D.J. Thiele and J.E Lemontt. 1985. Function and autoregulation of yeast copperthionein. Science. 228: 685-690.

Klaaverkamp, J.F., W.A. Macdonald, D.A. Duncan and R. Wagemann. 1984. Metallothionein and acclimation heavy metals in fish: a review. IN: Contaminant Effects on Fisheries. V.W. Cairns, P.V. Hodson and J.O Nriagu. (eds.). John Wiley and Sons, Inc. pp. 100-113.

Kogima, Y. and J.H.R. Kagi. 1978. Metallothionein. Trends Biochem. Sci. 3: 89-94.

Laemmli, U.K. 1970. Cleavage of structural proteins during the assembly of the head of bacteriophage T4. Nature 227: 680-685.

Olafson, R.W., S. Loya and R.G. Sim. 1980. Physiological parameters of prokaryotic metallothionein induction. Biochem. Biophys. Res. Comm. 95: 1495-1503.

Roch, M., J.A. McCarter, A.T. Matheson, M.J.R. Clark and R.W. Olafson. 1982. Hepatic metallothionein induction in rainbow trout (*Salmo gairdneri*) as an indicator of metal pollution in the Campbell River System. Can. J. Fish. Aquatic. Sci. 19: 1596-1601.

Roesijadi, G. 1981. The significance of low molecular weight, metallothionein-like proteins in marine invertebrates: current status. Mar. Environ. Res. 4: 167-179.

Roesijadi, G. 1982. Uptake and incorporation of mercury into mercury-binding proteins of gills of *Mytilus edulis* as a function of time. Mar. Biol. 66: 151-157.

Roesijadi, G. and A.S. Drum. 1982. Influence of meracaptoethanol on the isolation of mercury-binding proteins from the gills of *Mytilus edulis*. Comp. Biochem. Physiol. 71B: 455-459.

Signorella, A.P. and W.C. Hymer. 1984. An enzyme-linked immunosorbent assay for rat prolactin. Anal. Biochem. 136: 372-381.

Viarengo, A., M. Pertica, G. Mancinelli, S. Palermero, G. Zanicchi and M. Orunesu. 1981. Synthesis of Cu-binding proteins in different tissues of mussels exposed to the metal. Mar. Bull. Pollut. 12: 347-350.

Voller, A., D.E. Bidwell and A. Bartlett. 1979. The Enzyme Linked Immunosorbent Assay. Dynatech Laboratories, Inc., Alexandria. 125 pp.

Waalkes, M.P., J.S. Garvey and C.D. Klaasen. 1985. Comparison of methods of metallothionein quantification: cadmium radioassay, mercury radioassay, and radioimmunoassay. Toxicol. Appli. Pharmacol. 79: 524-527.

Webb, M. 1979. The metallothioneins. IN: The Chemistry, Biochemistry, and Biology of Cadmium. M. Webb (ed.). Elsevier North-Holland Biomedical Press, NY. pp. 195-266.

ENVIRONMENTAL INDUCTION OF TELEOST CYTOCHROME P-450 DETECTED BY IMMUNOCHEMICAL ANALYSIS

John J. Stegeman
Biology Department
Woods Hole Oceanographic Institution
Woods Hole, Massachusetts 02543

INTRODUCTION

Organic foreign compounds have been recognized as having worldwide distribution. Polychlorinated biphenyls, for example, are found in regions extending from the atmosphere (Atlas and Giam, 1981) to the deep sea (Stegeman *et al.*, 1986). In some estuaries and coastal waters there are local high concentrations of these and other compounds that possess biological activity. The significance of such compounds in the sea is not yet known, but may be disclosed by studies on mechanisms of their biological action and transformation. Many such compounds are transformed by oxidative metabolism to products that are either more reactive or less reactive than the parent compound.

Cytochromes P-450 comprise a family of enzymes that catalyze monooxygenase reactions important in oxidative biotransformation of many organic compounds. Cytochrome P-450 in fish occurs in multiple forms, several of which ave been purified (Klotz *et al.*, 1983; Williams and Buhler, 1983). One type of cytochrome P-450 is represented by cytochrome P-450E, purified from liver of the coastal marine teleost, scup (*Stenotomus chrysops*). Cytochrome P-450E is the form primarily responsible for transformation of some widely distributed carcinogenic polynuclear aromatic hydrocarbons such as benzo(a)pyrene (Klotz *et al.*, 1983). This transformation could initiate events leading to environmental tumorigenesis. Cytochrome P-450E is also inducible by many foreign compounds, including polynuclear aromatic hydrocarbons and certain chlorobiphenyl congeners. Increased levels of cytochrome P-450E, or its counterpart in other species, could indicate that animals are being exposed to biochemically significant levels of these and similar foreign compounds.

Immunochemical Detection

Monoclonal and polyclonal antibodies have been prepared against scup cytochrome P-450E. The specificity and cross-reactivity of these antibodies has been described (Park *et al.*, 1986 ; Kloepper-Sams *et al.*, 1987). The antibodies to cytochrome P-450E recognize only this protein and no other microsomal proteins in scup liver (Kloepper-Sams *et al.*, 1987). Further, both types of antibodies also recognize the counterpart to cytochrome P-450E in other vertebrates, from fish to mammals. These antibodies have now been shown to be useful for analysis of environmental effects, as expressed in the induction of cytochrome P-450E.

Two examples will be mentioned here. First, monoclonal anti-P-450E (MAB 1-12-3) was used in western blot (immunoblot) analysis of hepatic cytochrome P-450E induction in scup from local Massachusetts waters (Kloepper-Sams *et al.*, 1987). In that study, animals sampled near New Bedford Harbor were found to have 50% or more of the total microsomal cytochrome P-450 present as cytochrome P-450E. New Bedford Harbor is a site characterized by very high PCB content in the sediments (Weaver, 1984). Fish taken from a site east of Martha's Vineyard, where contamination by compounds such as PCB's is believed to be substantially less than that near New Bedford Harbor, had very low levels of cytochrome P-450E.

The second example (Stegeman *et al.*, 1986) involves the use of these specific antibodies to detect effects in a species other than scup. The cross-reactivity of anti-P-450E suggested that we might use such antibodies to evaluate environmental induction of a counterpart to cytochrome P-450E in other species. In this case, rattails (*Coryphaenoides armatus*) were sampled from the deep sea, at about 3000 m depth off eastern North America. Analysis of hepatic preparations had previously revealed the presence of native cytochrome P-450 and intact microsomal electron transport systems in *C. armatus* retrieved from depth. Subsequent analysis of *C. armatus* from two locations revealed differences in the content of a protein detected by immunoblot with polyclonal anti-P-450E counterpart. The content of this protein correlated positively with the levels of catalytic activities associated PCB or PAH induction. These same activities are induced in many fish species (Stégeman, 1981). Furthermore, the levels of "cytochrome P-450E" in the rattails correlated positively with the levels of specific chlorobiphenyls measured in the liver of the *C. armatus*. The chlorobiphenyls included some known to induce a counterpart to cytochrome P-450E in mammals.

In addition to these studies, a similar immunological analysis of induced cytochrome P-450 has been described in English sole from Puget Sound (Varanasi *et al.*, 1986). In that study, antibodies to rainbow trout cytochrome P-450 LM_4, a counterpart to cytochrome P-450E, were used.

Two points are derived from these studies. First, it is apparent that animals in estuarine as well as deep-sea environments are showing effects of some compounds with activity like PCB's or polynuclear aromatic hydrocarbons. The extent or significance of biological change seen as cytochrome P-450E induction is not known. Second, it is evident that antibodies to cytochrome P-450E are useful in describing this biological change. Additional studies along the lines of those above could reveal the geographic and biological extent of such change and be useful in evaluating its significance.

ACKNOWLEDGEMENTS

Continuing support of the National Science Foundation (OCE83-10505) is gratefully acknowledged.

REFERENCES

Atlas, E.L. and C.S. Giam. 1981. Global transport of organic pollutants: Ambient levels of organic contaminants in the remote marine atmosphere. Science 211: 163-165.

Kloepper-Sams, P.J., S.S. Park, H.V. Gelboin and J.J. Stegeman. 1987. Specificity and cross-reactivity of monoclonal and polyclonal antibodies against cytochrome P-450E of the marine fish scup. Arch. Biochem. Biophys.

Klotz, A., J. Stegeman and C. Walsh. 1983. An aryl hydrocarbon hydroxylating hepatic cytochrome P-450 from the marine fish *Stenotomus chrysops*. Arch. Biochem. Biophys. 226: 578-592.

Park, S.S., H. Miller, A.V. Klotz, P.J. Kloepper-Sams, J.J. Stegeman and H. V. Gelboin. 1986. Monoclonal antibodies against cytochrome P-450E from the marine teleost *Stenotomus chrysops* (scup). Arch. Biochem. Biophys. 249: 339-350.

Stegeman, J.J. 1981. Polynuclear aromatic hydrocarbons and their metabolism in the marine environment. IN: Polycyclic Hydrocarbons and Cancer. H.V. Gelboin and P.O.P. Ts'o (eds.). Academic Press. 3(Chapter 1): 1-60.

Stegeman, J.J., P.J. Kloepper-Sams and J.W. Farrington. 1986. Monooxygenase induction and chlorobiphenyls in the deep sea fish *Coryphaenoides armatus*. Science 231: 1287-1289.

Varanasi, U.S., T.K. Collier, D.E. Williams and D.R. Buhler. 1986. Hepatic cytochrome P-450 isozymes and aryl hydrocarbon hydroxylase in English sole (*Parophrys vetulus*). Biochem. Pharmacol. 35: 2967-2971.

Weaver, G. 1984. PCB contamination in and around New Bedford Mass. Environ. Sci. Technol. 18: 22A-27A.

Williams, D. and D. Buhler. 1983. Comparitive properties of purified cytochrome P-448 from ß-naphthoflavone treated rats and rainbow trout. Comp. Biochem. Physiol. 75C: 25-32.

QUANTIFYING STOMACH CONTENTS USING IMMUNOASSAYS: A CRITIQUE

Robert J. Feller
Department of Biology
Marine Science Program
Belle W. Baruch Institute for
Marine Biology and Coastal Research
University of South Carolina
Columbia, SC 29208

Robert B. Ferguson
Belle W. Baruch Institute for
Marine Biology and Coastal Research
University of South Carolina
Columbia, SC 29208

INTRODUCTION

Immunological techniques have a long history of use in the qualitative examination of invertebrate diets, especially for insects (Calver, 1984). Typical dietary questions asked in the ecological application of serological methods include: does predator A ingest prey X, Y, and Z ?; what proportion of the predator population eats a particular prey organism ?; how long is a meal detectable ?; does selective predation occur (that is, does the predator eat prey X but not Y or Z) ? After using antisera to answer similar questions in marine food webs (Feller *et al.*, 1979), it has become apparent that ecologists will never be satisfied with answers to such qualitative questions alone.

CONCEPTUAL PROBLEMS

The original approach of Dempster (1960) was to estimate predation rates of various insects using the formula:

$$\frac{(\text{predator abundance}) \times (\text{proportion eating prey}) \times (\text{prey abundance})}{(\text{detection time of prey in gut})}$$

where the product of this predation rate and the time over which predation occurs yields the number of prey eaten during the time period of interest. The proportion eating prey was estimated as the percentage of positive serological tests of numerous predator stomachs. A critical assumption in the use of this formulation, that a positive test detects a meal consisting of one prey item, must be evaluated carefully before numerical quantification of prey is attempted. However, Dempster's basic formulation did provide the cornerstone for several subsequent `quantitative' studies of predation (Rothschild, 1966; Reynoldson and Davies, 1970; Ashby, 1974; Sunderland and Sutton, 1980; Adams, 1981).

Confronted with the task of developing models of material and energy flux in the marine environment (e.g., Schwinghammer *et al.*, 1986; Rowe *et al.*, 1986), it is necessary to quantify biomass transfers among various compartments of the food web. In many situations it is very difficult to identify prey even qualitatively in a predator's gut contents. Although various immunoassays have successfully revealed the identity of participants in these otherwise `invisible' food webs, satisfactory quantitative ecological interpretation of qualitative immunoassay results has yet to be achieved. Methodological constraints aside, this is a difficult task even under optimal conditions. An optimal scenario might, for instance, involve predators which feed only at certain times in an ambit of known size and which can easily be observed feeding. It is also preferable that the predator's ingesta be both readily identifiable and each prey component measurable as biomass or energy.

This idealized situation is always constrained by a variety of non-ideal, but nonetheless real, attributes of predatory organisms and their prey (Table 1). In terms of serological analysis, the most problematic of these include the unknown meal size (how many prey comprise a meal or were bites taken ?), variable gut retention time (different types of prey are detectable for different lengths of time), and finally the antiserum specificity (is the antiserum detecting the targeted prey or merely cross-reacting with similar prey ?). The question of meal size can be approached using laboratory feeding trials to estimate maximum meal size (Ashby, 1974), whereas detection times require standardization for the immunodetection system employed. Techniques with higher sensitivity can detect meals for longer periods of time than less sensitive methods (Greenstone, 1983; Calver *et al.*, 1986). Furthermore, gut residence time and, hence, detection time is temperature-dependent in poikilotherms and can vary with meal size (degree of satiation), prey type (hard-bodied versus soft-bodied prey), and whether feeding is continuous or periodic. Corrections for variable detection times are essential for interpreting results of immunoassays (Reynoldson and Davies, 1970), and have been elegantly conceptualized as contact times by Peterson and Bradley (1978).

Table 1. Problems associated with quantifying stomach contents data for A) standard visual methods of analysis and B) immunoassays. None of the problems is mutually exclusive.

A)
- selecting the appropriate measure: numerical importance, gravimetric, volumetric, energetic, frequency of occurrence, indices of fullness, index of relative importance (IRI), etc.
- variable prey detection times and differential rates of digestion
- high variance estimates of predator and prey abundance
- size of foraging area ill-defined for motile predators
- meal size (numbers or biomass) unknown and doubtful extrapolations from laboratory feeding experiments
- feeding periodicities or lack thereof

B)
- antiserum specificity and sensitivity may be inadequate (low titers or from digestive breakdown of antigenic determinants on target proteins)
- antisera not available for all potential prey taxa
- interference from multiple prey in gut or secondary ingesta (prey of prey)
- cannot detect cannibalism
- inability to differentiate among ingestion of living or dead prey or pieces of prey
- cross-reactions from untested prey (false positives)
- translation of protein concentrations to other currencies

It should be quite obvious that the quantification of predator-prey interactions can take a variety of forms (Kiritani and Dempster, 1973; Berg, 1979). No single method of quantification will satisfy all requirements. The index of relative importance (IRI) devised by Pinkas *et al.* (1971) incorporates volumetric, numerical, and frequency of occurrence data into a single measure of prey importance. However, this conceptually pleasing index can also be biased towards prey having long contact times if this particular variable is not taken into account.

The following section outlines an approach to this general problem of quantifying the diet of benthic omnivores in soft-substrate habitats. Parts of these methods have been tested, but more work is needed. We make no claim that any of these ideas will provide final solutions to the problem of quantification, for although the methods themselves are highly quantitative, the ecological interpretation of the data will probably always be equivocal.

QUANTIFICATION WITH ROCKET IMMUNOELECTROPHORESIS

For stomach contents which are visually unrecognizable, any quantitative analysis must first be preceded by qualitative assays to determine what, or which, prey may be present in the sample. This is typically done using the double immunodiffusion technique of Ouchterlony (Wang, 1982; Feller, 1984a). Dominance of different prey types may be measured as their frequency of occurrence in replicate predator gut samples. Once it is known which prey are present in or dominate a gut content sample (either of an individual predator or a composite of several gut contents), it is then possible to make certain quantitative measurements. For instance, it may be of interest to know what proportion of the proteins in a predator's gut is comprised of particular prey proteins. This measure of relative proportion or contribution is possible by using the rocket electrophoresis technique of Laurell (1966) which is designed to measure concentrations of specific proteins (Axelsen and Bock, 1972; Verbruggen, 1975; Laurell and McKay, 1981) (Figure 1). The stomach content antigens (soluble proteins remaining from the ingested prey) are moved electrophoretically from small wells into a horizontal slab of agarose gel containing a constant amount of antiserum specific for the prey type of interest. A precipitin line forms in the shape of a rocket and encloses an area on the gel which is proportional to the concentration of prey antigen and inversely proportional to the concentration of antiserum in the gel. The rocket electrophoresis technique is capable of detecting and quantifying proteins in the 0.1 to 10 ng range (Kenny and Foy, 1975). The total amount of soluble protein present in the predator gut sample can be measured with the standard Bio-Rad Protein Assay (Bio-Rad, 1977) which is sensitive to concentrations as low as 1 µg/ml. The concentration of total soluble protein in starved predators should serve as a control for potential bias of gut protein concentrations due to sloughing of soluble proteins from the predator's gut wall or other contamination

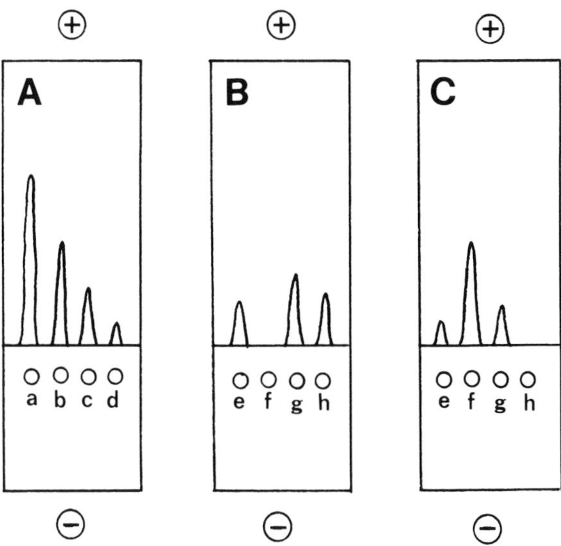

Figure 1. Diagram illustrating the use of rocket electrophoresis for quantitative gut contents analysis on three agarose-coated microscope slides. Upper portion of gels (A-C) contains antiserum to proteins of interest.
A - antiserum to BSA, bovine serum albumin.
B - antiserum to prey type X.
C - antiserum to prey type Z. Circular wells a - d contain decreasing concentrations of BSA protein standard; wells e - h contain equal volumes of gut contents from four different predators. Slide A serves as a protein concentration standard. All slides may be electrophoresed simultaneously. The gut contents in well f contain no antigens of prey type X, but antigens of prey type Z are present in this example.

unrelated to ingesta. Thus by knowing the concentrations of prey antigens in the sample, one can calculate the proportion of the total soluble protein concentration comprised by each prey type. One could state, for instance, that a shrimp stomach contained 5 mg/ml total soluble protein at the time of sampling, of which 40% was polychaete protein, 30% bivalve protein, etc. These proportions will necessarily be minimum estimates since any antiserum will react only with the antigenic proteins of the prey and not all of its soluble proteins. The sum of all prey protein concentrations may not equal the gut total if proteins of unknown origin are present.

This approach seems particularly attractive for compartmental models of energy flux, but of course to be meaningful, samples should be taken repeatedly through time so that one might adjust for feeding periodicities that could bias a sample from any single point in time. Variability in the detection times of each prey type should also be taken into account. Rocket electrophoresis has also been used as a rapid screening device for analyzing the stomach contents of larger numbers of predators for the frequency of occurrence of specific prey (Grisley and Boyle, 1985).

CALIBRATION

A further quantification might entail calibration of protein concentrations in a predator's gut contents with a dilution series of known numbers or biomass of uniformly-sized prey (and hence having known soluble protein concentrations). If the volume of gut contents is known and the effects of digestion are ignored (possible for predators collected immediately after they have fed), then prey protein concentrations in a gut content sample can be converted directly to numbers or biomass of standard prey. An obvious problem here is that one never knows exactly when a predator may have fed, especially if they feed continuously. Luckily many predators exhibit periodic or episodic feeding, in which case it is possible to sample them soon after the start of their gut loading process.

An extension of the standardization and calibration of prey proteins was attempted by Feller (1984b). In this study of shrimp feeding, the concentration of prey proteins in unit areas or volumes of sediment was used as a translation factor for interpreting the concentrations of prey proteins found in gut content samples. That is, concentrations of specific prey proteins found in a gut with rocket electrophoresis were equated to amounts of that protein found in so many unit areas of sediment (from grinding known amounts of sediment and analyzing them for specific prey proteins). The goal was to be able to state that a given predator's stomach contained an amount of prey protein equivalent to that found in so many unit areas of sediment, implying among other things that carrying capacity could be estimated from such data if daily rations were known. Again, however, it was only too easy to find problems with this approach since we seldom if ever know where predators feed in soft-bottoms. Additionally, the variances around estimates of sediment prey abundance and sediment protein concentrations were almost always very large, further complicating attempts to interpret otherwise quantitative data in an ecologically meaningful way.

CONCLUSIONS

The first point to remember about using immunoassays for gut contents analysis is that their use is usually called for when every other easier method fails. All the possible combinations of potential artifacts and departures from various necessary assumptions make ecological interpretation of quantitative measurements of prey proteins somewhat tenuous. Quantification of any immunoassay is fraught with uncertainty, even more so when one realizes that the problems inherent in standard visual or gravimetric methods also underlie all immunoassays. Whether antisera and immunoassays can ever be used reliably to quantify predator-prey interactions is largely a question of how much uncertainty one is willing to accept.

ACKNOWLEDGEMENTS

Concepts described in this paper were developed with the support of NSF grants OCE 81-10148 and OCE 85-21345 (Biological Oceanography Program) and grants from the South Carolina Sea Grant Consortium. This is contribution number 659 from the Belle W. Baruch Institute for Marine Biology and Coastal Research.

REFERENCES

Adams, J. 1981. Serological analysis of the diet of *Bdellocephala punctata*, a freshwater triclad. Oikos 36: 99-106.

Ashby, J.W. 1974. A study of arthropod predation of *Pieris rapae* L. using serological and exclusion techniques. J. Appl. Ecol. 11: 419-425.

Axelsen, N.H. and E. Bock. 1972. Identification and quantification of antigens and antibodies by means of quantitative immunoelectrophoresis. A survey of methods. J. Immunol. Methods 1: 109-121.

Berg, J. 1979. Discussion of methods of investigating the food of fishes, with reference to a preliminary study of the prey of *Gobiusculus flavescens* (Gobiidae). Mar. Biol. 50: 263-273.

Bio-Rad. 1977. Bio-Rad Protein Assay. Tech. Bull. 1051, April. Bio-Rad Laboratories, Richmond, Calif. 3 pp.

Calver, M.C. 1984. A review of ecological applications of immunological techniques for diet analysis. Aust. J. Ecol. 9: 19-25.

Calver, M.C., J.N. Matthiessen, G.P. Hall, J.S. Bradley and J.H. Lillywhite. 1986. Immunological determination of predators of the bush fly, *Musca vetustissima* Walker (Diptera: Muscidae), in south-western Australia. Bull. Ent. Res. 76: 133-139.

Dempster, J.P. 1960. A quantitative study of the predators on the eggs and larvae of the broom beetle, *Phytodecta olivaca* Forster, using the precipitin test. J. Anim. Ecol. 29: 149-167.

Feller, R.J. 1984a. Dietary immunoassay of *Ilyanassa obsoleta*, the eastern mud snail. Biol. Bull. 166: 96-102.

Feller, R.J. 1984b. Serological tracers of meiofaunal food webs. Hydrobiologia 118: 119-125.

Feller, R.J., G.L. Taghon, E.D. Gallagher, G.E. Kenny and P.A. Jumars. 1979. Immunological methods for food web analysis in a soft-bottom benthic community. Mar. Biol. 54: 64-75.

Greenstone, M.H. 1983. Site-specificity and site tenacity in a wolf spider: a serological dietary analysis. Oecologia, Berlin. 56: 79-83.

Grisley, M.S. and P.R. Boyle. 1985. A new application of serological techniques to gut content analysis. J. Exp. Mar. Biol. Ecol. 90: 1-9.

Kenny, G.E. and H.M. Foy. 1975. Detection and quantitation of circulating polysaccharide in pneumococcal pneumonia by immunoelectroosmophoresis (counterelectrophoresis) and rocket electrophoresis. Microbiology 1975. Am. Soc. Microbiol. Wash. D.C. 1: 97-102.

Kiritani, K. and J.P. Dempster. 1973. Different approaches to the quantitative evaluation of natural enemies. J. Appl. Ecol. 10: 323-330.

Laurell, C.-B. 1966. Quantitative estimation of proteins by electrophoresis in agarose gel containing antibodies. Anal. Biochem. 15: 45-52.

Laurell, C.-B. and E.J. McKay. 1981. Electroimmunoassay. Methods in Enzymology 73: 339-369.

Peterson, C.H. and B.P. Bradley. 1978. Estimating the diet of a sluggish predator from field observations. J. Fish. Res. Board Can. 35: 136-141.

Pinkas, L., M.S. Oliphant and I.L.K. Iverson. 1971. Food habits of albacore, blue fin tuna, and bonito in California waters. Fish. Bull. Calif. 152: 1-105.

Reynoldson, T.B. and R.W. Davies. 1970. Food niche and coexistence in lake-dwelling triclads. J. Anim. Ecol. 39: 599-617.

Rothschild, G.H.L. 1966. A study of a natural population of *Conomelus anceps* (Germar) (Homoptera: Delphacidae) including observations on predation using the precipitin test. J. Anim. Ecol. 35:413-434.

Rowe, G.T., N. Merrett, J. Shepherd, G. Needler, B. Hargrave and M. Marietta. 1986. Estimates of direct biological transport of radioactive waste in the deep sea with special reference to organic carbon budgets. Oceanologica Acta 9: 199-208.

Schwinghamer, P., B. Hargrave, D. Peer and C.M. Hawkins. 1986. Partitioning of production and respiration among size groups of organisms in an intertidal benthic community. Mar. Ecol. Prog. Ser. 31:131-142.

Sunderland, K.D. and S.L. Sutton. 1980. A serological study of arthropod predation on wood lice in a dune grassland ecosystem. J. Anim. Ecol. 49: 987-1004.

Verbruggen, R. 1975. Quantitative immunoelectrophoretic methods: a literature survey. Clin. Chem. NY. 21: 5-43.

Wang, A.-C. 1982. Methods of immune diffusion, immuno-electrophoresis, precipitation, and agglutination. IN: Antibody as a Tool, the Applications of Immunochemistry. J.J. Marchalonis and G.W. Warr. (eds.). John Wiley & Sons, Chichester. pp. 139-161.

EUPHAUSIID PREDATION ON LARVAL ANCHOVY AT TWO CONTRASTING SITES OFF CALIFORNIA DETERMINED WITH AN ELISPOT IMMUNOASSAY

Gail H. Theilacker
Southwest Fisheries Center La Jolla Laboratory
National Marine Fisheries Service, NOAA
PO Box 271
La Jolla, CA 93038

INTRODUCTION

Understanding the processes affecting recruitment is a fundamental objective in fishery research. Starvation and predation are believed to be the major factors influencing survival of young fish (Hunter, 1984), and mortality due to these two factors must be quantified to understand how they control recruitment of young fish to a fish stock.

Evidence supporting the occurrence of starving larval fish in the ocean was mainly circumstantial (May, 1974; Jones and Hall, 1974; Lasker, 1975) until recently, when the presence of starving ocean-caught larval northern anchovy, *Engraulis mordax*, was documented using histological criteria (O`Connell, 1980). Subsequently, using similar criteria, Theilacker (1986) quantified starvation-induced mortality for ocean-caught larval jack mackerel, *Trachurus symmetricus*, and showed that starvation was a significant source of mortality for first-feeding fish.

Limited information is available on predation of larval fishes. Field evidence of predation by crustaceans has been mainly circumstantial (Westernhagen and Rosenthal, 1976; Alvariño, 1980; Frank and Leggett, 1982; Bailey and Yen, 1983; Brewer *et al.*, 1984), yet many crustaceans eat larval fish under laboratory conditions (reviewed by Hunter, 1984). Little field information exists for carnivorous Crustacea because they macerate their prey and prey remains cannot be visually identified. However, predation on larval fishes by pelagic cnidarians and ctenophores, gelatinous predators, was documented in the field and, in some cases, quantified (Moller, 1984; Purcell, 1981; 1984). Furthermore several field studies have demonstrated that planktivorous fishes also consume fish eggs and larvae (reviewed by Hunter, 1984).

Off California, the euphausiid crustacean *Euphausia pacifica* co-occurs temporally and spatially with northern anchovy larvae. In February and March, during northern anchovy spawning season, field collections containing 5000 adult euphausiids per m^3 are common (Brinton and Wyllie, 1976). In the laboratory, adult *E. pacifica* can consume up to 25 larvae per day (Theilacker and Lasker, 1974). If this feeding rate occurs in the ocean, euphausiids could be an important predator of larval anchovy.

These results are of a preliminary field study conducted to test the feasibility of using a newly developed immunoassay to assess predation on larval anchovy by euphausiids. *E. pacifica was sampled* from two contrasting sites off California. The inshore site was within an area of intense anchovy spawning, and the offshore site was outside of the main spawning area. An ELISPOT immunoassay was used to detect the presence of northern anchovy yolk protein in the euphausiids' stomachs (Theilacker *et al.*, 1986). Evidence obtained shows that euphausiids eat anchovy eggs and/or larvae; quantification of the euphausiid consumption rate in the field; and suggests that the predation by euphausiids may be significant.

This preliminary study proved the feasibility of using the ELISPOT technique for field samples. A more comprehensive spatial and temporal estimation of anchovy egg and larval mortality due to predation by euphausiids will be made next year. The proposed research is discussed.

METHODS

Field Collection

Inshore collections were made in March 1985 at two sites, 8 miles apart, 50 miles off the coast of southern California where anchovy spawning was intense. The offshore collection was made in February 1986, 100 miles off the southern California coast where anchovy spawning was minimal. These collections were made aboard the NOAA Ship David Starr Jordan. Between 2200 and 0200, 5 minute oblique tows were taken to 50 m, the depth where euphausiids and larval northern anchovy co-occur for about 10 h at night (Ahlstrom, 1959; Brinton, 1976). A Bongo frame was used, fitted with a 505 µm mesh net and plastic cod ends. After carefully diluting the live euphausiids into several buckets containing 14°C surface seawater, swimming animals were selected, blotted and placed individually into

1 ml Eppendorf tubes set in crushed ice. The sorting process, from seawater to cooled tube, took about 1 min per animal. Groups of animals were frozen every 10 min in liquid nitrogen and stored at -80° C.

In addition to the euphausiid collections, quantitative, oblique plankton tows were taken with a 505 µm mesh Bongo to a depth of 200 m at the inshore and offshore sites. Abundances of anchovy eggs, yolk-sac larvae, and adult (12-22 mm) and juvenile (6 - 11 mm) euphausiids were estimated from these samples.

Laboratory immunoassay

To test for the presence of anchovy yolk protein (which cannot be visually detected or identified) in the stomachs of field-collected *E. pacifica*, an immunoassay (Enzyme-Linked Immunospot; ELISPOT) was used that detected nannogram quantities of anchovy egg-yolk protein in stomach of laboratory-fed euphausiids. The details of the ELISPOT assay are given by Theilacker *et al.* (1986). The individual, field-collected animals were prepared for the assay by dissecting the stomach, hepatopancreas, and hindgut, teasing the tissues in buffer, and dotting the extract, about 10 µl, onto nitrocellulose paper (0.45 µm pore size). For the ELISPOT procedure, after the antigen (unknown gut contents; potentially, anchovy yolk protein) was immobilized on nitrocellulose, the antigen was then reacted with the primary antibody to anchovy yolk protein which was produced in rabbits. The resulting antigen-antibody complex was treated with a secondary antibody which had been tagged with the enzyme alkaline phosphatase. This secondary antibody was produced in goats by immunizing them with rabbit sera. The complex on th nitrocellulose was visualized with a dye, and positive phosphatas activity produced a blue-black reaction.

Control euphausiids, fed algae in the laboratory for 24-48 h a 14° C to clear their guts of non-algal prey, were dissected and ru with each test. A test consisted of one control animal with 6- unknowns; as many as 6 tests were run simultaneously with this assay Dotted gut contents from field animals that developed color visuall darker than the gut contents of the control animals were counted a positive. In this application, the immunoassay was used to detec presence or absence of yolk in the predator's stomach. That is, eve though the protein quantity is measurable with the assay, the number of larvae eaten cannot be quantified because the absolute amount of yolk contained in an individual egg and yold-sac larva descreases with age.

RESULTS

A total of 114 adult *E. pacifica* were taken at the two inshore collections. The fraction positive at the first site was 43/55 (78%) and at the second site was 47/59 (80%). Offshore, the fraction positive was 5/65 (7%).

At the two inshore sites euphausiid abundance, estimated as the number of individuals under 10 m^2 sea surface, was 192 and 542 adults and 526 and 927 juveniles; offshore, the number of individuals was 3376 adults and 399 juveniles (Table 1).

Estimates of anchovy egg and larval abundance inshore (estimated, as above, for numbers/10 m^2 sea surface and summed over the duration, 6 days, that eggs and larvae are eaten by predatory euphausiids, see Theilacker and Lasker, 1974) were 18207 eggs and 3200 yolk-sac larvae; estimates for the offshore site were 1942 eggs and 240 yolk-sac larvae (Table 1).

To estimate daily consumption by the euphausiid population, additional field and laboratory information was needed. Brinton's (1976) field data on vertical distribution of *E. pacifica* showed that for 10 hours at night 87% of the animals occur in the upper 50 meters of the sea surface, which is the depth distribution for northern anchovy eggs and larvae (Ahlstrom, 1959). Earlier laboratory studies revealed that juvenile *E. pacifica* feed at one third the rate of adults (Theilacker and Lasker, 1974). Thus to calculate total predation by euphausiids the number of juveniles was adjusted by 0.3 (Table 1; footnote 2). The final information needed, euphausiid digestion times at *in situ* temperatures, was taken from the results of a digestion experiment where 12 adult euphausiids that had eaten 1 or 2 anchovy larvae were subsequently fed copepods and algae for 1 to 6 h. More experiments are needed, but preliminary results showed that yolk protein was undetectable in the euphausiid guts after 4 h digestion. For this study, a digestion time of 4 h was assumed to calculate euphausiid feeding rates.

A Poisson distribution was used to calculate the mean number of anchovy eggs and larvae in a euphausiid's stomach because the mean of a Poisson distribution can be easily estimated directly from the frequency of "0" counts, or negative assays ($\mu = -ln\ f(0)$). As mentioned earlier, the immunoassay technique determines presence or absence of anchovy yolk protein and does not provide information on how many prey were consumed. Thus the positive assays combine counts of one or more eggs or larvae, and the negative assays identify the euphausiids that had not eaten anchovies in the last 4 h, the assumed digestion time. Using the Poisson distribution, the mean number of eggs or larvae eaten per 4 h is equal to *-ln* of the frequency of negative assays. Because the euphausiids and anchovy co-occur for 10 h per day, the number of anchovy eggs and larvae eaten per day is 2.5 μ (Table 1; footnote 2). This relation was used to estimate the impact of euphausiid predation on the anchovy population at the two contrasting sites given the above laboratory and field information (Table 1).

Table 1.

	ESTIMATED ABUNDANCE/10m²		PREDATION BY EUPHAUSIIDS		
	ANCHOVY (eggs/larvae)	EUPHAUSIIDS (juv./adults)	ELISPOT (fraction +)	EST. FEEDING RATE/10h (indiv./pop.)	% ANCHOVY POPULATION EATEN
$Inshore_1$	18,207/3200	526/192	43/55	3.8[1]/1,156[2]	5
$Inshore_2$		927/542	47/59	4.0/2,854	13
$Offshore_1$	1,942/240	399/3,376	5/65	0.2/608	28

1 Poisson distribution $f(0) = e^{-\mu}$; $\mu = -\ln f(0)$ = mean number eaten/4h

2 Number eggs or larvae eaten/10h = (number adult euphausiids + 0.3 number juveniles) x 2.5μ x 0.87

The egg and larval (yolk-sac) mortality rates due to euphausiid predation at the inshore sites were 0.05 and 0.13/d, with an average of 0.09 (percent of anchovy eaten; Table 1). At the offshore station, daily anchovy mortality due to euphausiid predation was 0.28. The total instantaneous mortality rate (Z) calculated from the above egg (N_1) and larval (N_2) data for the inshore site was 0.6/d and for the offshore site was 0.7/d ($Z = ln\ N_1 - ln\ N_2)/(t2-t1)$). Thus the results from this preliminary study show that the difference in mortality between the inshore and offshore sites can be accounted for by differences in rates of euphausiid predation.

DISCUSSION

This preliminary study shows that invertebrate predation rates can be estimated by using the ELISPOT immunoassay together with additional laboratory and field data. Refining recruitment estimates may be possible by combining information on predation mortality together with estimates of total larval fish mortality and starvation-induced mortality.

Efforts to make a comprehensive spatial and temporal evaluation of anchovy mortality due to predation by euphausiids are being planned. Three two-week cruises, beginning in November, 1986, will be conducted to sample the northern anchovy habitat off California before, during and after peak spawning. A net with a larger mesh, 2 mm, will be used to collect the animals. This net allows eggs and larvae to pass through, precluding biases caused by euphausiids feeding in the cod-end (Nicol, 1984). Short, 5 minute tows were used in the preliminary study reported here to decrease the probability of cod-end feeding. In the laboratory, additional euphausiid digestion experiments will be conducted with adult and juvenile animals.

ACKNOWLEDGEMENTS

I appreciate the assistance of Marlene Johnson with the feeding experiments and immunoassays. Roger Leong spawned the anchovies, Alec MacCall suggested the statistical treatment, Reuben Lasker and John Hunter commented on the manuscript, and an anonymous reviewer made additional constructive comments.

REFERENCES

Ahlstrom, E.H. 1959. Vertical distribution of pelagic fish eggs and larvae off California and Baja California, U.S. Wildl. Serv. Fish. Bull. 60: 107-146.

Alvariño, A. 1980. The relation between the distribution of zoopalnkton predators and anchovy larvae. Calif. Coop. Oceanic Fish. invest. Rep. 21: 150-160.

Bailey, K.M. and J. Yen. 1983. Predation by a carnivorous marine copepod, *Euchaeta elongata* Esterly, on eggs and larvae of the Pacific hake, *Merluccius productus*. J. Plankton Res. 5: 71-82.

Brewer, G.D., F.S. Kleppel and M. Dempsey. 1984. Apparent predation on ichthyoplankton by zooplankton and fishes in nearshore waters of southern California. Mar. Biol. 80: 17-28.

Brinton, E. 1976. Population biology of *Euphausia pacifica* off southern California. Fish. Bull. U.S. 74: 733-762.

Brinton, E. and J.G. Wyllie. 1976. Distributional atlas of euphausiid growth stages off southern California, 1953 through 1956. CalCOFI (Calif. coop. ocean. Fish. Invest.) Atlas No. 24.

Frank, K.T. and W.C. Leggett. 1982. Coastal water mass replacement: its effect on zooplankton dynamics and the predator-prey complex associated with larval capelin (*Mallotus villosus*). Can. J. Fish. Aquat. Sci. 39: 991-1003.

Hunter, J.R. 1984. Inferences regarding predation on the early life stages of cod and other fishes. *Flodevigen rapportser*. 1, The propagation of cod, *Gadus morhua* L.: 533-562.

Jones, R. and W.B. Hall. 1974. Some observations on the population dynamics of the larval stage in common gadoids. IN: Early life history of fish, J.H.S. Blaxter (ed.). Springer-Verlag. Berl. pp. 87-102.

Lasker, R. 1975. Field criteria for survival of anchovy larvae: the relation between inshore chlorophyll maximum layers and successful first feeding. Fish. Bull. U.S. 73: 453-462.

May, R.C. 1974. Larval mortality in marine fishes and the critical period concept. IN: The early life history of fish. J.H.S. Blaxter (ed.). Springer-Verlag, NY. pp 3-19.

Moller, H. 1984. Reduction of a larval herring population by jellyfish predator. Science 224: 621-622.

Nicol, S. 1984. Cod end feeding by the euphausiid *Meganyctiptianes norvegica*. Mar. Biol. 80: 29-33.

O'Connell, C.P. 1980. Percent of starving northern anchovy larvae (*Engraulis mordax*) in the sea as estimated by histological methods. Fish. Bull. U.S. 78: 475-489.

Purcell, J.E. 1981. Feeding ecology of *Rhizophysa eysenhardti*, a siphonophore predator of fish larvae. Limnol. Oceanogr. 26: 424-432

Purcell, J.E. 1984. Predation on larval fish by the Portuguese man of war, *Physalia physalis*. Mar. Ecol. Prog. Ser. 19: 189-191.

Theilacker, G.H. 1986. Starvation-induced mortality of young sea-caught jack mackerel, *Tranchurus symmetricus*, determined with histological and morphological methods. Fish. Bull. U.S. 84: 1-17.

Theilacker, G.H., A.S. Kimball and J.S. Trimmer. 1986. Use of an ELISPOT immunoassay to detect euphausiid predation on larval anchovy. Mar. Ecol. Prog. Ser. 30: 127-131.

Theilacker, G.H. and R. Lasker. 1974. Laboratory studies of predation by euphausiid shrimps on fish larvae. IN: The early life history of fish. J.H.S. Blaxter (ed.). Springer-Verlag, Berlin. pp. 287-299.

Westernhagen, H. von and H. Rosenthal. 1976. Predator-prey relationship between pacific herring, *Clupea harengus* pallasi, larvae and a predatory hyperiid amphipod, *Hyperoche medusarum*. Fish. Bull. 74: 669-674.

ENZYME IMMUNOASSAY UTILIZING MONOCLONAL ANTIBODIES FOR IDENTIFICATION OF EUROPEAN EEL VIRUS (EEV), AN AQUATIC BIRNAVIRUS

V. Lipipun, P. Caswell-Reno
P. Reno and B. Nicholson

Department of Microbiology and
Center for Marine Studies
University of Maine
Orono, Maine 04469

INTRODUCTION

European eel virus (EEV) is the etiological agent of branchio-nephritis, one of the most serious diseases of eels in Asian fish culture (Sano *et al.*, 1981). This virus was first isolated from the European eel (*Anguilla anguilla*) and termed EEV; later, it was also isolated from the Japanese eel (*A. japonica*). Subsequently, EEV has been isolated from both healthy and diseased eels throughout Japan and Taiwan (Kou *et al.*, 1983; Sano *et al.*, 1981). Also, a virus similar, and possibly identical, to EEV has been isolated from *Tilapia* in Taiwan (Kou *et al.*, 1983). Experimental infection with EEV produced high levels of mortality in eels and *Tilapia* sp. but not in rainbow trout.

EEV is a member of the newly recognized virus family *Birnaviridae*, which also includes infectious pancreatic necrosis virus (IPNV) and related viruses. The birnaviruses are medium sized viruses (approximately 59 nm in diameter) characterized by a single layer capsid structure and a bi-segmented, double-stranded RNA (dsRNA) genome (Dobos *et al.*, 1979). Most of the aquatic birna-viruses cross-react to some extent in neutralization tests with polyclonal antisera. However, at least 9 separate serotypes have been identified based on the relative degree of cross-neutralization (Hill and Way, 1983). All of the viral isolates from the United States represent a single serotype designated as either West Buxton or VR299. Four serotypes (Jasper, Canada 1, Canada 2 and Canada 3) have been identified from Canada and four serotypes (Sp, Ab, Te, and He) have been reported in Europe. To date, all birnavirus isolates from Asia represent either the West Buxton or Ab serotype, although most isolates appear to belong to the latter serotype.

Recently, monoclonal antibodies (MAb) have been developed in an effort to investigate the antigenic and evolutionary relationships of these aquatic birnaviruses in greater depth and to provide more specific and standardized reference antisera. These MAbs have been used in neutralization and enzyme-linked immunosorbent assays (ELISA) to demonstrate that some epitopes are apparently widely dispersed among these viruses whereas other epitopes are serotype or even strain specific (Caswell-Reno *et al.*, 1986).

This report describes the development of additional birnavirus MAbs and the use of these MAbs in an immunodot test for general birnavirus screening and specific strain identification of EEV.

MATERIALS AND METHODS

Virus propagation and purification

EEV, Ab, and Sp birnaviruses were propagated in CHSE-214 cell cultures (Lannan *et al.*, 1984) at 20°C in Eagle's Minimum Essential Medium (MEM) supplemented with 10% fetal bovine serum as described previously (Nicholson *et al.*, 1979). Viruses were concentrated by polyethylene glycol (PEG) precipitation, extracted with freon and purified by isopycnic CsCl gradient centrifugation according to the method of Macdonald and Yamamoto (1977).

Hybridoma production

Eight week old BALB/c mice were immunized with an initial injection of purified EEV (5 x 10^8 tissue culture infectious doses - 50%, $TCID_{50}$) in complete Freunds adjuvant followed one month later by a second intraperitoneal injection (10^9 $TCID_{50}$) without adjuvant and an intravenous injection (10^9 $TCID_{50}$). Four days later, mice with high antibody titers as measured by ELISA (Caswell-Reno *et al.*, 1986) were used as a source of spleen cells.

The fusion protocol followed that of Caswell-Reno *et al.* (1986). Briefly, spleen cells were fused with Sp2/0-Ag 14 mouse myeloma cells at a ratio of 10:1 with PEG 1600 in serum free growth medium (SF-GM) consisting of Dulbecco's modified Eagle medium with D-glucose (4.5 mg/ml), NCTC 109 (10%), bovine pancreatic insulin (0.2 µg/ml), oxalacetic acid (1 mM) and pyruvic acid (0.45 mM). Fused cells were resuspended at a concentration of 2.5 x 10^5 /ml in GM supplemented with 15% fetal bovine serum, hypoxanthine (0.1 mM), aminopterin (4 x 10^{-4} mM) and thymidine (1.6 x 10^{-2} mM) (GM-HAT). The cell suspension was then dispensed in 0.1 ml aliquots in each well of 96 well microtitration plates and incubated at 37° C in a 10% CO_2 atmosphere. The cultures were maintained in GM-HAT for approximately three weeks and then adapted to HAT-free GM. Culture fluids were tested periodically in an ELISA (Caswell-Reno *et al.*, 1986). Hybridomas producing specific antibody were expanded, cloned at least two times, and frozen in liquid nitrogen. The monoclonal antibodies (MAb) used in this investigation were obtained from supernatants of these hybridoma cultures.

Determination of immunoglobulin isotype

MAb isotypes were determined in an ELISA using heavy and light chain class and subclass specific reagents (Boehringer Mannheim Biochemicals, Indianapolis, Indiana).

Immunodot test

Supernatent fluids (2 µl) from virus-infected CHSE-214 cell cultures ($\geq 10^7$ $TCID_{50}$/ml) were "spotted" onto nitrocellulose paper strips (Bio-Rad Laboratories, Richmond, CA), air dried, washed four times with tris buffered saline (TBS), and non-specific reactive sites blocked by soaking in 3% bovine serum albumin (BSA) and 10% calf-serum (CS) for 1 h. Individual strips were then incubated for 3 h with each MAb, washed with TBS, blocked for 15 min with 3% BSA - 10% CS, and incubated for 2 h with goat anti-mouse IgG antibody conjugated with horse radish peroxidase (Hyclone Laboratories, Inc., Logan, Utah). After washing, the strips were incubated for 15 min in the substrate (0.5 mg 4-chloro-naphthol and 10 µl of 3% H_2O_2/ml TBS). A positive reaction consisted of a dark spot where the virus had been applied to the paper.

RESULTS

Production of hybridomas

Approximately 68% of the wells of microtitration plates seeded with PEG fused cells yielded hybridomas of which approximately 7% produced sustained levels of EEV specific antibodies detected by ELISA screening. Ultimately, six hybridomas were selected for further characterization and cloned at least twice. Three of these MAbs, designated E2, E3 and E5, were used in this investigation. The isotypes of these MAbs were determined to be: E2 = IgG_{2a}, E3 = IgG_{2b}, and E5 = IgG_1. All MAbs contained k light chains.

Reactivity in immunodot assay of EEV MAbs with selected aquatic birnaviruses

The ability of these three EEV MAbs to react in an immunodot assay with EEV and two other aquatic birnaviruses was evaluated. The Ab virus isolate was chosen because it is the type strain of the Ab serotype, which includes EEV. The Sp virus was also used as a representative of the most common serotype in Europe. The results are shown in Figure 1 and Table 1. In all cases the source of virus was culture fluids from infected CHSE-214 cells. MAb E5 reacted with all three birnaviruses used here; indeed, preliminary evidence (unpublished data) indicates that MAb E5 recognizes an epitope shared by all aquatic birnaviruses. MAb E2 reacted only with the two Ab serotype viruses, Ab and EEV. MAb E3 was strain specific and reacted only with EEV.

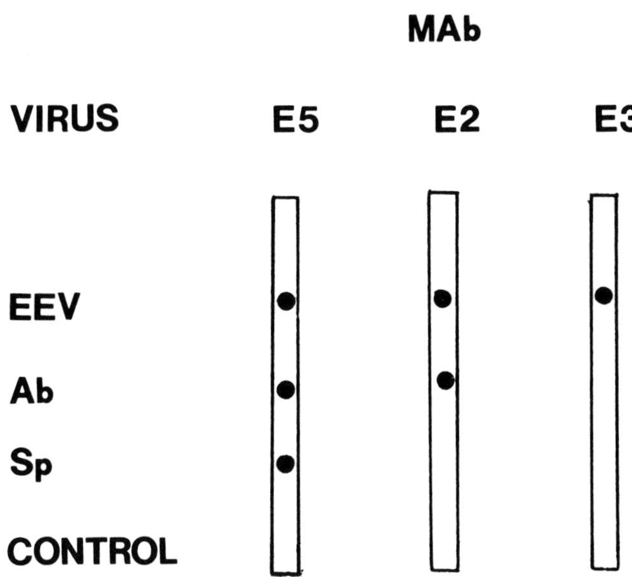

Figure 1. Diagram of reactivity pattern in immunodot test with monoclonal antibodies prepared to EEV. Reactions of three different antibodies (E5, E2, E3) with the homologous virus strain (EEV), the homologous serotype (Ab), and a heterologous serotype (Sp).

Table 1. Reactivity in Immunodot Assay of EEV MAbs with Selected Aquatic Birnaviruses.

	MAb		
Virus	E5	E2	E3
EEV	+	+	+
Ab	+	+	-
Sp	+	-	-

The sensitivity of the assay was determined by evaluating the ability of various MAb preparations to give a positive reaction with different concentrations of birnaviruses. In all cases, virus preparations from infected cell cultures exhibiting extensive cytopathic effect (CPE) and a virus titer of $\geq 10^4$ $TCID_{50}/\mu l$ gave clearly visible positive results.

DISCUSSION

Recently, a simplified immunoblot assay employing polyclonal rabbit antisera was reported for identification of several salmonid fish viruses, including the birnavirus IPNV (McAllister and Schill, 1986). The results presented here demonstrate the usefulness of a simpler immunodot test utilizing monoclonal antibodies (MAb) for the identification of aquatic birnaviruses. As with standard immunodot tests, this assay is relatively rapid, simple, inexpensive and does not require sophisticated instrumentation; also, virus identification is possible using unpurified and unconcentrated virus taken directly from supernatent fluids of infected cell cultures used for virus isolation. However, the use of a panel of MAbs as described here provides several advantages over polyclonal antisera including standardization of antibodies and the ability to identify specific virus serotypes and strains. MAb E5 reacts with all nine major birnavirus type strains (unpublished data) and, therefore, can be used for general virus screening to identify viral isolates as members of the aquatic *Birnaviridae*. MAbs E2 and E3 can be used to specifically identify an isolate as serotype Ab and strain EEV, respectively. This immunodot assay using MAbs should find widespread application in the diagnosis of viral diseases of fish, particularly in Asia where eel aquaculture is a major industry. Also, these MAbs provide additional specific probes for studying the antigenic structure and evolutionary relationships of the aquatic birnaviruses in general.

ACKNOWLEDGEMENTS

This research was supported by research grants from the National Science Foundation (INT-810447), the National Oceanic and Atmospheric Administration (NOAA) University of Maine Sea Grant College Program, and the Maine Agriculture Experiment Station.

REFERENCES

Caswell-Reno, P., P.W. Reno and B.L. Nicholson. 1986. Monoclonal antibodies to infectious pancreatic necrosis virus: analysis of viral epitopes and comparison of different isolates. J. Gen. Virol. 67: 2193-2205

Dobos, P., B.J. Hill, R. Hallet, D.T.C. Kells, H. Bect and D. Teninges. 1979. Biophysical and biochemical characterization of five animal viruses with bisegmented double stranded RNA genomes. Journal of Virology. 32: 593-605

Hill, B.J. and K. Way. 1983. Serological classification of fish and shellfish birnaviruses. Abstracts of First International Conference of the European Association of Fish Pathologists. Plymouth, England. pp. 10.

Kou, G.H., S.N. Chen, Y. Yeno, R.P. Hewdrick and J.L. Fryer. 1982. Investigation of viral infections in cultured fishes in Taiwan. Abs. ROC-Japan Symposium on Fish Diseases, Tungkang, Taiwan, ROC. 42-43.

Lannan, C.N., J.R. Winton and J.L. Fryer. 1984. Fish cell lines: establishment and characterization of nine cell lines from salmonids. In Vitro. 20: 671-676.

Macdonald, R.D. and T. Yamamoto. 1977. The structure of infectious pancreatic necrosis virus RNA. Journal of General Virology. 34: 235-247.

McAllister, P.E. and W.B. Schill. 1986. Immunoblot assay: a rapid and sensitive method for identification of salmonid fish viruses. Journal of Wildlife Diseases. 22: 468-474.

Nicholson, B.L., G.W. Thorne, C. Janicki and A. Hanson. 1979. Studies on a host range variant from different isolates of infectious pancreatic necrosis virus (IPNV). Journal of Fish Diseases. 2: 367-379.

Sano, T., N. Okamoto and T. Nishimura. 1981. A new viral epizootic of *Anguilla japonica* Temminck and Schlegel. Journal of Fish Diseases. 4: 127-139.

MONOCLONAL ANTIBODIES TO ATLANTIC SALMON (*SALMO SALAR*) IMMUNOGLOBULIN: PRODUCTION AND PRELIMINARY CHARACTERIZATION

C.E. Moody, P.W. Reno and A.E. Gagliardi
Department of Microbiology
University of Maine
Orono, Maine 04469

INTRODUCTION

The predominant humoral immunoglobulin (Ig) of teleosts is a high molecular weight tetrameric IgM-like molecular (Cobel, 1975). Limited studies with certain species suggest that some molecular weight (MW) size heterogeneity may be present. One report suggests that rainbow trout (*Salmo gairdneri*) may have two species of high molecular weight Ig (Hodgkins *et al.*, 1967) and several reports support the presence of size heterogeneity in goldfish immunoglobulin (Trumps, 1970; Everhart, 1972). Moreover, some species of fish have been reported to possess a low molecular weight humoral Ig which may be an IgM-like monomer (Clem and McClean, 1975). Despite reports that fish have lymphocyte populations analogous to the T and B cells of homeothermic vertebrates (Etlinger *et al.*, 1976), other reports have demonstrated abundant surface Ig on most, if not all, lymphocytes of fish by membrane immunofluorescence using polyclonal antisera to fish Ig (Warr *et al.*, 1979). As perhaps the most primitive vertebrate to possess an adaptive immune system, it is of some phylogenetic interest to determine the complexity of cell types and humoral components involved in fish immune responses. Furthermore, from a practical standpoint successful immunoprophylaxis of hatchery-reared populations is ultimately dependent on an understanding of these components and their role in protection against infectious disease.

The method of lymphocyte fusion described in 1975 by Köhler and Milstein provides a method to overcome the limitation of polyclonal antisera. Mouse myeloma tumor cells are fused to spleen cells derived from a mouse previously immunized with an antigen. A portion of the hybrid cells combine the desired parental traits: continual growth in vitro and production of antibody specific to the immunizing antigen. This technique offers several advantages: 1) each hybrid clone produces only one antibody, 2) an unlimited supply of antibody is available, 3) impure antigens lead to pure antibody reagents, and 4) all specificities can be rescued. Since its introduction this technology of monoclonal antibody (MAb) production has been widely used in studies of cell surface markers, for antigen purification, and in disease diagnosis.

Studies using such reagents to catfish (Lobb and Clem, 1984) and carp (Secombes *et al.*, 1983) IgM demonstrate that only a subpopulation of peripheral lymphocytes express membrane-bound IgM, supporting a dichotomy between T and B cells. Others have reported a heterogeneity of light chains in catfish using monoclonal antibodies, possibly corresponding to kappa and lambda light chains of higher vertebrates (Lobb *et al.*, 1984). Since it seems reasonable to conclude that monoclonal antibodies will provide more definitive information with regard to the teleost immune system, we chose to produce such antibodies to the Ig of a commercially important salmonid species, Atlantic salmon (*Salmo salar*).

MATERIALS AND METHODS

Preparation of Atlantic salmon Ig

Normal Atlantic salmon serum was precipitated three times with 50% saturated ammonium sulfate (SAS), pH 7.6. The precipitate was collected by centrifugation at 1400 x g for 10 min and the final pellet was resuspended in one half the original volume in phosphate buffered saline (PBS).

Bio Gel A-5m (Bio-Rad Laboratories, Richmond, CA), exclusion limit 5000 kilodalton (Kd), equilibrated in PBS, pH 7.5, was packed in a glass column 41 x 2.5 cm. The column was maintained at 4°C and an ascending flow rate of 20 ml/h was achieved using a peristaltic pump. Protein samples were applied to the column in a volume equivalent to 1% of the total gel bed volume. An additional 2 ml of 10% sucrose in PBS was added immediately after each sample to aid in sharper sample application. The column was calibrated using purified aldolase, catalase, ferritin and thyroglublin (Pharmacia Fine Chemicals, Piscataway, NJ). Both normal whole Atlantic salmon serum and SAS precipitated serum were applied to the column and collected in 5 ml fractions. The first eluted peak from the column was presumed to contain mostly AsIg and was pooled and concentrated using an Amicon model 52 stirred cell concentrator under 5-10 psi nitrogen pressure. Immunoelectrophoresis (IEP) and sodium dodecyl sulfate-polyacrylamide gel electrophoresis (SDS-PAGE) analysis of the fractions generated by Bio Gel A-5 m gel filtration demonstrated that this column was not producing purified AsIg. Therefore, SAS precipitated Bio Gel A-5 m fractions were applied to a Sephadex G-200 column. Eluted material from the void volume was found to contain purified proteins by IEP and SDS-PAGE.

SDS-PAGE Analysis of AsIg.

The technique was the method of Laemmli (1970). Samples of salmon serum and fractions of purified AsIg were diluted 1:3 and boiled for 5 min. Five µl of standards and 20-30 µl of salmon serum or AsIg samples were applied to each lane. Gels were run at 250-300 volts and were visualized using a silver staining kit (Bio-Rad).

Production of Hybridomas

Five-week-old female BALB/c mice were given five injections over a two month period of 0.15-0.25 mg Atlantic salmon Ig that had been partially purified (ppAsIg) by sieve gel chromatography on the A-5m column. The first four injections were administered intra-peritoneally with adjuvant (the first with complete Freund's, the next three with incomplete Freund's adjuvant), followed by a final intravenous injection without adjuvant.

The fusion protocol was a modification of that of Kwan, Yelton and Scharff (1980). The spleen was removed, cleaned of blood and connective tissue and transferred to sterile, 60 mm plastic dishes containing 2 ml of serum free growth medium (SF-GM) consisting of Dulbecco's modified MEM (DMEM) with D-glucose (4.5 mg/ml), gentamicin (50 µg/ml), fungizone (2.5 µg/ml), NCTC 109 (10%), bovine pancreatic insulin (0.2 µ/ml), ocxalacetic acid (1 mM) and pyruvic acid (0.45 mM). Spleen tissues were forced through a 100 mesh tissue screen. The spleen cell suspension was placed in a sterile centrifuge tube and clumps of tissue were allowed to settle for 5 min. For fusion, spleen leukocytes were mixed with washed SP2/0-Ag 14 mouse myelomas cells at a ratio of 10:1 in a 50 ml centrifuge tube. The mixture was centrifuged for 10 min at 225 X g at room temperature, the supernatant removed, and 1 ml of 40% (v/v) polyethylene glycol, MW 4000, in SF-GM was added slowly to the cell pellet at 37° C. The cells were resuspended and centrifuged for 3 min at 225 x g; the cell pellet then was allowed exposure to PEG for a total of seven min at 30° C. Ten ml of SF-GM at room temperature were added slowly to this cell pellet with gentle agitation. The cells were centrifuged and resuspended to a concentration of 5 x 10^5 cells/ml in GM supplemented with 15% FBS, hypoxanthine (0.1 mM), aminopterin (4 x 10^{-4} mM) and thymidine (1.6 x 10^{-2} mM)(HAT medium). Cells were seeded in 96 well microtiter plates at densities of 5 x 10^3 to 5 x 10^4 myeloma cells per well. The cultures were replenished at 3-4 day intervals with 50 µl of fresh HAT medium.

Enzyme Linked Immunosorbent Assay (ELISA)

An indirect ELISA was used to screen for MAb production. Immulon II microtitration plates (Dynatech Laboratories, Alexandria, VA) were coated with 0.1 µg ppAsIg protein/well in 0.05 M carbonate-bicarbonate buffer, pH 9.6, overnight at 4°C in a humid chamber. Excess Ag was removed with washing buffer comprised of PBS plus 0.05% tween 20 (PBST). Hybridoma supernatant (0.2 ml) was added to each well and the plate was incubated for 2 h at 4°C. Supernatant collected from SP2/0 cultures and mouse anti-whole Atlantic salmon serum were used as negative and positive controls. The plate was washed again with PBST and 0.2 ml of a 1:1500 dilution of alkaline phosphatase conjugated goat anti-mouse Ig (Cooper Biomedical, Inc., Malvern, PA) was added to each well and incubated for 2 h at 4°C. After a final wash with PBST, 0.2 ml of 1 mg/ml p-nitrophenyl phosphate disodium (Sigma Chemical Co., St. Louis, MO) dissolved in 10% diethanolamine buffer pH 9.8 was added to all wells and incubated for 45-60 min at room temp. The reaction was terminated by adding 0.05 ml 3 M NaOH. The absorbance of each well was measured at 410 mm. Wells with readings greater than or equal to twice the negative control were considered positive for Ab specific for ppAsIg. Positive wells were then cloned twice by the method of Oi and Herzenberg (1980).

Determination of immunoglobulin isotype

Monoclonal antibody isotypes were determined in an ELISA using heavy and light chain class and subclass specific reagents (Zymed Laboratories, Inc., South San Francisco, CA, and Boehringer, Mannheim, Indianapolis, IN).

RESULTS

Purification of AsIg

AsIg was presumed to be in the first peak of the Bio Gel A-5m column. This peak had a MW of 890 Kd. On testing in IEP against

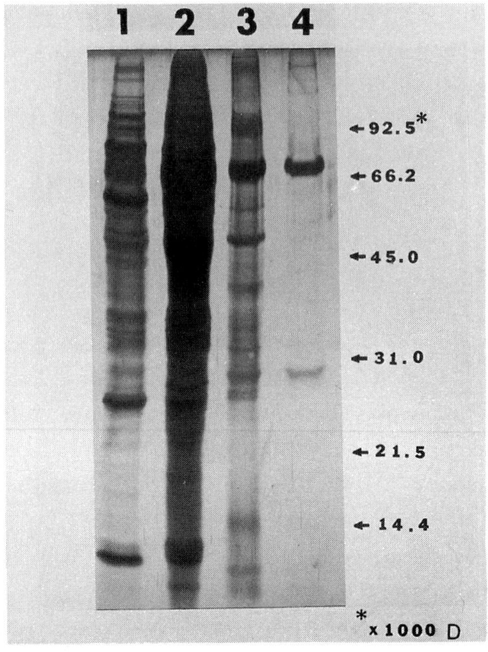

Figure 1. SDS-PAGE of Atlantic salmon serum at each stage of purification. Lane 1: one to twenty dilution of whole serum. Lane 2: Ammonium sulfate precipitated serum. Lane 3: partially purified serum after A-5m column chromatography. Lane 4: Atlantic salmon Ig after additional Sephadex G-200 column chromatography. MWs and positions of protein standards phosphorylase B, bovine serum albumin, ovalbumin, carbonic anhydrase, soybean trypsin inhibitor and lysozyme are indicated in order of descending MW.

*NA = not available

polyclonal mouse anti-Atlantic salmon sera, one line of precipitation was observed. Protein from this fraction was used to immunize BALB/c mice. However, later SDS-page analysis of this preparation revealed the presence of numerous contaminating proteins (Figure 1, lane 3). However, after passage through Sephadex G-200 column, SDS-PAGE analysis revealed two proteins of 72 and 31 Kd, presumably purified heavy and light chains of AsIg. This purified preparation was utilized in subsequent ELISA's testing for MAb production.

Hybridomas and MAb production

Of 984 total wells seeded at various concentrations, 61.1% produced hybridomas. However, only 1% were found to produce MAb reactive with purified AsIg in an indirect ELISA after repeated cloning. Two of these, designated SS-1 and SS-2 were chosen for these initial studies. Both were found to be of the IgM class with a kappa light chain isotype (Table 1).

Table 1. Immunoglobulin Class and Subclass of SS-1 and SS-2

MAb	IgA	IgG1	IgG2a	IgG2b	IgG3	IgM	k	l
				(P/N Ratio)*				
SS-1	0.46	0.74	0.45	0.40	0.41	3.42	2.41	0.52
SS-2	0.47	0.81	0.51	0.54	0.44	5.83	4.16	0.56

*Indirect ELISA. P/N ration = OD of MAb tested/OD of negative control.
P/N $\geq$ 2.0 is considered positive.

Reactivity of MAbs SS-1 and SS-2 with vertebrate serum.

Dilutions of whole Atlantic salmon serum ranging in estimated protein concentrations of 0.05 to 25 µg protein/well were tested by indirect ELISA using SS-1. The dose response curve obtained from this experiment generated positive results with P/N ratios all exceeding 2.0 for all salmon serum concentrations tested. The dilution of 5 µg/well protein/well was selected as an optimal working value and was used for all subsequent cross reactivity studies with vertebrate sera. The results evaluating reactivity of SS-1 and SS-2 with different Atlantic salmon sera (Table 2) detected no significant difference among the samples tested. Consequently, a working dilution of 5 µg protein/well was used as the concentration to test other invertebrate sera for reactivity with both MAbs.

Sera were tested from seven different poikilothermic species and from three homeothermic species. The data in Table 3 indicate the ability of MAb from both clones to recognize a common determinant in serum of many of the species tested. No significant difference in reactivity of SS-1 and SS-2 with sera tested was apparent. It should be noted that P/N ratios represent an average of three separate

Table 2. Elisa reactions of SS-1 and SS-2 tested against whole Atlantic salmon serum from different sources

Atlantic salmon[1]	SS-1	SS-2
	(P/N ratio)[6]	
grilse (1)[2]	2.39	2.38
grilse (2)[2]	2.63	2.66
salt water[3]	2.89	2.63
fresh water[4]	3.41	2.72
OPI[5]	3.42	2.67

[1] Serum samples tested at 5 µg protein/well.

[2] Serum collected from two individual, hatchery reared, stressed grilse maintained in fresh water.

[3] Pooled serum from several 2 year old salmon that had been acclimated to living in salt water.

[4] Pooled serum from several 2 year old salmon that were maintained in fresh water.

[5] Pooled serum from healthy Atlantic salmon maintained in salt water at Ocean Products Incorporated, Eastport, ME.

[6] P/N ratio = OD of MAb tested by indirect ELISA/OD of negative control. Each value is an average of three trials. P/N $\geq$ 2.0 is considered positive.

Table 3. Cross-reactivity of SS-1 and SS-2 with serum from other vertebrates.

	SS-1	SS-2
Species	P/N ratio*	P/N ratio
Atlantic salmon	2.39	2.39
Goldfish	4.69	3.87
Brook trout	3.67	3.09
Flounder	3.12	2.87
Pollack	3.01	3.36
Sculpin	2.64	2.72
Hagfish	1.74	1.70
Lamprey	1.64	1.88
Horse	1.62	2.04
Guinea pig	1.58	1.85
Bovine (fetal)	1.05	1.15

*Indirect ELISA. P/N ratio = OD of vertebrate serum tested/OD of negative control. Each value represents the average of three trials. P/N ratio $\geq$ 2.0 is considered positive. Serum samples were plated at 5 μg protein/well.

trials and each serum tested was collected from only one individual of each species. Nonetheless both MAbs were consistently positive with teleost serum and negative with agnathan and homeotherm serum.

DISCUSSION

The purpose of the present work was to isolate and partially characterize AsIg, in order to produce MAbs to AsIg. This would provide a potential tool for diagnosis of diseases in hatchery-reared salmonids and could be utilized as a more sophisticated method to analyze the role of Ig in other immune functions of salmon. Preliminary results characterizing AsIg and two MAbs are reported here.

The first peak eluted from the A-5m column was presumed to contain AsIg. The MW of the AsIg was estimated to be 890 Kd. Although there is no other available information regarding the MW of AsIg, this estimate is close to that previously reported for Ig of other salmonid species (Table 4). The structure for this IgM-like Ig

Table 4. Molecular weight estimations of immunoglobulin from several salmonid species.

Species	IgM4 MW (Kd)	MW H-chain and L-chain (Kd)	References
Chum salmon *Onchorhynchus keta*	730	72	Kobayachi *et al.*, 1982
Coho salmon *Onchorhynchus kisutch*	750	75 26	Cisar and Fryer, 1974
Rainbow trout *Salmo gairdneri*	610	60 NA*	Elcombe *et al.*, 1985
Brown trout *Salmo trutta*	670	NA	Ingram and Alexander, 1979
Atlantic salmon *Salmo salar*	890	72 31	Present study

has been determined in *Osteichthyes* to be a tetramer whereas in homeotherms, IgM is a pentameric molecule with a MW of approximately 900 Kd. The MW of the heavy and light chains of AsIg were determined to be approximately 72 and 31 Kd respectively, based on their relative mobilities in SDS-PAGE. The reason for the range of reported MW of various Igs is not known, but may be due to differences in glycosylation. The purity of the AsIg isolated from the A-5m column was originally assessed by means of IEP using a polyvalent mouse antiserum to Atlantic salmon serum proteins. Although the preparation revealed a single precipitin line, later analysis using SDS-PAGE revealed numerous proteins still present in this preparation (Figure 1, lane 3). It is possible that technical problems such as improper packing of this column could account for its inability to satisfactorily separate the Ig from other proteins present in SAS precipitated Atlantic salmon serum. In addition, it may be that there are a lot of high MW proteins in Atlantic salmon serum, making it more difficult to isolate only AsIg. However, it is interesting to note that when Ingram and Alexander (1979) attempted to purify Ig from the brown trout using Sepharose 6B sieve gel chromatography, the Ig obtained from the first and second peaks eluted was not free of other serum proteins. When this preparation was refractionated through the same column, it eluted as a single peak which was predominantly brown trout Ig. The recycling of Ig containing fractions over Sephadex G-200, adopted from Ingram and Alexander (1979), proved to be an effective way of purifying AsIg.

Hybridomas were constructed by PEG-mediated cell fusion of SP2/OAg 14 myeloma cells and splenocytes of BALB/c mice immunized with partially purified AsIg. The secretion of mouse anti-AsIg Ab by hybrid cell cultures was tested by absorption of the Ab to AsIg Ab in an indirect ELISA. Of 601 populations of hybridomas which grew, six hybrid cell cultures producing MAb directed towards AsIg were isolated. Two of these hybridomas were selected for further analysis including characterization of their class and subclass, binding properties to heavy and/or light chains of AsIg, specificity for other vertebrate Ig and their binding affinity to Atlantic salmon peripheral blood cells. Both of these hybridomas designated SS-1 and SS-2, were found to produce mu isotype heavy chains and kappa light chains which reacted specifically with AsIg in an indirect ELISA.

The finding that MAbs SS-1 and SS-2 were able to detect Ig in all the Atlantic salmon tested (refer to Table 3), indicates they can be a useful tool for monitoring Ab levels in this species whether it be to monitor the effectiveness of different routes of immunization or simply as a method to detect the presence of specific Ig to a pathogenic microorganism. It is important to reiterate that the serum tested, represent a sampling of only a limited number of fish.

It would be useful to screen a larger number of different age groups of Atlantic salmon. The results obtained by examining the specificity of MAbs SS-1 and SS-2 for other vertebrate serum proteins are somewhat surprising because they appear to be directed against a determinant common in other poikilotherms. It is of note that the goldfish and brook trout serum produced the strongest reactions in the ELISA. This information further suggests the MAbs are potentially reacting with these vertebrates Igs because both these species had been hyperimmunized with SRBC's and *Vibrio anguillarum* respectively. There exist at least two possible explanations for the somewhat broad specificity exhibited by MAbs SS-1 and SS-2 for different vertebrate sera. It has been reported that Vh_a-related determinants are shared by Igs of a wide range of vertebrates from sharks to man and thus seem to be the epitopes which have been conserved during vertebrate evolution. Results showed that an allotypic marker found on the V1-domain of rabbit Ig and shared by man (Knight *et al.*, 1975) is also expressed by at least a subset of Ig Vh-domains from evolutionarily distinct vertebrate species representing primitive classes. This distribution demonstrates a strong phylogenetic conservation of Vh-regions and may explain why strong inter-species cross reactions are often observed. Numerous studies have reported cross reactivities of anti-Ig with Ig of different species. For example, Eshhar *et al.* (1983) have also found that MAb to purified Vh cross reacts with heavy chains from a wide range of vertebrate species.

In view of these findings, it is therefore possible the MAbs SS-1 ad SS-2 are directed against the Vh-regions or some other highly conserved region of AsIg, which would account for their reactivity with a common determinant present in these diverse vertebrate species. An additional explanation for cross reactivity of MAbs SS-1 and SS-2 with sera from different species may be that these MAbs could be directed towards carbohydrate moiteties. Yamaga *et al.* (1978) determined that some their anti-rainbow trout Ig Abs were directed against carbohydrate determinants in trout Ig. Whatever the basis for the cross-reactivity we have observed, it appears to be limited to other species of teleosts.

Preliminary efforts to stain Atlantic salmon peripheral blood cells with SS-1 and SS-2 ascitic fluid by indirect immunofluorescence using FITC conjugated anti-mouse IgMs indicate that a subpopulation of the cells present contained surface determinants recognized by these MAbs. Further analysis of the cell populations and the characterization and production of additional MAbs are presently in progress.

ACKNOWLEDGEMENTS

This work was supported by research awards from the National Oceanic and Atmospheric Administration (NOAA) University of Maine Sea Grant Program and the Maine Agriculture Experiment Station (Publication No. 1157).

REFERENCES

Cisar, J.O. and J.L. Fryer. 1974. Characterization of anti *Aeromonas salmonicida* antibodies from coho salmon. Infec. Immun. 9: 236.

Clem, L.W. and W.E. McClean. 1975. Phylogeny of immunoglobulin structure and function: VII. Monomeric and tetrameric immunoglobulin of the margate, a marine teleost fish. Immunol. 29: 791.

Cobel, M.J. 1975. The immune responses in fish: a review. J. Fish. Biol. 7: 539.

Elcombe, B.M., C.J. Chang, C.J. Taves and J.L. Winkelhake. 1985. Evolution of antibody structure and effector functions: Comparative hemolytic activities of monomeric and tetrameric IgM from rainbow trout, *Salmo gairdnerii*. Comp. Biochem. Physiol. 80B: 697.

Eshhar, Z., O. Gigi and Y. Ben-Neriah. 1983. Monoclonal anti-vh_a antibodies recognize a common Vh determinant expressed on immunoglobulin heavy chains from various species. Eur. J. Immunol. 13: 533.

Etlinger, H.M., H.O. Hodgkins and J.M. Chiller. 1976. Evolution of the lymphoid system. II. Evidence for lymphoid heterogeneity in rainbow trout revealed by the organ distribution of mitogenic responses. J. Immunol. 116: 1547.

Everhart, D.L. 1972. Antibody response of goldfish to a protein immunogen and to haptenic determinants. Immunol. 29: 791.

Hodgkins, H.O., R.S. Weiser and C.S. Ridgway. 1967. The nature of antibodies and the immune response in rainbow trout (*Salmo gairdneri*). J. Immunol. 99: 534.

Ingram, G.A. and J.B. Alexander. 1979. The immunoglobulin of the brown trout, *Salmo trutta* and its concentration in the serum of antigen-stimulated and non-stimulated fish. J. Fish. Biol. 14: 249.

Knight, K.L., T.R. Malek and S. Dray. 1975. Human immunoglobulins with a 1 allotypic determinants of rabbit immunoglobulin heavy chains. Nature 253: 216.

Kobayashi, K., A. Hara, K. Takano and H. Hirai. 1982. Studies on subunit components of immunoglobulin M from a bony fish, the chum salmon (*Oncorhynchus keta*). Mol. Immunol. 19: 95.

Köhler, G. and C. Milstein. 1975. Continuous cultures of fused cells secreting antibody of predefined specificity. Nature 256: 495.

Kwan, S.P., D.E. Yelton and M.D. Scharff. 1980. Production of monoclonal antibodies. IN: Genetic Engineering. J.K. Setlow and A. Hollaender (eds.). Plenum Publishing Corp. NY. 2: 31.

Laemmli, U.K. 1970. Cleavage of structural proteins during the assembly of bacteriophage T_4. Nature 227: 680.

Lobb, C.J. and L.W. Clem. 1984. Fish lymphocytes differ in the expression of surface immunoglobulin. Dev. Comp. Immunol. 6: 473.

Lobb, C.J., M.D. Olson and L.W. Clem. 1984. Immunoglobulin light chain classes in a teleost fish. J. Immunol. 132: 1917.

Oi, V.T. and L.A. Herzenberg. 1980. Immunoglobulin producing hybrid cell lines. IN: Selected Methods in Cellular Immunology. B.B. Mishell and S.M. Shiigi (eds.). W.H. Freeman and Co., San Franscisco, CA. 351 pp.

Secombes, C.J., J.J. Van Gronigen and E. Egberts. 1983. Separation of lymphocyte subpopulations in carp *Cyprinus carpio* L. by monoclonal antibodies: immunohistochemical studies. Immunol. 48: 165.

Trumps, G.N. 1970. Goldfish immunoglobulins and antibodies to bovine serum albumin. J. Immunol. 104: 1267.

Warr, G.W., D. DeLuca and B.R. Griffin. 1979. Membrane immunoglobulin is present on thymic and splenic lymphocytes of the trout, *Salmo gairdneri*. J. Immunol. 123: 910.

Yamaga, K.M., R.T. Kubo and H.M. Etlinger. 1978. Studies on the question of conventional immunoglobulin on thymocytes from primitive vertebrates I. Presence of anti-carbohydrate antibodies in rabbit anti-trout Ig sera. J. Immunol. 120: 2068.

IMMUNOCHEMICAL EVALUATION OF THE INFECTIOUS PANCREATIC NECROSIS (IPN) CARRIER STATE IN SALMONIDS AND OTHER SPECIES

Paul W. Reno
Department of Microbiology and
Migratory Fish Research Institute
University of Maine
Orono, Maine 04469

INTRODUCTION

One of the most serious and pervasive diseases of cultured salmonids is infectious pancreatic necrosis (IPN) which is caused by an unusual virus of the same name (IPNV). The disease is especially prevalent on the West coast of the U.S. and the Canadian Maritime Provinces, but is also found in Asia, the British Isles and Europe. More than fifty species of fish and shellfish have been determined to harbor the virus (Dorson, 1983), although overt disease is not a necessary consequence of infection. When disease does occur, however, it is lethal to a large proportion of the affected population; this has been documented in many species including Atlantic salmon, *Salmo salar;* Pacific salmon, *Oncorhyncus* sp.; striped bass, *Morone saxitilis;* menhaden, *Brevoortia tyrranus;* rainbow trout, *Salmo gairdneri;* brook trout, *Salvelinis fontinalis;* and eels, *Anguilla* sp. Isolation of the virus from asymptomatic animals has been accomplished with at least twenty other aquatic animals, including freshwaster, estuarine, and marine vertebrates and invertebrates. Among the more intriguing species infected are shrimp, hard clam, American and European oyster, blue crab, minnows, milkfish and tilapia (Dorson,1983).

One of the factors which contributes substantially to the pervasive nature of this disease is the fact that species that are susceptible to IPN disease and those which are not, appear to be prone to exist in a dynamic balance with the virus. This phenomenon is generally alluded to as a "carrier state" and consists of an infection without visual or microscopic evidence of disease. Carriers generally arise when survivors of IPN disease episodes sustain the virus in relatively large quantities within their visceral organs (Frantsi and Savan, 1971; Reno,1976). Virus is often shed from the fish in bodily fluids, especially feces, urine, and sex products - milt and ovarian fluid - and can thus be readily transmitted to susceptibles in the proximate aqueous environment. The finding that a diverse array of marine, estuarine, and freshwater animals harbor the virus, coupled with the fact that the virus is quite stable in ocean water and freshwater fishes (Reno, 1976) and the severe epizootics which arise from infection make the elucidation of the basic characteristics of the IPN carrier state quite important. This paper reports the basic virologic and serologic characteristics of the carrier state in trout, and demonstrates the utility of immunochemical methods in defining viral carrier states in fish and potentially other aquatic animals.

MATERIALS AND METHODS

Fish

The fish used in these studies were brook trout, *Salvelinus fontinalis*, which were obtained from Cobequid Provincial Fish Hatchery of the Department of Fisheries and Oceans in Nova Scotia, Canada. Samples of feces from fish at this hatchery prior to their use in these studies indicated an IPN carrier rate of at least 20%. Young-of-the-year, one plus, and two plus fish were transported to the isolation facilities of the Ontario Veterinary College at Guelph, Ontario, where they were held for approximately a year for examination of the duration and extent of the IPN carrier state. The fish were held in 75 liter basins supplied with flowing, dechlorinated Guelph city water which had a temperature range of 6-13°C. Fish were fed pelletized trout feed (Martin Feed Mills, Elmira, Ontario) to about two percent body weight per day; photoperiod was adjusted to coincide with the external photoperiod.

Virus and cell culture methods

The virus used in this study was an isolate derived from brook trout from New Brunswick, Canada which was determined to be identical biophysically and serolgically to the isolate from Cobequid. Virus was grown at a temperature of 20°C on RTG-2 continuous cell line derived from rainbow trout gonad (Wolf and Quimby, 1962). Cells were grown and maintained in Eagle's Minimal Essential Medium supplemented with 10% fetal bovine serum in closed vessels to obviate the need for carbon dioxide atmosphere. Cells were prepared in 25 cm^2 flasks or 96 well microtitration plates.

Preparation of anti-IPNV antiserum

Virus was purified by the method of Kelly and Loh (1972) by the use of polyethylene glycol precipitation followed by freon extraction and isopyknic cesium chloride density gradient centrifugation. Approximately 10^8 infectious virus in complete Freund's Adjuvant was injected intramuscularly into three New Zealand white rabbits of approximately three kg weight. Seven weeks later, the animals were reinjected with the same amount of virus in incomplete Freund's Adjuvant and exsanguinated two weeks later. The antisera had a neutralizing titer of 10^6 against homologous virus after absorption with RTG-2 cells. Rabbit immunoglobulins were purified to homogeneity by ammonium sulfate precipitation, seiving on G-200, and DEAE chromatography. The experiment should also be done using serum taken from the same animal that was immunized but taken before immunization was begun. This would ensure that the reactions obtained are a result of immunization and not something the animal was exposed to in his earlier history.

Immunoperoxidase testing

Goat anti-rabbit Ig obtained from Gibco (Grand Island, NY) was conjugated with horseradish peroxidase type VI (Sigma, St. Louis, MO) according to the method of Nakane *et al.*, (1974) and the test was performed by the technique of Weir *et al.*, (1974). Paraffin embedded tissues were processed to remove endogenous peroxidase with acidic methanol after deparaffinizing and rehydration. Rabbit antiserum or normal serum was added for one hour at 37°C, washed with buffered saline and treated with peroxidase labelled anti-rabbit serum. After washing thoroughly, diamino-benzidine substrate (1 µg/ml in McIlvane's buffer, pH 5.0) was added and incubated for 30 minutes. After washing in buffer, the sections were dehydrated through ethanol and mounted in Permount.

Infectious center assay

Organs were aseptically removed, minced and trypsinized in 0.025% trypsin for one hour at 22°C. The dispersed cells were washed and suspended in MEM + 5% FBS. Cell counts were performed on a hemocytometer and dilutions of the cells were mixed with equal volumes of melted 1.5% agarose in growth medium and plated in a thin layer over confluent monolayers of RTG-2 cells. After five days plaques were counted and the number of infectious centers was calculated.

RESULTS

Isolation of virus from carrier fish

Of 150 fish tested by routine cell culture methods, 148 were found to harbor infectious virus in at least one of their visceral organs. Titers in the various organs varied greatly, but the organs from which virus was most frequently isolated were the kidney, spleen, and pyloric ceca. The amount of infectious ranged from $10^{1.45}$ to 10^{6} ,but averaged about 10^{3} per gram of tissue. Generally, not all organs in a given individual were found to contain detectable infectious IPN, and the serum always appeared to be free of virus, although it contained low levels of neutralizing antibody. In order to determine how many tissue cells were harboring infectious virus, infectious center assays were performed. In spite of the fact that all fish were carriers and that substantial amounts of infectious IPN was detected in the various organs, the amount of virus detected by infectious center assays was low (Table 1). While titers per gram of tissue (containing approximately 10^{9} cells) were high, the number of cells within the organs which were shown to contain virus was quite low, ranging from 8 - 6234 infectious centers per 10^{6} tissue cells. This is calculated to be from 0.001 to 0.6 % of the cells infected.

Table 1. Proportion of tissue cells from IPN carrier Brook Trout wi detectable IPN virus by the infectious center assay (PFU/1 cells) or quantal assay (TCID50/g)

	FISH #									
ORGAN	1		2		3		4		5	
	TCID*	PFU	TCID	PFU	TCID	PFU	TCID	PFU	TCID	PFU
SPLEEN	3.7	-	-	-	1.45	-	1.45	-	3.45	75
KIDNEY	4.45	-	-	-	2.45	-	1.70	-	4.2	42
P.CAECA	4.45	63	2.95	12	2.45	-	-	-	2.95	8
GONAD	5.45	6056	-	-	-	-	-	-	-	-
LIVER	2.25	68	-	1.95	-	-	-	-	2.95	14

*Expressed as $\log^{10}$ of TCID50/g tissue

Immunoperoxidase assays

In order to determine if the immunoperoxidase test would detect IPN virus in tissues, preliminary tests were attempted in cell cultures infected with virus. Results indicated that the reaction was easily detected and was specific for IPN virus. The same procedure was used with tissues from carrier brook trout in an attempt to correlate the extent of infectious virus and viral antigen in the various tissues. Figure 1 demonstrates that IPN antigen is pervasive throughout the spleen tissue, with virtually every cell indicating some degree of antigen presence. Nearly identical reactions were found in kidney (Figure 2?) and liver (data not shown) - all cells seemed to contain IPN antigen. However, gastrointestinal tract tissue, especially the intestine and intestinal portion of the pyloric caeca were shown to contain antigen only in endothelial, mucosal, and submucosal tissues but not in the muscularis layers (Figures 3 and 5). Similarly, gonadal tissue exhibited evidence of IPN antigen in the sex cells themselves, but showed no indication of antigen in the supporting fibrous tissue (Figure 4).

When the strength of the immunoperoxidase reaction was compared to the IPN virus titer in the tissues of specific trout, it was seen that the immunoperoxidase test detected antigen even in organs and fish which were determined not to harbor virus (Table 2). In fact, fish from other locales which were determined to be virus free by cell culture methods were required to provide negative controls for the assay, since even the two "noncarriers" were found to be positive.

Table 2. Comparison of IPNV titers and immunoperoxidase reactions in tissues of carrier trout.

	FISH					
ORGAN	1		2		3	
	IPNV	IPASE	IPNV	IPASE	IPNV	IPASE
SPLEEN	2.70*	+ **	4.45	++	NEG	++
KIDNEY	3.2		4.70		NEG	
renal		-		++		-
hemat.		+		++		+/-
PYLORIC CECA	4.20		6.95		NEG	
pancreas		++		++		+/-
caeca		+		++		+
LIVER	2.20	+	4.20	-	NEG	+
GONADS	NEG	-	NEG	-	NEG	++
INTESTINE	2.20		4.95		NEG	
mucosa		++		++		+
muscle		-		-		-

* log10 of TCID50/g
** - = No reaction; +/- = Slight reaction; + = Moderate reaction; ++ = Strong reaction

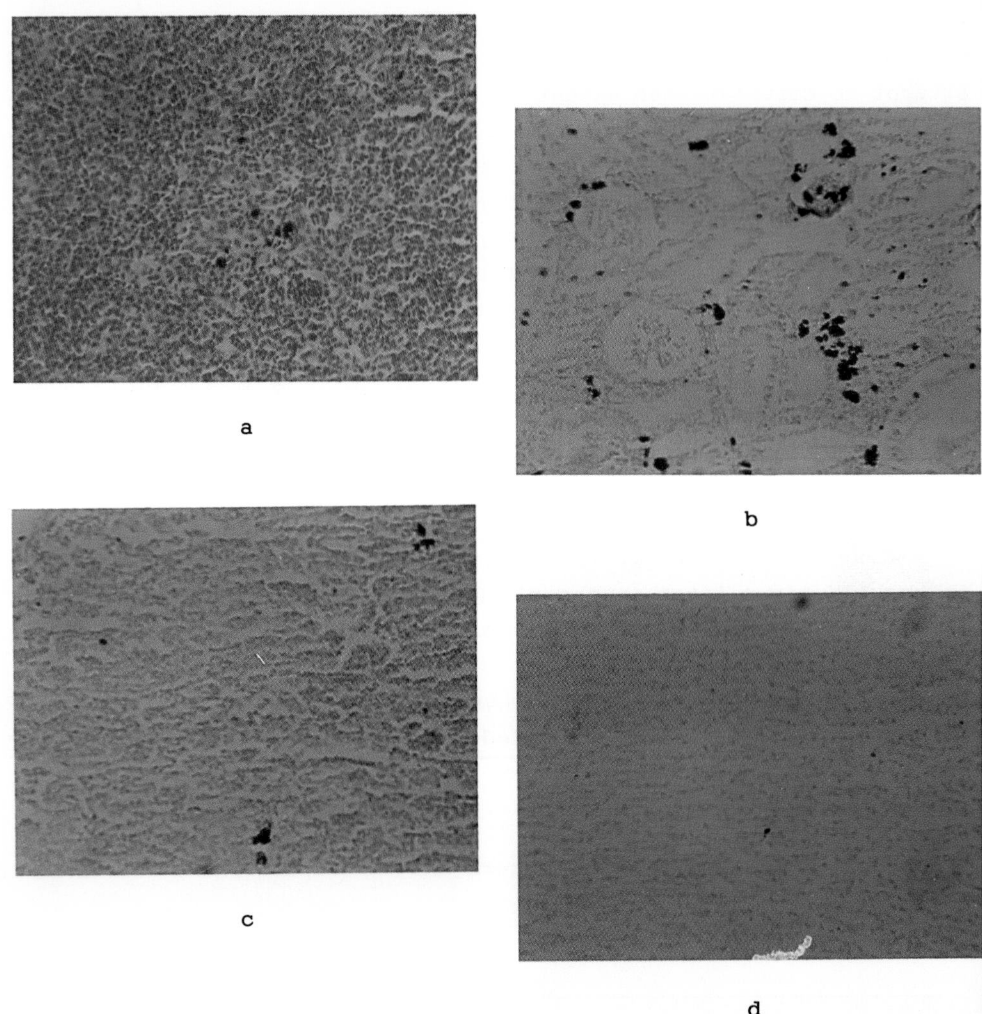

Figure 1. Immunoperoxidase test on spleen from IPN carrier brook trout. a) hemotoxylin and eosin stain. b) treated with normal rabbit serum. c) treated with anti-IPN serum.

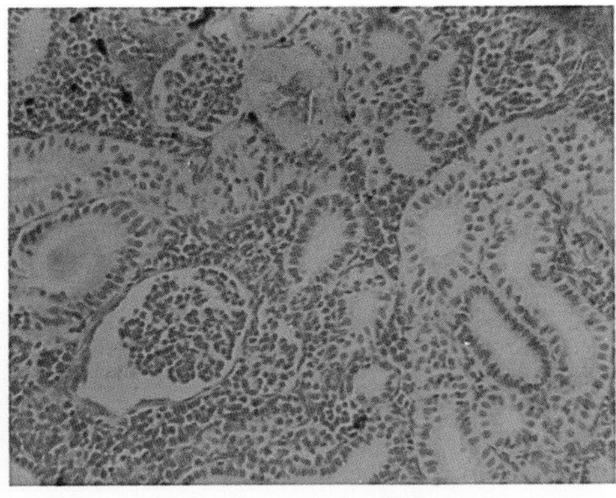

a

b

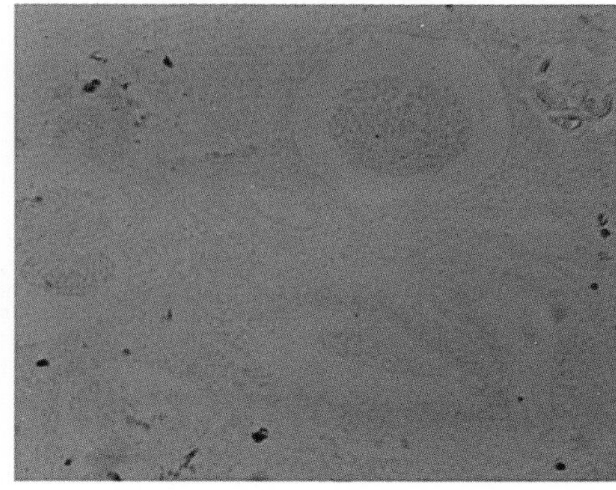

c

Figure 2. Immunoperoxidase test on kidney from IPN carrier brook trout. a) hematoxylin and eosin stain . b) treated with anti-IPN serum. c) treated with anti-IPN serum absorbed with IPN virus.

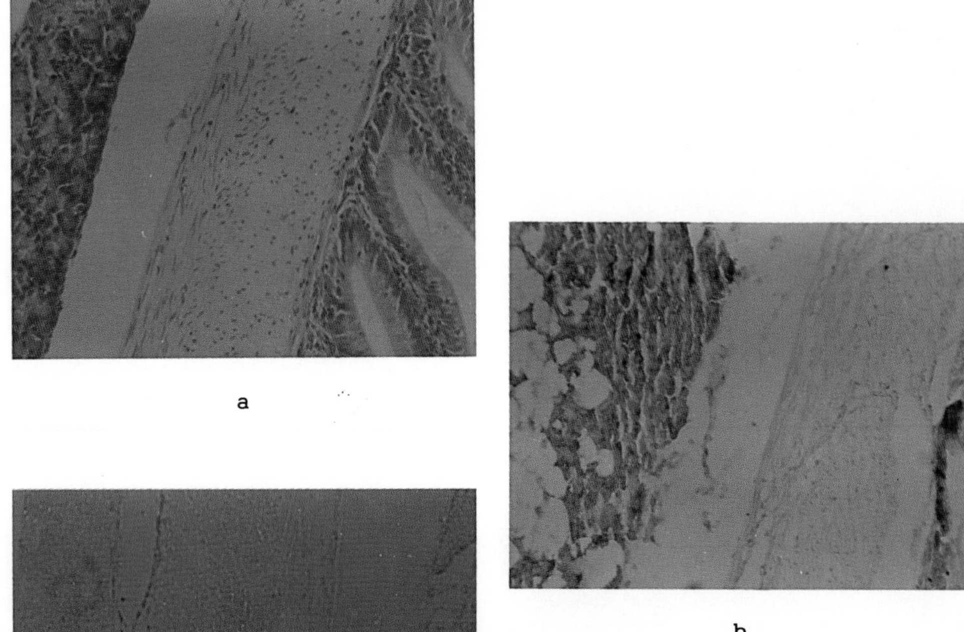

a

b

c

Figure 3. Immunoperoxidase test on pyloric caeca from IPN carrier brook trout. a) hematoxylin and eosin. b) treated with normal rabbit serum. c) treated with anti-IPN serum.

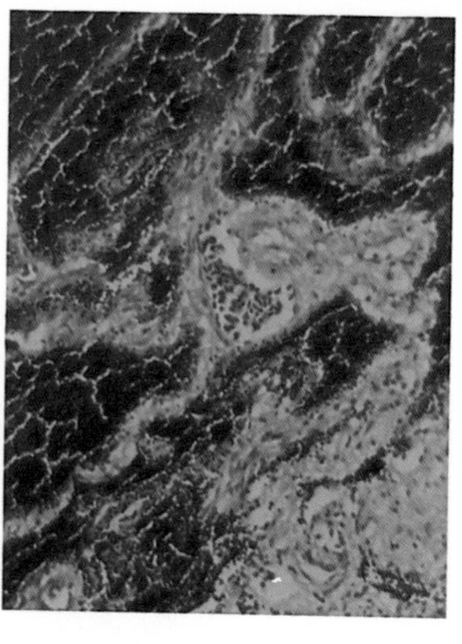

a

b

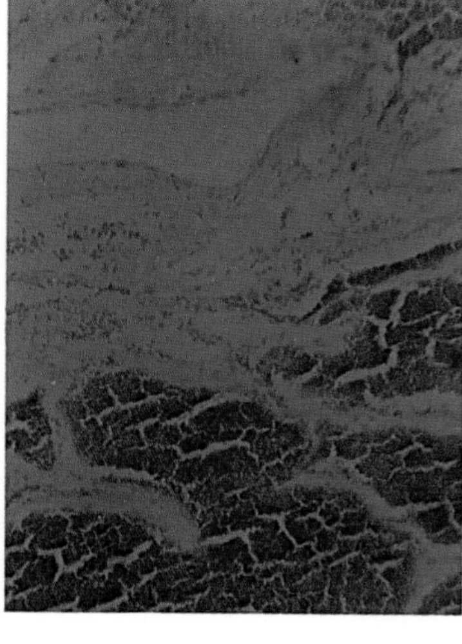

c

Figure 4. Immunoperoxidase test on gonadal (testicular) tissue from IPN carrier brook trout. a) hematoxylin and eosin stain. b) treated with normal rabbit serum. c) treated with anti-IPN serum.

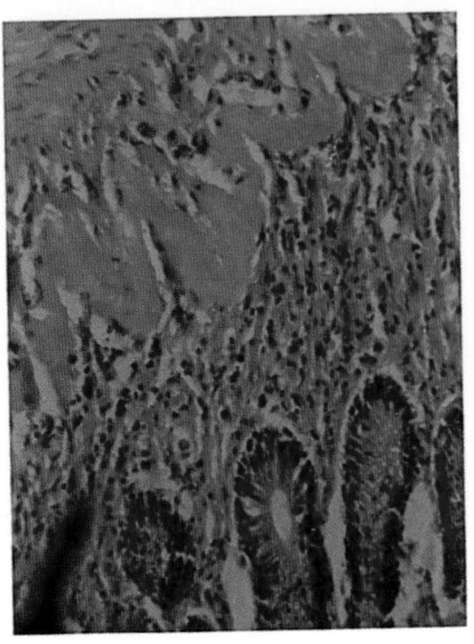

a

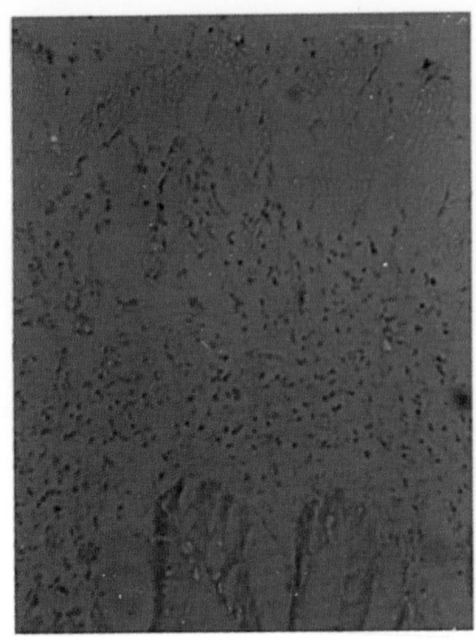

b

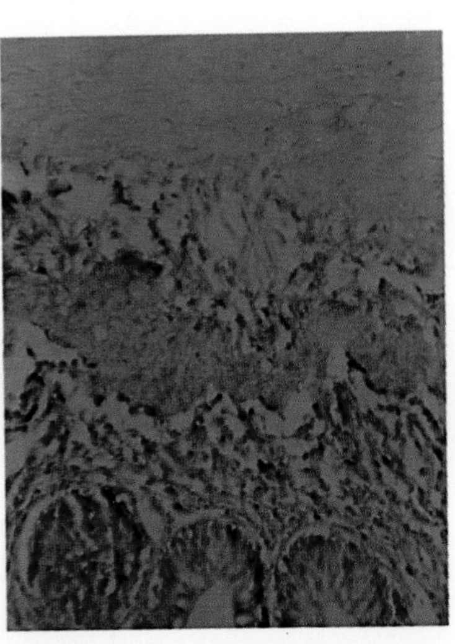

c

Figure 5. Immunoperoxidase test on intestine of IPN carrier brook trout. a) hematoxylin and eosin. b) treated with normal rabbit serum. c) treated with anti-IPN serum.

DISCUSSION

This study was designed to evaluate the IPN carrier state in terms of the distribution and quantitation of virus within the visceral tissues, and to determine whether the distribution of viral antigen correlated with the infectious virus localization. Other investigators have also detected infectious virus in virtually all of the visceral organs, but not in the serum or brain (Dorson, 1983). This distribution of virus might have been anticipated in light of the fact that the disease cycle of IPN in most salmonids is generally conceded to be defined by an initial infection either at the time of spawning, or during vitellogenesis, with disease episodes occurring several weeks after feeding commences. Survivors of epizootics, as has been demonstrated here, harbor the virus in their bodies for long periods, presumably for the duration of their lives. If there is a general viremic phase during disease,then it can be assumed that virus would be generally distributed throughout the body; the findings reported here, however, do not support this contention.

In most cases, infectious virus was detected only in three to four of the organs tested, indicating that there is a lack of dipersion or survival of virus in the organs. It is important to establish whether the virus is, in fact, not present in these organs, or is the lack of sensitivity due to the fact that the assay itself is not sufficient enough to detect low numbers of virus copies per organ. Since fish in the throes of disease are approximately 10 g in weight and contain upwards of 10^6-10^7 infectious units of virus per gram of fish, then total virus loads are in the range of 10^8 infectious units per fish. When survivors of IPN epizootics attain adulthood and weigh in the range of 100-1000 g, the mean virus content per gram is in the range of 10^3-10^4 per gram. This indicates a total virus load of about 10^8 infectious units - approximately the same level as found in the diseased fish. This would not indicate that the virus had necessarily replicated during the intervening time period, but may simply have remained quiescent. However, the immunoperoxidase data reported here indicates that the distribution of viral antigen is quite pervasive, and in fact, is detected in virtually every tissue in every organ with the exception of smooth muscle tissue and supporting fibrous tissue. This would tend to indicate that viral replication had taken place but that the bulk of protein produced was not packaged into infectious virus. The nature of the proteins produced and detected by the polyclonal antiserum used here can be more meticulously elucidated by the use of monoclonal antibodies. For instance, we could determine which of the three virus structural proteins were being produced and at what levels by employing several of the monoclonal antibodies to IPN viral proteins which we have already produced (Caswell-Reno *et al.*, 1986). Since defective interfering particles are known to be produced in vivo, the proteins detected here may be from these incomplete virions and may be responsible for the generation of the carrier state as seen in these fishes.

In addition to the study of the fundimental aspects of the IPN carrier state, the immunoperoxidase technique lends itself well to diagnostic applications. Nicholson and Henshal (1978) have demonstrated IPN virus in cell culture with the indirect immunoperoxidase technique even prior to cytopathic effects. It appears that the technique as used here is capable of detecting IPN viral antigen in fish which were found to be negative by cell culture methods (Table 2). This may be a result of small amounts of infectious virus undetectable with the present assay system, or because no infectious virus was present. However, since viral antigens may be present in the absence of infectious virus the immunoperoxidase technique can possibly evaluate whether a fish had been exposed to IPN virus early in its life, but is carrying virus at undetectable levels in adulthood. This would result in more effective capacity to regulate the spread of IPN in fish and shellfish by enabling diagnosticians to determine with high sensitivity that individuals have been exposed to IPN virus, an accomplishment not now achievable with cell culture methods.

REFERENCES

Caswell-Reno, P., P.W. Reno, and B.L. Nicholson. 1986. Monoclonal Antibodies to Infectious Pancreatic Necrosis Virus (IPNV): Analysis of Viral Epitopes and Comparison of Different Isolates. J. Gen. Virol. 67: 2193-2205.

Dorson, M. 1983. Infectious Pancreatic Necrosis. IN: Antigens of Fish Pathogens. D. Anderson, M. Dorson, and P. Dubourget, (eds.). Collection Fond. M. Merieux.

Frantsi, C. and M. Savan. 1971. Infectious Pancretic Necrosis: Comparative Frequencies of Isolation from Feces and Organs of Brook Trout, *Salvelinus fontinalis*. J. Fish. Res. Bd. Can. 26: 1064-1065.

Kelly, R. and P. Loh. 1972. Electron Microscopical and Biochemical Characterization of Infectious Pancreatic Necrosis Virus. J. Virol. 10: 824-834.

Nakane, P. and A. Kawaoi. 1974. Peroxidase-labelled Antibody, a New Method of Conjugation. J. Histo. Cyto. 22: 1084-1091.

Nicholson, B.L. and E. Henchal. 1978. Rapid Identification of Infectious Pancreatic Necrosis virus in Infected Cell Cultures by Immunoperoxidase Techniques. J. Wild. Dis. 14: 465-469.

Reno, P.W. 1976. Qualitative and Quantitative Aspects of Infectious Pancreatic Necrosis Carrier State. Ph. D. Thesis, University of Guelph, Ontario.

Weir, E.E., T. Pretlow, A. Pitts, E. Williams. 1974. A More Sensitive and Specific Histochemical Peroxidase Stain for the Localization of Cellular Antigen by the Enzyme-antibody Conjugate Method. J. Histo. Cyto. 22: 51.

Wolf, K. and M.C. Quimby. 1962. Established Eurythermic Line of Fish Cells In Vitro. Sci. 135: 1065-1066.

REFERENCES

Achtman, M. and G. Pluschke. 1986. Clonal analysis of descent and virulence among selected *Escherichia coli*. Ann. Rev. Microbiol. 40: 185-210.

Adams, J. 1981. Serological analysis of the diet of *Bdellocephala punctata*, a freshwater triclad. Oikos 36: 99-106.

Agnew, W.S., S.R. Levinson, J.S. Brabson and M.A. Raftery. 1978. Purification of the Tetrodotoxin-Binding Component Associated with the Voltage-Sensitive Sodium Channel from *Electrophorus electric* vs Electroplax Membranes. Proc. Natl. Acad. Sci. USA. 75: 2606.

Ahlstrom, E.H. 1959. Vertical distribution of pelagic fish eggs and larvae off California and Baja California, U.S. Wildl. Serv. Fish. Bull. 60: 107-146.

Akazawa, T., Takabe and H. Kobayashi. 1984. Molecular evolution of ribulose-1,5-bisphosphate carboxylase/oxygenase (Rubisco). Trends Biochem Sci. 9: 380-383.

Alam, M.I., C.P. Hsu and Y. Shimizu. 1979. Comparisons of toxins in three isolates of *Gonyaulax tamarensis* (Dinophyceae). J. Phycol. 15: 106-110.

Alberte, R.S., and A.L. Friedman. 1986. A diatom-specific signature: use of immunological techniques. Abstract. EOS. 67: 1056.

Alberte, R.S., A.M. Wood, T.A. Kursar and R.R.L. Guillard. 1984. Novel phycoerythrins in marine *Synechococcus* spp.: characterization and evolutionary and ecological implications. Plant Physiol. 75: 732-739.

Alexander, A.G. and G.E. Kenny. 1980. Characterization of the strain-specific and common surface antigens of *Mycoplasma arginini*. Infection and Immunity 29: 442-451.

Alexander, R.B., W.B. Isaacs and E.R. Barrack. 1985. Immunogold probes for electron microscopy: evaluation of staining by fluorescence microscopy. J. Histochem. Cytochem. 33(10): 995-1000.

Alvariño, A. 1980. The relation between the distribution of zoopalnkton predators and anchovy larvae. Calif. Coop. Oceanic Fish. invest. Rep. 21: 150-160.

Anderson, D.M., A.W. White and D.G. Baden. 1985. Eds. Toxic Dinoflagellates. Elsevier Science Publishers, NY. 561 pp.

Anderson, N.G. and N.L. Anderson. 1978. Analytical Techniques for cell fractions. XXI. Two-dimensional analysis of serum and tissue proteins: Multiple isoelectric focusing. Anal. Biochem. 85: 331-340.

Anderson, N.L. and N.G. Anderson. 1978. Analytical techniques for cell fractions. XXII. Two-dimensional analysis of serum and tissue proteins: Multiple gradient-slab electrophoresis. Ibid., pp 334-354.

Andrews, D.W. 1984. Use of monoclonal antibody immunoaffinity column to purify subsets of human HLA-DR antigens. IN: Methods in Enzymology. J.J. Langone, H. VanVunakis and G. DiSabato. (eds.). 108: 600.

Arnold, S.J. 1983. Morphology, performance, and fitness. Amer. Zool. 23: 347-361.

Arnold, S.J. 1986. Laboratory and field approaches to the study of adaptation. IN: *Predator-prey relationships*. M.E. Feder and G.V. Lauder (eds.). Chicago/London. University of Chicago Press. pp. 157-177.

Arnold, S.J. and M.J. Wade. 1984. On the measurement of natural and sexual selection: theory. Evolution. 38: 709-719.

Arnold, S.J. and M.J. Wade. 1984. On the measurement of natural and sexual selection: applications. Evolution. 38: 720-734.

Ashby, J.W. 1974. A study of arthropod predation of *Pieris rapae* L. using serological and exclusion techniques. J. Appl. Ecol. 11: 419-425.

Association of Official Analytical Chemists. 1984. IN: Official Methods of Analysis. S. Williams (ed.). 14th Edition, Washington, D.C. p. 344.

Atlas, E.L. and C.S. Giam. 1981. Global transport of organic pollutants: Ambient levels of organic contaminants in the remote marine atmosphere. Science 211: 163-165.

Avise, J.C. 1986. Mitochondrial DNA and the evolutionary genetics of higher animals. Phil. Trans. R. Soc. Lond. B. 312: 325-342.

Avise, J.C., E. Bermingham, L.G. Kessler and N.C. Saunders. 1984. Characterization of mitochondrial DNA variability in a hybrid swarm between subspecies of bluegill sunfish (*Lepomis macrochirus*). Evolution. 38: 931-934.

Axelsen, N.H. and E. Bock. 1972. Identification and quantification of antigens and antibodies by means of quantitative immunoelectrophoresis. A survey of methods. Journal of Immunological Methods. 1: 109-121.

Baden, D.G. 1983. Marine food-borne dinoflagellate toxins. Int. Rev. Cytol. 80: pp. 99-150.

Baden, D.G., G.M. Bikhazi, S.J. Decker, F.F. Foldes and I. Leung, 1984. Neuromuscular blocking action of two brevetoxins from the Florida red tide organism *Ptychodiscus brevis*. Toxicon. 22: pp. 75-84.

Baden, D.G. and T.J. Mende. 1982. Toxicity of two toxins from the Florida red tide dinoflagelate *Ptychodiscus brevis*. Toxicon. 20: 457-461.

Baden D.G., T.J. Mende and L. Brand. 1985. Cross-reactivity in immunoassays directed against toxins isolated from *Ptychodiscus brevis*. IN: Toxic Dinoflagellates. D.M Anderson, A.W. White and D.G. Baden (eds.). Elsevier Science Publishers, NY. pp. 363-368.

Baden, D.G., T.J. Mende, W. Lichter and L. Wellham. 1981. Crystallization and toxicology of T34: A major toxin from Florida's red tide organism (*Ptychodiscus brevis*). Toxicon. 19: pp. 455-462.

Baden, D.G., T.J. Mende, J. Walling and D.R. Schultz. 1984. Specific antibodies directed against toxins of *Ptychodiscus brevis* (Florida's red tide dinoflagellate). Toxicon 22: 783-790.

Bagshaw, J.C. and A.H. Warner. 1986. Restriction fragment length polymorphism in isolated populations of *Artemia*. Am. Soc. Biol. Chem. Fed. Proc. 45: 1883.

Bailey, K.M. and J. Yen. 1983. Predation by a carnivorous marine copepod, *Euchaeta elongata* Esterly, on eggs and larvae of the Pacific hake, *Merluccius productus*. J. Plankton Res. 5: 71-82.

Baker, K. H. and A.L. Mills. 1982. Determination of the number of respiring *Thiobacillus ferrooxidans* cells in water samples by using combined fluorescent antibody-2-(p-iodophenyl)-3-(p-nitrophenyl)-5- phenyltetrazolium chloride staining. Appl. Environ. Microbiol. 43: 338-344.

Balch, W.M. submitted. Studies of nitrate transport by marine phytoplankton using $^{36}Cl\text{-}ClO_3$ as a transport analogue. I. Physiological Findings. J. Phycol.

Balch, W.M., C. Garside and E.H. Renger. in press. Studies of nitrate transport by marine phytoplankton using $^{36}Cl\text{-}ClO_3$ as a transport analogue. II Field Observations. Deep Sea Res.

Barker, H.A. 1935. The culture and physiology of marine dinoflagellates. Arch. Mikrobiol. 6: 157-81.

Barlow, R.G. and R.S. Alberte. 1985. Photosynthetic characteristics of phycoerythrin-containing marine *Synechococcus* spp. I. Responses to growth photon flux density. Mar. Biol. 86: 63-74.

Bates, H.A., R. Kostriken and H. Rapoport. 1978. A Chemical Assay for Saxitoxin. Improvements and Modifications. J. Agric. Food Chem. 26: 252.

Bates, H.A. and H. Rapoport. 1975. A Chemical Assay for Saxitoxin, the Paralytic Shellfish Poison.. J. Agric. Food Chem. 23: 237.

Bauwe, H. 1984. Photosynthetic enzyme activities and immunofluorescence studies on the localization of ribulose 1,5-bisphosphate carboxylase/oxygenase in leaves of C3, C4 and C3-C4 intermediate species of *Flaveria* (Asteraceae). Biochem. Physiol. Pflanzen 179: 253-268.

Bayer, E.A. and M. Wilcheck. 1978. The avidin-biotin complex as a tool in molecular biology. TIBS. 3: N257.

Becker, W.A. 1984. Manual of Quantitative Genetics. Pullman. Academic Enterprises.

Benbough, J. E. and K.L. Martin. 1976. An indirect radiolabelled antibody staining technique for the rapid detection and identification of bacteria. Jour. Appl. Bact. 41: 47-58.

Bermingham, E. and J.C. Avise. 1986. Molecular zoogeography of freshwater fishes in the southeastern United States. Genetics. 113: 939-965.

Berg, J. 1979. Discussion of methods of investigating the food of fishes, with reference to a preliminary study of the prey of *Gobiusculus flavescens* (Gobiidae). Mar. Biol. 50: 263-273.

Bigger, C.H., P.L. Jokeil, W.H. Hildemann and I.S. Johnston. 1982. Characterization of alloimmune memory in a sponge. J. Immunol. 129: 1570-1572.

Bio-Rad. 1977. Bio-Rad Protein Assay. Tech. Bull. 1051, April. Bio-Rad Laboratories, Richmond, Calif. 3 pp.

Bjorkman, O. 1968. Carboxydismutase activity in shade-adapted and sun-adapted species of higher plants. Physiol. Plant. 21: 481-485.

Blasco, D., J.J. MacIsaac, T.T. Packard and R.C. Dugdale. 1984. Relationship between nitrate reductase and nitrate uptake in phytoplankton in the Peru upwelling region. Limnol. Oceanogr. 29: 275-286.

Block Engineering Inc., Framingham, MA. Dye Tagged Reagent., US. Patent #4166105. Thomas Hirschfeld

Blue, M.L., J.F. Daley, H. Levine and S.F. Schlossman. 1986. Discrete stages of human thymocyte activation and maturation in vitro: correlation between phenotype and function. Eur. J. Immunol. 16: 771-777.

Boag, P.T. 1983. The heritability of external morphology in Darwin's ground finches *Geospiza* on Isla Daphne Major Galapagos Ecuador. Evolution. 37: 877-894.

Bonin, D.J., M.R. Droop, S.Y. Maestrini and M-C. Bonin. 1986. Physiological features of six micro-algae to be used as indicators of seawater quality. Cryptogamie, Algologie. 7: 23-83.

Bonné-Tamir, B., M.J. Johnson, A. Natali, D.C. Wallace and L.L. Cavalli-Sforza. 1986. Human mitochondrial DNA types in two Israeli populations--a comparative study at the DNA level. Am. J. Hum. Gen. 38: 341-351.

Bohlool, B.B. and E.L. Schmidt. 1980. The immunofluorescence approach in microbial ecology. IN: Advances in Microbial Ecology. M. Alexander (ed.). 4: 203-241.

Bohmer, R.M. and N.J. King. 1984. Immuno-gold labeling for flow cytometric analysis. J. Immunol. Methods. 74(1): 49-57.

Boreham, P.F.L. and C.E. Ohiagu. 1978. The use of serology in evaluating invertebrate prey-predator relationships: a review. Bulletin of Entomological Research. 68: 171-194.

Bose, R.J. and J.E. Reid. 1979. Evidence for the Heterogeneity of Paralytic Shellfish Toxin in Clam and Mussel Samples Gathered near Prince Rupert, British Columbia. IN: Toxic Dinoflagellate Blooms. D.L. Taylor and H.H. Seliger. (eds.). Elsevier North Holland, Inc., New York, NY. pp. 399-402.

Bowes, G., W.L. Ogren and R.H. Hageman. 1972. Light saturation, photosynthesis rate, RuDP carboxylase activity, and specific leaf weight in soybeans grown under different light intensities. Crop Sci. 12: 77-79.

Boyer, G.L., C.F. Wichmann, J. Mosser, E.J. Schantz and H.K. Schnoes. 1979. Toxins Isolated from Bay of Fundy Scallops. IN: Toxic Dinoflagellate Blooms. D.L. Taylor and H.H. Seliger. (eds.). Elsevier North Holland, Inc., New York, NY. pp. 373-376.

Braarud, T. 1951. Taxonomical studies of marine dinoflagellates. Nytt. Mag. Nat. Vidensk. 88: 43-48.

Braarud, T. 1961. Cultivation of marine organisms as means of understanding environmental influences on populations. IN: Oceanography. M. Sears (ed.). Publ. No. 67. Am. Assoc. Adv. Sci. Washington D.C. pp. 271-298.

Brand, L.E. 1981. Genetic variability in reproduction rates in marine phytoplankton populations. Evolution. 35: 1117-1127.

Brand, L.E. 1982. Genetic variability and spatial patterns of genetic differentiation in the reproductive rates of the marine coccolithphores *Emiliana huxleyi* and *Gephyrocapsa oceanica*. Limnol. Oceanogr. 27: 236-245.

Brand, L.E. 1984. The salinity tolerance of forty-six marine phytoplankton isolates. Est. Coast. Shelf Sci. 18: 543-556.

Brand, L.E. 1985. Low genetic variability in reproduction rates in populations of *Prorocentrum micans* ehrenb. (Dinophyceae) over Georges Bank. J. Exp. Mar. Biol. Ecol. 88: 55-65.

Brand, L.E. and R.R.L. Guillard. 1981. The effects of continuous light and light intensity on the reproduction rates of twenty-two species of marine phytoplankton. J. Exp. Mar. Biol. Ecol. 50: 119-132.

Brand, L.E., R.R.L. Guillard and L.S. Murphy. 1981. A method for the rapid and precise determination of acclimated phytoplankton reproduction rates. J. Plankt. Res. 3: 193-201.

Brand, L.E., L.S. Murphy, R.R.L. Guillard and H-t. Lee. 1981. Genetic variability and differentiation in the temperature niche component of the diatom *Thalassiosira pseudonana*. Mar. Biol. 62: 103-110.

Brand, L.E., W.G. Sunda and R.R.L. Guillard. 1983. Limitation of marine phytoplankton reproductive rates by zinc, manganese, and iron. Limnol. Oceanogr. 28: 1182-1198.

Brand, L.E., W.G. Sunda and R.R.L. Guillard. 1986. Reduction of marine phytoplankton reproduction rates by copper and cadmium. J. Exp. Mar. Biol. Ecol. 96: 225-250.

Brandon, R.N. 1982. The levels of selection. IN: P.S.A. 1982, Vol. I. P. Asquith and T. Nickles. (eds.). East Lansing. Philosophy of Sci. Assoc. pp. 315-322.

Bravdo, B. and J.E. Pallas, Jr. 1981. Photosynthesis photo-respiration and RuBP carboxylase/oxygenase activity in selected peanut genotypes. Photosynthetica 16: 36:42.

Brewer, G.D., F.S. Kleppel and M. Dempsey. 1984. Apparent predation on ichthyoplankton by zooplankton and fishes in nearshore waters of southern California. Mar. Biol. 80: 17-28.

Brigati, D.J., D. Myerson, J.J. Leary, B. Fpalholz, S.Z. Travis, C.K.Y. Fong, G.D. Hfiung and D.C. Ward. 1983. Detection of viral genomes in cultures cells and parafin-imbedded tissue sections using biotin-labelled hybridization probes. Virology. 126: 32-50.

Brinton, E. 1976. Population biology of *Euphausia pacifica* off southern California. Fish. Bull. U.S. 74: 733-762.

Brinton, E. and J.G. Wyllie. 1976. Distributional atlas of euphausiid growth stages off southern California, 1953 through 1956. CalCOFI (Calif. coop. ocean. Fish. Invest.) Atlas No. 24.

Buckland, R.M. 1986. Strong signals from streptavidin-biotin. Nature. 320: 557.

Buscema, M. and G. Van de Vyver. 1984. Cellular aspects of alloimmune reactions in sponges of the genus *Axinella* II. *Axinella verrucosa* and *Axinella danicornis*. J. Expt. Zool. 229: 19-32.

Caldwell, H.D., C.C. Kuo and G.E. Kenny. 1975. Antigenic analysis of Chlamydiae by two-dimensional immunoelectrophoresis I. Antigenic heterogeneity between *C. trachomatis* and *C. psittacti*. The Journal of Immunology 115: 963-968.

Caldwell, H.D., C.C. Kuo and G.E. Kenny. 1975b. Antigenic analysis of Chlamydiae by two-dimensional immunoelectrophoresis. II. A Trachoma-LGV specific antigen. The Journal of Immunology. 115: 969-975.

Calver, M.C. 1984. A review of ecological applications of immunological techniques for diet analysis. Aust. J. Ecol. 9: 19-25.

Calver, M.C., J.N. Matthiessen, G.P. Hall, J.S. Bradley and J.H. Lillywhite. 1986. Immunological determination of predators of the bush fly, *Musca vetustissima* Walker (Diptera: Muscidae), in south-western Australia. Bull. Ent. Res. 76: 133-139.

Campbell, D.H., J.S. Garvey, N.E. Cremer and D.H. Sussdorf. 1964. Methods in immunology. W.A. Benjamin, Inc. NY. pp. 22-23, 41-42.

Campbell, L. 1985. Investigations of marine, phycoerythrin-containing *Synechococcus* spp. (Cyanobacteria): Distribution of serogroups and growth rate measurements. 186 pp. Ph.D. Dissertation, State University of New York, Stony Brook, NY.

Campbell, L. and E.J. Carpenter. 1986. Characterization of phycoerythrin-containing *Synechococcus* populations by immunofluorescence. submitted.

Campbell, L., E.J. Carpenter and V.J. Iacono. 1983. Identification and enumeration of marine Chroococcoid cyanobacteria by immunofluorescence. Appl. Environ. Microbiol. 46: 553-559.

Campbell, W.H. 1985. The biochemistry of higher plant nitrate reductase. IN: Nitrogen Fixation and CO_2 Metabolism. P.W. Ludden and J.E. Burris. (eds.). Elsevier Science Publishing Co., Inc., NY pp. 143-151.

Campbell, W.H. and J.L. Remmler. 1986. Regulation of corn (*Zea Mays*) leaf nitrate reductase: I. Immunochemical methods for analysis of the enzymes protein component. Plant Physiol. 80: 435-441.

Campbell, W.H. and K.G. Ripp. 1984. An ELISA for higher nitrate reductase. Ann. New York. Acad. Sci. 435: 123-151.

Carlson, R.E., M.L. Lever, B.W. Lee and P.E. Guire. 1984. Development of Immunoassays for Paralytic Shellfish Poisoning: A Radioimmunoassay for Saxitoxin. IN: Seafood Toxin. E.P. Ragelis (ed.). ACS Symposium Series, Washington D.C. 262: 181.

Carlsson, J., H. Drevin and R. Axen. 1978. Protein thiolation and reversible protein-protein conjugation: N-succinimydyl 3 (2-pyridyldithio) propionate, a new heterobifunctional reagent. Biochem. J. 173: 723-737.

Carlucci, A.F. and D. Pramer. 1957. Factors influencing the plate method for determining abundance of bacteria in sea water. Proc. Soc. Exp. Biol. Med. 96: 392-394.

Carpenter, E.J. and R.R.L. Guillard. 1971. Intraspecific differences in nitrate half-saturation constants for three species of marine phytoplankton. Ecology. 52: 183-185.

Caswell-Reno, P., P.W. Reno, and B.L. Nicholson.1986. Monoclonal Antibodies to Infectious Pancreatic Necrosis Virus (IPNV): Analysis of Viral Epitopes and Comparison of Different Isolates. J. Gen. Virol. 67: 2193-2205.

Catterall, W.A. 1980. Neurotoxins that Act on Voltage-Sensitive Sodium Channels in Excitable Membranes. Ann. Rev. Pharmacol. Toxicol. 20: 15.

Catterall, W.A. and M.A. Risk. 1980. Toxin T_{46} from *Ptychodiscus brevis* enhances activation of voltage-sensitive sodium channels by veratridine. Mol. Pharmacol. 19: pp. 345-348.

Caugant, D.A., B.R. Levin, I. Orskov, F. Orskov, C.V. Eden and R.K. Selander. 1985. Genetic diversity in relation to serotype in *Escherichia coli*. Inf. and Immun. 49: 407-413.

Cembella, A.D. and F.J.R. Taylor. 1985. Biochemical variability within the *Protogonyaulax tamarensis/catenella* species complex. IN: Toxic Dinoflagellates. D.M. Anderson, A.W. White and D.G. Baden. (eds.) North Holland/New York. Elsevier. pp. 55-60.

Cembella, A.D. and F.J.R. Taylor. 1986. Electrophoretic variability within the *Protogonyaulax tamarensis/catenella* species complex: pyridine linked dehydrogenases. Biochem. Syst. and Ecol. 14: 311-323.

Cembella, A.D., J.J. Sullivan, G.L. Boyer, F.J.R. Taylor and R.J. Andersen. 1986. Variation in paralytic shellfish toxin composition within the *Protogonyaulax tamarensis/catenella* species complex: Red Tide dinoflagellates. Bioch. Syst. and Ecol. (in press).

Chen, C.H. 1985. Tandem conjugates of phycobiliproteins. Phycobiliprotein Symposium, Seattle WA, September 9th-10th.

Chess, L., R.P. MacDermott and S. F. Schlossman. 1974. Immunologic functions of isolated human lymphocyte subpopulations. I. Quantitative isolation of human T and B cells and response to mitogens. J. Immunol. 113: 1113-1121.

Chu, F.S. and T.S.L. Fan. 1985. Indirect Enzyme-Linked Immunosorbent Assay for Saxitoxin in Shellfish. J. Assoc. Off. Anal. Chem. 68: 13.

Cisar, J.O. and J.L. Fryer. 1974. Characterization of anti *Aeromonas salmonicida* antibodies from coho salmon. Infec. Immun. 9: 236.

Clem, L.W. and W. E. McClean. 1975. Phylogeny of immunoglobulin structure and function: VII. Monomeric and tetrameric immunoglobulin of the margate, a marine teleost fish. Immunol. 29: 791.

Cleveland, J.S. and M.J. Perry (in press) Quantum yield, relative specific absorption, and fluorescence in nitrogen-limited *Chaetoceros gracilis*. Mar. Biol.

Clevenger, C.V. and A.L. Epstein. 1984. Identification of a nuclear protein component of interchromatin granules using a monoclonal antibody and immunogold electron microscopy. Exp. Cell Res. 151: 194-207.

Clinical Chemistry. 1982. Part II, 28(4): pp 737-1092.

Clinical Chemistry. 1984. Part I, 30(2): pp 1897-2108.

Cobel, M.J. 1975. The immune responses in fish: a review. J. Fish. Biol. 7: 539.

Cochran, W.G. and G.M. Cox. 1957. *Experimental Designs*, 2nd ed. New York/London. Wiley.

Cohen, J.E. 1978. Food webs and niche space. Princeton University Press, Princeton.

Collatz, G.J, M. Badger, C. Smith and J.A. Berry 1979. A radioimmune assay for RuP_2 carboxylase protein. Carnegie Inst Yearbook. 78: 171-175.

Collos, Y. and G. Slawyk. 1976. Significance of cellular nitrate content in natural populations of marine phytoplankton growing in shipboard culture. Mar. Biol. 34: 27-32.

Collos, Y. and G. Slawyk. 1977. Nitrate reductase activity as a function of *in situ* nitrate uptake and environmental factors of euphotic zone profiles. J. Exp. Mar. Biol. Ecol. 29: 119-130.

Colwell, D.E., S.M. Michalek and J.R. McGee. 1986. Method for generating high frequency of hybridomas producing monoclonal antibodies. IN: Methods in Enzymology. J.J. Langone and H. VanVunakis (eds.). Academic Press, NY. 121: 42.

Coons, H.A., J.H. Creech and N.R. Jones. 1941. Immunological properties of an antibody containing a fluorescent group. Proc. Soc. Exptl. Biol. Med. 47: 200.

Cousins, R.J. 1985. Absorption, trasport, and hepatic metabolism of copper and zinc: special reference to metallothionein and ceruloplasmin. Physiol. Rev. 65: 238-309.

Cramer, E., K.B. Pryzwansky, J.L. Villeval, U. Testa and J. Breton-Gorius. 1985. Ultrastructural localization of lactoferrin and myeloperoxidase in human neutrophils by immunogold. Blood. 65(2): 423-32.

Critctlow, R.E. and S.C. Stearns. 1982. The structure of food webs. American Naturalist. 120: 478-479.

Crowle, A.J. 1973. Immunodiffusion, 2nd Edition, Academic Press Inc., New York. pp. 1-373.

Crowle, A.J., G.J. Revis and K. Jarrett. 1972. Preparatory electroimmunodiffusion for making precipitins to selected native antigens. Immunological Communications. 1: 325-336.

Cucci, T.L., S.E. Shumway, R.C. Newell, R. Selvin, R.R.L. Guillard and C.M. Yentsch. 1985. Flow cytometry: a new method for characterization of differential ingestion, digestion and egestion by suspension feeders. Mar. Ecol. Prog. Ser. 24: 201-204.

Dagg, M.J. 1974. Loss of prey body contents during feeding by an aquatic predator. Ecology. 55: 903-906.

Dahle, A.B. and M. Laake. 1981. Diversity dynamics of marine bacteria: immunofluorescence stain on membrane filters. J. Appl. Microbiol. 43: 169-176.

Dahle, A.B. and M. Laake. 1982. Diversity dynamics of marine bacteria studied by immunofluorescent staining on membrane filters. Appl. Environ. Microbiol. 43: 169-176.

Daley, R.J. and S. R. Brown 1973. Chlorophyll, nitrogen, and photosynthetic patterns during growth and senescence of two blue green algae. J. Phycol. 9: 395-401.

Davies, R.W. 1969. The production of antisera for detecting specific triclad antigens in the gut contents of predators. Oikos. 20: 248-260.

Davies R.W., F.J. Wrona and R.P. Everett. 1978. A serological study of prey selection by *Nephelopsis obscura* Verrill (Hirudinoidea). Canadian Journal of Zoology. 56: 587-591.

Davies, R.W., F.J. Wrona, L. Linton and J. Wilkialis. 1981. Inter- and intra specific analyses of the food niches of two sympatric species of *Eprobellidae* (Hirudinoidea) in Alberta, Canada. Oikos. 37: 105-111.

Davio, S.R. 1985. Neutralization of Saxitoxin by Anti-Saxitoxin Rabbit Serum. Toxicon 23: 669.

Davis, B.D., R. Dulbecco, H.N. Eisen, H.S. Ginsberg and W.B. Wood. 1973. Microbiology, 2nd ed. Harper and Row Publishers, Inc., Hagerstown, MD.

Dawkins, R. 1982. *The Extended Phenotype: The Gene as a Unit of Selection*. San Francisco. Freeman.

deMacario, E.C., M.J. Wolin and A.J.L. Macario. 1981. Immunology of Archaebacteria that produce methane gas. Science 214: 74-75.

DeMey, J. 1983. Colloidal Gold Probes. IN: Immunocytochemistry. J.M. Polak and S. Van Noorden (eds.). Wright-PSG, Bristol, London, Boston. pp. 82-112.

DeMey,J., A.M. Lambert, A.S. Bajer, M. Moeremans and M. DeBrabander. 1982. Visualization of microtubules in interphase and mitotic plant cells of *Haemanthus endosperm* with the immuno-gold staining method. Proc. Natl. Acad. Sci. USA. 79(6): 1898-902.

Dempster, J.P. 1960. A quantitative study of the predators on the eggs and larvae of the broom beetle, *Phytodecta olivaca* Forster, using the precipitin test. J. Anim. Ecol. 29: 149-167.

DeWaele, M., J. DeMey, M. Moeremans, L. Broodtaerts, L. Smet and B. VanCamp. 1982. Colloidal gold as a marker for the light microscope detection of leukocyte cell surface antigens with monoclonal antibodies. J. Clin. Immunol. 2(3 Suppl): 24S-31S.

DeWaele. M., J. DeMey, M. Moeremans, M. DeBrabander and B. VanCamp. 1982. Immunogold staining method for the light microscopic detection of leukocyte cell surface antigens with monoclonal antibodies: its application to the enumeration of lymphocyte subpopulations. J. Histochem. Cytochem. 31(3): 376-381.

DeWaele, M., J. DeMey, W. Renmans, C. Labeur, P. Reynaert and B. VanCamp. 1986. An immunogold-silver staining method for detection of cell-surface antigens in light microscopy. J. Histochem. Cytochem. 34(7): 935-939.

DeWaele, M., J. DeMey, P. Reynaert, M.F. Dehou, W. Gepts and B. VanCamp. 1986. Detection of cell surface antigens in cryostat sections with immunogold-silver staining. Am. J. Clin. Pathol. 85(5): 573-578.

Dobos, P., B.J. Hill, R. Hallet, D.T.C. Kells, H. Bect and D. Teninges. 1979. Biophysical and biochemical characterization of five animal viruses with bisegmented double stranded RNA genomes. Journal of Virology. 32: 593-605

Dobzhansky, Th. 1970. Genetics of the Evolutionary Process. New York. Columbia Univ. Press.

Dorson, M. 1983. Infectious Pancreatic Necrosis. IN: Antigens of Fish Pathogens. D. Anderson, M. Dorson, and P. Dubourget, (eds.). Collection Fond. M. Merieux.

Dortch, Q., S.I. Ahmed and T.T. Packard. 1979. Nitrate reductase and glutamate dehydrogenase activities in *Skeletonema costatum* as measures of nitrogen assimilation rates. J. Plank. Res. 1: 169-185.

Dortch, Q. and H. Maske. 1982. Dark uptake of nitrate and nitrate reductase activity of a red-tide population off Peru. Mar. Ecol. Prog. Ser. 9: 299-303.

Drebes, G. 1977. Sexuality. IN: The Biology of Diatoms. D. Werner. (ed.). Berkeley/Los Angeles. Univ. Calif. Press. pp. 250-283.

Duarte, A.J.S., C.B. Carpenter and T.B. Strom. 1982. Expression of T cell differentiation antigens and Ia on rat cytotoxic T lymphocytes. Jour. Immunol. 128: 580

Ducklow, H.W. 1984. Geographical ecology of marine bacteria: physical and biological variability at the mesoscale. IN: Current Prespectives in Microbial Ecology. M.J. Klug and C.A. Reddy. (eds.). Washington D.C. Am. Soc. Microbiol. pp. 22-31.

Dugdale, R.C. and J.J. Goering. 1967. Uptake of new and regenerated forms of nitrogen in primary productivity. Limnol. Oceanogr. 12: 196-206.

Dugdale, R.C. and F.P. Wilkerson. 1986. The use of ^{15}N to measure nitrogen uptake in eutrophic oceans: experimental considerations. Limnol. Oceanogr. 31: 673-689.

Dunlap, J. and J.W. Hastings. 1981. The biological clock in *Gonyaulax* controls luciferase activity by regulating turnover. J. Biol. Chem. 256: 10509-10518.

Dunn-Coleman, N.S., J. Smarrelli, Jr. and R.H. Garrett. 1984. Nitrate assimilation in eukaryotic cells. IN: Internatl. Rev. Cytol. G.H. Bourne, J.F. Danelli and K.W. Joen (eds.). Academic Press, NY. 92: 1-50.

Eckert, R. and G.T. Reynolds. 1967. The subcellular origin of bioluminescence in *Noctiluca miliaris.* J. Gen. Physiol. 50: 1429-1458.

Eckert, R., and T. Sibaoka. 1968. The flash triggering action potential of the luminescent dinoflagellate *Noctiluca*. J. Gen. Physiol. 52: 258-282.

Eckman, J.E. 1983. Hydrodynamic processes affecting benthic recruitment. Limnology and Oceanography. 28: 241-257.

Endler, J. 1977. Geographic Variation, Speciation, and Clines. Princeton. Princeton Univ. Press.

Endler, J. 1986. Natural Selection in the Wild. Princeton. Princeton Univ. Press.

Eppley, R.W. 1978. Nitrate reductase in marine phytoplankton. IN: Handbook of Phycological Methods. J.A. Hellebust and J.S. Craigii. (eds.). Cambridge University Press. pp. 217-223.

Eppley, R.W., J.L. Coatsworth and L. Solorzano. 1969. Studies of nitrate reductase in marine phytoplankton. Limnol. Oceanogr. 14: 194-205.

Eppley, R.W., R.W. Holmes and J.D.H. Strickland. 1967. Sinking rates of marine phytoplankton measured with a fluorometer. J. Exp. Mar. Biol. Ecol. 1: 191-208.

Eppley, R.W. and B.J. Peterson. 1979. Particulate organic matter flux and planktonic new production in the deep ocean. Nature 28: 677-680.

Eppley, R.W. and E. Renger. submitted. New production above the nitracline: plankton utilize nanomolar concentrations of nitrate in nutrient-depleted surface waters. Oceanographie Tropicale.

Eppley, R.W., E.H. Renger, W.G. Harrison and J.J. Cullen. 1979. Ammonium distribution in southern California coastal waters and its role in the growth of phytoplankton. Limnol. Oceanogr. 24: 495-509.

Eppley, R. W., J.N. Rogers and J.J. McCarthy. 1969. Half-saturation constants for uptake of nitrogen and ammonium by marine phytoplankton. Limnol. Oceanogr. 14: 912-20.

Escaig, J. 1982. New instruments which facilitate rapid freezing at 83°K and 6°K. J. Microscopy. 126: 221-229.

Evans, H.J. and A. Nason. 1953. Plant Physiol. 28: 233-254.

Falconer, D.S. 1981. *Introduction to Quantitative Genetics*, 2nd ed. London. Longman.

Falkowski, P. 1983. Enzymology of nitrogen assimilation. IN: Nitrogen in the Marine Environment, E.J. Carpenter and D.G. Capone (eds.). Academic Press, NY. pp. 839-868.

Feller, R. J. 1984. Dietary immunoassay of *Ilyanassa obsoleta*, the Eastern mud snail. Biological Bulletin. 166: 96-102.

Feller, R.J. 1984. Serological tracers of meiofaunal food webs. Hydrobiologia 118: 119-125.

Feller, R.J. 1986. Immunological detection of *Mercenaria mercenaria* in a predator and preparation of size-class specific antibodies. The Veliger. 28: 341-347.

Feller, R.J. and E.D. Gallagher. 1982. Antigenic similarities among estuarine soft-bottom benthic taxa. Oecologia. 52: 305-310.

Feller, R.J., G.L. Taghon, E.D. Gallagher, G.E. Kenny and P.A. Jumars. 1979. Immunological methods for food web analysis in a soft-bottom benthic community. Marine Biology. 54: 61-74.

Feller, R.J., G. Zagursky and E. A. Day. 1985. Deep-sea food web analysis using cross-reacting antisera. Deep-Sea Research. 32: 485-497.

Fichter, B.L. and W.P. Stephen. 1981. Time related decay in prey antigens ingested by the predator *Podius maculiventris* (Hemiptera, Pentatomidae) as detected by ELISA. Oecologia. 51: 404-407.

Fichter, B.L. and W.P. Stephen. 1984. Time-related decay of prey antigens ingested by arboreal spiders as detected by ELISA. Environmental Entomology. 13: 1583-1587.

Fisher, N.S. 1977. On the differential sensitivity of estuarine and open-ocean diatoms to exotic chemical stress. Am. Nat. 111: 871-895.

Fisher, N.S. and D. Frood. 1980. Heavy metals and marine diatoms: influence of dissolved organic compounds on toxicity and selection for metal tolerance among four species. Mar. Biol. 59: 85-93.

Fisher, N.S., L.B. Graham, E.J. Carpenter, and C.F. Wurster. 1973. Geographic differences in phytoplankton sensitivity to PCBs. Nature. 241: 548-549.

Fisher, R.A. 1918. The correlation between relatives on the supposition of Mendelian inheritance. Trans. Roy. Soc. Edinb. 52: 399-433.

Fisher, R.A. 1958. The genetical theory of natural selection, 2nd. ed. New York. Dover.

Fliermans, C.B. and E.L. Schmidt. 1975. Autoradiography and immunofluorescence combined for autecological study of single cell activity with *Nitrobacter* as a model system. Appl. Microbiol. 30: 676-684.

Fliermans, C.B. and E.L. Schmidt. 1977. Immunofluorescence for autecological study of a unicellular bluegreen alga. J. Phycol. 13: 364-368.

Fogel, M., R. Schmitter and J.W. Hastings. 1972. On the physical identity of scintillons: Bioluminescent particles in *Gonyaulax polyedra*. J. Cell. Sci. 11: 305-317.

Fourcroy, D. Klein-Eude and C. Lambert. 1985. Phytochrome control of gene expression in radish seedlings. II. Far-red light mediated appearance of ribulose-1,5-bisphosphate carboxylase and the mRNA for its small subunit. Pl. Sc. Lett. 37: 235-244.

Frank, K.T. and W.C. Leggett. 1982. Coastal water mass replacement: its effect on zooplankton dynamics and the predator-prey complex associated with larval capelin (*Mallotus villosus*). Can. J. Fish. Aquat. Sci. 39: 991-1003.

Frantsi, C. and M. Savan. 1971. Infectious Pancretic Necrosis: Comparative Frequencies of Isolation from Feces and Organs of Brook Trout, *Salvelinus fontinalis*. J. Fish. Res. Bd. Can. 26: 1064-1065.

Freyssinet, G., M. Freyssinet and D.E. Buetow. 1983. Immunodetection of ribulose-1,5-bisphosphate carboxylase in mutants of *Euglena gracilis* impaired in photosynthesis. Plant Sci. Lett. 32: 61-72.

Friedman, A.L. and R.S. Alberte. 1984. A diatom light-harvesting pigment-protein complex. Pl. Physiol. 76: 483-489.

Friedman, A.L. and R.S. Alberte. 1986. Biogenesis and light regulation of the major light harvesting chlorophyll-protein of diatoms. Pl. Physiol. 80: 43-51.

Friedman, A.L. and R.S. Alberte. (submitted) Phylogenetic distribution of the major diatom light-harvesting pigment-protein determined by immunological methods.

Friedrich, J.W. and R.C. Huffaker. 1980. Photosynthesis, leaf resistances, and ribulose-1,5-biphosphate carboxylase degradation in senescing barley leaves. Plant Physiol. 65: 1103-1107.

Friefelder, D. 1982. Immunological Methods. W.H. Freeman & Co., NY. Chapter 1. p. 323.

Fujiwara, K. and T.D. Pollard. 1980. J. Cell. Biol. 87(2): 222A.

Funkhouser, E.A. and C.S. Ramadoss. 1980. Synthesis of nitrate reductase in *Chlorella*. II. Evidence for synthesis in ammonia-grown cells. Plant Physiol. 65: 944-948.

Gallagher, E.D. 1983. The mechanisms of benthic succession. Ph.D. Dissertation, University of Washington, Seattle, WA. 263 pp.

Gallagher, E.D., P.A. Jumars and D.D. Trueblood. 1983. Facilitation of soft-bottom benthic succession by tube-builders. Ecology. 64: 1200-1216.

Gallagher, E.D., P.A. Jumars and P. Goodin. Competition among the pioneers in soft-bottom benthic succession. In preparation.

Gallagher, J.C. 1980. Population genetics of *Skeletonema costatum* (Bacillariophyceae) in Narragansett Bay. J. Phycol. 16: 464-474.

Gallagher, J.C. 1982. Physiological variation and electrophoretic banding patterns of genetically different seasonal populations of *Skeletonema costatum* (Bacillariophyceae).

Gallagher, J.C. and R.S. Alberte. 1985. Photosynthetic and cellular photoadaptive characteristics of three ecotypes of the marine diatom, *Skeletonema costatum* (Grev.) Cleve. J. Exp. Mar. Biol. Ecol. 94: 233-250.

Gallagher, J.C., A.M. Wood and R.S. Alberte. 1984. Ecotypic differentiation in a marine diatom. I. Influence of light intensity on the photosynthetic apparatus. Mar. Biol. 82: 121-134.

Gallagher, J.P. and P. Shinnick-Gallagher. 1980. Effect of *Gymnodinium breve* toxin in the rat phrenic nerve diaphragm preparation. Br. J. Pharmacol. 69: pp. 367-374.

Garvey, J.S., R.J. Vander Mallie and C.C. Chang. 1982. Radio immunoassay of metallothioneins. IN: Methods in Enzymology. J.J. Langone and H. Van Vunakis (eds.). Academic Press, NY. pp. 121-138.

Garside, Chris. 1982. A chemiluminescent technique for the determination of nanomolar concentrations of nitrate, nitrate and nitrite or nitrite alone in seawater. Mar. Chem. 11: 159-167.

Garside, Chris. 1985. The vertical distribution of nitrate in open ocean surface water. Deep Sea Research. 32: 723-732.

George, S.G. and B.J.S. Pirie. 1980. Metabolism of zinc in the mussel, *Mytilus edulis* (L.): a combined ultrastructural and biochemical study. J.Mar. Biol. Ass. U.K. 60: 575-590.

Gershey, R.M., R.A. Neve, D.L. Musgrave and P.B. Reichardt. 1979. A Colorimetric Method for Determination of Saxitoxin. J. Fish Res. Board Can. 34: 559.

Gervais, A.J. and J.L. MacLean. 1985. Management. IN: Toxic Dinoflagellates. D.M. Anderson, A.W. White and D.G. Baden (eds.). Elsevier Science Publishers, NY. pp. 530-533.

Giles, N. and R.S. Phillips. 1985. A note on the production and use of antisera for the detection of part-digested sticklebacks in predator stomach-content samples. J. Fish. Biol. 27: 827-829.

Glazer, N.A. and L. Stryer. 1984. Phycofluor probes. TIBS. 9: 483.

Glibert, P.M., T.M. Kana, R.J. Olsen, D.L. Kirchman and R.S. Alberte. 1986. Clonal comparisons of growth and photosynthetic responses to nitrogen availability in marine *Synechococcus* spp. J. Exp. Mar. Biol. Ecol. 101: 199-208.

Glover, H.E. 1985. The physiology and ecology of the marine cyanobacterial genus *Synechococcus*. IN: Advances in aquatic microbiology. H.W. Jannasch and P.J. LeB Williams (eds.). Academic Press, NY. 4: 49-107.

Glover, H.E., L. Campbell and B. Prezelin. 1986. Contribution of *Synechococcus* spp. to size-fractioned primary productivity in three water masses in the Northwest Atlantic. Mar. Biol. 91: 193-203.

Glover, H.E., M.D. Keller and R.R.L. Guillard. 1986. Light quality and oceanic ultraphytoplankters. Nature. 319: 142-143.

Glover, H.E., M.D. Keller and R.W. Spinrad. 1987. The effects of light quality and intensity on photosynthesis and growth of marine eukaryotic and prokaryotic phytoplankton clones. J. Exp. Mar. Biol. Ecol. (in press).

Goldman, J.C. and E.J. Carpenter. 1974. A kinetic approach to the effect of temperature on algal growth. Limnol. Oceanogr. 19: 756-766.

Gompper, R. and W. Hagele. 1966. N-Sulfonyl-iminodithiokohlensaureester und N-Sulfonyl-dithiourethane. Chem. Ber. 99: 2885.

Gooch, J.L. 1975. Mechanisms of evolution and population genetics. IN: Marine Ecology, Vol. II, Pt. 1. O. Kinne. (ed.). Chichester/New York/Brisbane/Toronto. Wiley. pp. 349-409.

Grant, P.R. 1983. Inheritance of size and shape in a population of Darwin's finches, *Geospiza conirostris.* Proc. Roy. Soc. Lond. B. 220: 219-236.

Grassle, J.F. and J.P. Grassle. 1974. Opportunistic life histories and genetic systems in marine benthic polychaetes. Journal of Marine Research 32: 253-284.

Grassle, J.R. and J.P. Grassle. 1977. Temporal adaptation in sibling species of *Capitella*. IN: Ecology of marine benthos, B.C. Coull, Ed., University of South Carolina, Columbia, South Carolina. pp. 177-189.

Gray, J.C. and R.G.O. Kekwick. 1974. An immunological investigation of the structure and function of ribulose-1,5-bisphosphate carboxylase. Eur. J. Biochem. 44: 481-489.

Greenstone, M.H. 1983. Site-specificity and site tenacity in a wolf spider: a serological dietary analysis. Oecologia, Berlin. 56: 79-83.

Grisley, M.S. and P.R. Boyle. 1985. A new application of serological techniques to gut content analysis. J. Exp. Mar. Biol. Ecol. 90: 1-9.

Guillard, R.R.L. 1968. B12 Specificity of marine centric diatoms. J. Phycol. 4: 59-64.

Guillard, R.R.L. 1975. Culture of phytoplankton for feeding marine invertebrates. IN: Culture of marine Invertebrate Animals. W.L. Smith and M.H. Chanley. (eds.). Plenum Publishing Co. NY. pp. 29-60.

Guillard, R.R.L., E.J. Carpenter and B.E.F. Reimann. 1974. *Skeletonema menzelii* sp. nov., a new diatom from the Western Atlantic Ocean. Phycologia. 13: 131-138.

Guillard, R.R.L. and P. Kilham. 1977. The ecology of marine planktonic diatoms. IN: The Biology of Diatoms. D. Werner. (ed.). Berkeley/Los Angeles. Univ. Calif. Press. pp. 372-469.

Guillard, R.R.L., P. Kilham, and T. A. Jackson. 1973. Kinetics of silicon-limited growth in the marine diatom *Thalassiosira pseudonana* Hasle and Heimdal (=*Cyclotella nana* Hustedt). J. Phycol. 9:233-237.

Guillard, R.R.L. and J. Ryther. 1962. Studies of marine planktonic diatoms I. *Cyclotella nana* Hustedt and *Detonula confervacea* (Cleve) Gran. Can. J. Microbiol. 8: 229-239.

Haldane, J.B.S. 1924. A mathematical theory of natural and artificial selection. Trans. Camb. Phil. Soc. 23: 19-40.

Haldane, J.B.S. 1954. The measurement of natural selection. Proc. IX. Intl. Cong. Genet. 1: 480-487.

Hall, S. and P.B. Reichardt. 1984. Cryptic Paralytic Shellfish Toxins. IN: Seafood Toxins. E.P. Ragelis (ed.). ACS symposium Series, Washington D.C. 262: 113-122.

Hall, S. and Y. Shimizu. 1985. Toxin analysis and assay methods. IN: Toxic Dinoflagellates. D.M. Anderson, A.W. White and D.G. Baden, (eds.). Elsevier Science Publishers, NY. pp. 545-548.

Hamer, D.H., D.J. Thiele and J.E Lemontt. 1985. Function and autoregulation of yeast copperthionein. Science. 228: 685-690.

Hanson, R.S. and J.A. Phillips. 1981. Chemical composition. IN: Manual of Methods for General Bacteriology. P. Gerhardt, (editor in chief). Amer. Soc. Microbiol., Washington, D.C. pp. 328-364.

Harding, J.P. and Tebbles, N. 1963. Speciation in the sea. London. The Systematics Association.

Hardy, R.R., K. Hayakawa, D.R. Parks, L.A. Herzenberg and L.A. Herzenberg. 1984. Murine B-cell differentiation lineages. J. Exp. Med. 159(4): 1169-1188.

Hargraves, P. and R.R.L. Guillard. 1974. Structural and physiological observations on some small marine diatoms. Phycologia. 13: 163-172.

Harris, G.P. 1980. Spatial and temporal scales in phytoplankton ecology. Mechanisms, methods, models, and management. Can. J. Fish. Aq. Sci. 37: 877-900.

Harris, G. P. 1986. Phytoplankton Ecology. London/New York. Chapman and Hall.

Harrison, W.G. 1975. Assay of nitrate reductase from plasmolyzed marine phytoplankton. Annual Report of the Food Chain Group. Scripps Institute of Oceanography, La Jolla, CA. 473-483.

Hartl, D.L. and D.E. Dykhuizen. 1984. The population genetics of *Escherichia coli*. Ann. Rev. Genet. 18: 31-68.

Hastings, J.W. 1986. Bioluminescence in bacteria and dinoflagellates. IN: Light Emission in Plants and Bacteria. J. Amesz Govindjee and D.C. Fork. (eds.). Academic Press, NY. pp. 363-398.

Hattersley, P.W., L. Watson and C.B. Osmond (1977) *In situ* immuno- fluorescent labelling of ribulose-1,5-bisphosphate carboxylase in leaves of C3 and C4 plants. Aust. J. Plant Physiol. 4: 523-539.

Hausman, K., 1978. Extrusive organelles in Protists. Int. Rev. Cytol. 52: 198-276.

Hayward, J. 1968. Studies on the growth of *Phaeodactylum tricornutum*. IV. Comparison of different isolates. J. Mar. Biol. Assoc. U.K. 48: 657-666.

Healy, J.A. and T.F. Cross. 1975. Immunoelectroosmophoresis for serological identification of predators of the sheep tick *Ixodes ricinus*. Oikos. 26: 97-101.

Hercend, T., J. Griffin, A. Bensussan, R. Schmidt, M.A. Edson, A. Brennan, C. Murray, J.F. Daley, S.F. Schlossman and J. Ritz. 1985. Generation of monoclonal antibodies to a human natural killer clone. J. Clin. Invest. 75: 932.

Hildemann, W.H., C.H. Bigger and I.S. Johnston. 1979. Histo-compatibility reactions and allogeneic polymorphism among invertebrates. Transplantation Proceedings 11: 1136-1142.

Hildemann, W.H., C.H. Bigger, P.L. Jokiel and I.S. Johnston. 1980. Characteristics of immune memory in invertebrates. IN: J. Manning, (ed.). Phylogeny of immunological memory. Elsevier/North-Holland Biomedical Press, Amsterdam, pp. 9-14.

Hill, B.J. and K. Way. 1983. Serological classification of fish and shellfish birnaviruses. Abstracts of First International Conference of the European Association of Fish Pathologists. Plymouth, England. pp. 10.

Hobbie, J.E., R.J. Daley and S. Jasper. 1977. Use of Nuclepore filters for counting bacteria by fluorescence microscopy. Appli. Environ. Microbiol. 33: 1225-1228.

Hochman, A., A., Nissanyk, D. Wynne, B. Kaplan and T. Berman. in press. Nitrate reductase: an improved assay method for phytoplankton. Limnol. Oceanogr.

Hodgkins, H.O., R.S. Weiser and C.S. Ridgway. 1967. The nature of antibodies and the immune response in rainbow trout (*Salmo gairdneri*). J. Immunol. 99: 534.

Hofmann, K., G. Titus, J.A,. Montibeller and F.A. Finn. 1982. Avidin binding of carboxyl substituted biotin andanalyogues. Biochem. 21: 978-984.

Hokama, Y. 1985. A rapid simplified enzyme immunoassay stick test for the detection of ciguatoxin and related polyethers from fish tissues. Toxicon. 23: 939-946.

Hokama, Y, M.A. Abad and L.H. Kimura. 1983. A rapid enzyme immunoassay for the detection of ciguatoxin in contaminated fish tissues. Toxicon 21: 817-824.

Hokama, Y, A.H. Banner, and D. Boyland. 1977. A radioimmunoassay for the detection of ciguatoxin. Toxicon 15: 317-325.

Hokama, Y., L.H. Kimura, M.A. Abad, L. Yokochi, P.J. Scheuer, M. Nukina, T. Yasumoto, D.G. Baden and Y. Shimizu. 1984. An enzyme immunoassay for the detection of ciguatoxin and competitive inhibition by related natural polyether toxins. IN: Seafood Toxins, ACS Symposium Series. E.P. Ragelis (ed.). American Chemical Society, Washington, D.C., 262: pp 307-320.

Hokama, Y, L.H. Kimura, and J.T. Miyahara. 1985. Immunological approaches to understanding marine toxins. IN: Aquaculture, Public Health Regulatory Management Aspects. C.E. Kimble (ed.). Zarcon Press, Zarcon Corp., Silver Spring, MD. pp 80-96.

Hokama, Y., A.M. Osugi, S.A.A. Honda and M. Matsuo. 1985. Monoclonal antibodies in the detection of ciguatoxin and related toxic polyethers in fish tissues by a rapid stick test. IN: Proceed., 5th Int. Coral Reef Cong., Tahiti, French Polynesia. 4: 449-455.

Holborow, E.J., G.D. Johnson and S. Chantler. 1982. Use of international reference preparations for immunofluorescence. IN: Immunofluorescence Technology, Selected Theoretical and Clinical Aspects, G. Wick, K.N. Traill and K. Schauenstein, (eds.). Elsevier Biomedical Press, New York. pp. 1-10.

Holm-Hansen, O., C.J. Lorenzen, W.R Holmes and J.D.H. Strickland. 1965. Fluorometric determination of chlorophyll. J. Cons. Int. Exp. Mer. 30: 3-15.

Horan, P.K., S.E. Slezak and G. Poste. 1986. Improved flow cytometric analysis of leukocyte subsets: simultaneous identification of five cell subsets using two color immunofluorescence. Proc. Natl. Acad. Sci. 83: 8361-8365.

Huang, J.M.C., C.H. Wu and D.G. Baden. 1984. Depolarizing action of a red tide dinoflagellate brevetoxin on axonal membranes. J. Pharmacol. exp. Therapeut. 229: pp. 615-621.

Hunter, J.R. 1984. Inferences regarding predation on the early life stages of cod and other fishes. *Flodevigen rapportser.* 1, The propagation of cod, *Gadus morhua* L.: 533-562.

Hurkman, W.J. 1979. Ultrastructural changes of chloroplast in attached and detached, aging primary wheat leaves. Am. J. Bot. 66: 64-70.

Hurn, B.A.L. and S.M. Chantler. 1980. Production of reagent antibodies. Methods in Enzymology. 70: 105- 142.

Hutchinson, G.E. 1967. A Treatise on Limnology. Vol. 2. John Wiley & Sons, Inc.

Ingram, G.A. and J.B. Alexander. 1979. The immunoglobulin of the brown trout, *Salmo trutta* and its concentration in the serum of antigen-stimulated and non-stimulated fish. J. Fish. Biol. 14: 249.

Iturriaga, R., J.H. Morrow and D.A. Kiefer. 1986. Application of microphotometry to the study of optical properties of marine particulates. EOS. 67:973.

Jacobberger, J.W., D. Fogleman and J.M. Lehman. 1986. Analysis of intracellular antigens by flow cytometry. Cytometry. 7: 356-364.

Jenkins, W.J. and J.C. Goldman. 1984. Seasonal oxygen cycling and primary production in the Sargasso Sea. J. Mar. Res. 43: 465-491.

Jensen, A., B. Rystad and S. Melsom. 1974. Heavy metal tolerance of marine phytoplankton. I. The tolerance of three algal species to zinc in coastal seawater. J. Exp. Mar. Biol. Ecol. 15: 145-157.

Johnson, C.H., S. Inoue, A. Flint and J.W. Hastings. 1985. Compartmentation of algal bioluminescence: autofluorescence of bioluminescent particles in the dinoflagellate *Gonyaulax* as studied with image intensified video microscopy and flow cytometry. J. Cell Biol. 100: 1435-1446.

Johnson, C.H., J.F. Roeber and J.W. Hastings. 1984. Circadian changes in enzyme concentration account for rhythm of enzyme activity in *Gonyaulax*. Science. 223: 1428-1430.

Johnson, H.M., P.A. Frey, R. Angelotti, J.E. Campbell and K.H. Lewis. 1964. Haptenic Properties of Paralytic Shellfish Poison Conjugated to Proteins by Formaldehyde Treatment. 1964. Proc. Soc. Exp. Biol. Med. 117: 425.

Johnson, H.M. and G. Mulberry. 1966. Paralytic Shellfish Poison: Serological Assay by Passive Haemagglutination and Bentonite Flocculations. Nature 211: 747.

Johnson, P.W. and J. McN. Sieburth. 1979. Chroococcoid cyanobacteria in the sea: A ubiquitous and diverse phototrophic biomass. Limnol. Oceanogr. 24: 928-935.

Jones, G.J., B. Palenik and F.M.M. Morel. 1987. Trace metal reduction by phytoplankton: the roll of plasmalemma redox enzymes. J. Phycol. (in press)

Jones, G.J., T.D. Waite and J.D. Smith. 1985. Light-dependent reduction of copper (II) and its effect on cell-mediated, thiolo-dependent superoxide production. Biochem. Biophys. Res. Comm. 128: 1031-1036.

Jones, R. and W.B. Hall. 1974. Some observations on the population dynamics of the larval stage in common gadoids. IN: Early life history of fish, J.H.S. Blaxter (ed.). Springer-Verlag. Berl. pp. 87-102.

Kalisz, S. 1986. Variable selection on the timing of germination in *Collinsia verna* (Scrophulariaceae). Evolution. 40: 479-491.

Kao, C.Y. 1981. Tetrodotoxin, Saxitoxin, Chiriquitoxin: New Perspectives on Ionic Channels. Fed. Proc. 40: 30.

Kawamura, A., Jr., and Y. Aoyama. 1983. Immunofluorescence in Medical Science. University of Tokyo Press, Springer-Verlag, New York. 262 pp.

Kearney, J.F., A. Radbruch, B. Liesegang and K. Rajewsky. 1978. A new mouse myeloma cell line that has lost immunoglobulin expression, but permits the construction of antibody-secreting hybrid cell lines. J. Immunol. 123: 1548-1550.

Keller, M.D. and R.R.L Guillard. 1985. Factors significant to marine dinoflagellate culture. IN: Toxic Dinoflagellates. D.M. Anderson, A.W. White and D.G. Baden (eds.). Elsevier Science Publishing Co., Inc. pp. 113-116.

Kelly, R. and P. Loh. 1972. Electron Microscopical and Biochemical Characterization of Infectious Pancreatic Necrosis Virus. J. Virol. 10: 824-834.

Kennett, R.H., T.J. McKearn and K.B. Bechton. Editors. 1980. Monoclonal Antibodies. Hybridomas: a new dimension in biological analysis.

Kenny, G.E. and H.M. Foy. 1975. Detection and quantitation of circulating polysaccharide in pneumococcal pneumonia by immunoelectroosmophoresis (counterelectrophoresis) and rocket electrophoresis. IN: Microbiology 1975, D. Schlessinger, Ed. American Society for Microbiology, Washington, D.C.pp. 92-102.

Kilham, S.S. 1975. Kinetics of silicon-limited growth in the freshwater diatom *Asterionella formosa*. J. Phycol. 11: 396-399.

Kilham, S.S., C.L. Kott and D. Tilman. 1977. Phosphate and silicate kinetics for the Lake Michigan diatom *Diatoma elongatum*. J. Great Lakes Res. 3:93-99.

Kilpatrick, K.A. 1985. The development of a method to measure marine cyanobacterial phycoerythrin extracted in solvents. M.S. thesis, Texas A & M University. 74 pp.

Kimura, M. 1983. The Neutral Theory of Molecular Evolution. New York. Cambridge Univ. Press.

Kimura, L.H., M. A. Abad, and Y. Hokama. 1982. Evaluation of the radioimmunoassay (RIA) for detection of ciguatoxin (CTX) in fish tissues. J. Fish. Biol. 21: 671-680.

Kimura, L.H., Y. Hokama, M. A. Abad, M. Oyama and J.T. Miyahara. 1982. Comparison of three different assays for the assessment of ciguatoxin in fish diseases: Radioimmunoassay, mouse bioassay and *in vitro* guinea pig atrium assay. Toxicon 20: 907-912.

Kiritani, K. and J.P. Dempster. 1973. Different approaches to the quantitative evaluation of natural enemies. J. Appl. Ecol. 10: 323-330.

Kiss, J.Z., A.C. Vasconcelos and R.E. Triemer. 1986. Paramylon synthesis and chloroplast structure associated with nutrient levels in *Euglena* (Euglenophyceae). J. Phycol. 22: 327-333.

Klaaverkamp, J.F., W.A. Macdonald, D.A. Duncan and R. Wagemann. 1984. Metallothionein and acclimation heavy metals in fish: a review. IN: Contaminant Effects on Fisheries. V.W. Cairns, P.V. Hodson and J.O Nriagu. (eds.). John Wiley and Sons, Inc. pp. 100-113.

Kloepper-Sams, P.J., S.S. Park, H.V. Gelboin and J.J. Stegeman. Specificity and cross-reactivity of monoclonal and polyclonal antibodies against cytochrome P-450E of the marine fish scup. Arch. Biochem. Biophys. in press.

Klotz, A., J. Stegeman and C. Walsh. 1983. An aryl hydrocarbon hydroxylating hepatic cytochrome P-450 from the marine fish *Stenotomus chrysops*. Arch. Biochem. Biophys. 226: 578-592.

Knight, K.L., T.R. Malek and S. Dray. 1975. Human immunoglobulins with a 1 allotypic determinants of rabbit immunoglobulin heavy chains. Nature 253: 216.

Kobayashi, K., A. Hara, K. Takano and H. Hirai. 1982. Studies on subunit components of immunoglobulin M from a bony fish, the chum salmon (*Oncorhynchus keta*). Mol. Immunol. 19: 95.

Koehn, R.E., V.E. Ghazarossian, E.J. Schantz, H.K. Schnoes and F.M. Strong. 1981. Derivatives of Saxitoxin. Bioorg. Chem. 10: 412.

Kogima, Y. and J.H.R. Kagi. 1978. Metallothionein. Trends Biochem. Sci. 3: 89-94.

Koestler, R.J., J.J. Lee, J. Reidy, R.P. Sheryll and X. Xenophontos. 1985. Cytological investigation of digestion and re-establishment of symbiosis in the larger benthic foraminifera *Amphistegina lessonii*. Endocyt. C. Res. 2: 21-54.

Köhler, G. and C. Milstein. 1975. Continuous cultures of fused cells secreting antibody of predefined specificity. Nature. 256: 495-497.

Kolenkine, X. 1971. Les histocompatibilitiés intraspécifique et interspécifique chez les hydres d'eau douce. Arch. Zool. Exp. Géner. 112: 63-70.

Kou, G.H., S.N. Chen, Y. Yeno, R.P. Hewdrick and J.L. Fryer. 1982. Investigation of viral infections in cultured fishes in Taiwan. Abs. ROC-Japan Symposium on Fish Diseases, Tungkang, Taiwan, ROC. 42-43.

Krasnow, R., J. Dunlap, W. Taylor and J.W. Hastings. 1980. Circadian spontaneous bioluminsecent glow and flashing of *Gonyaulax polyedra*. J. Comp. Physiol. 138: 19-26.

Kreitman, M. 1983. Nucleotide polymorphism at the alcohol dehydrogenase locus of *Drosophila melanogaster*. Nature. 304: 412-417.

Kreitman, M. and M. Aguade. 1986. Genetic uniformity in two populations of *Drosophila melanogaster* as revealed by the filter hybridization of four nucleotide-recognizing restriction enzyme digests. Proc. Nat. Acad. Sci. U.S.A. 83: 3562-3566.

Krøll, J. 1973. Line immunoelectrophoresis. Scandinavian Journal of Immunology. 2 (supplement no. 1): 61-67.

Krøll, J. 1973. Rocket-line immunoelectrophoresis. IN: Scandinavian Journal of Immunology. 2 (Supplement no. 1): pp. 83-87.

Krøll, J. 1981. Production of specific antisera by immunization with precipitin lines. Methods in Enzymology. 73: 52-57.

Krøll, J. 1981. Quantitation of antigens by densitometric scanning of immunoelectrophoretic precipitates. Methods in Enzymology. 73: 370-383.

Ku, M.S.B., M.R. Schmitt and G.E. Edwards. 1979. Quantitative determination of RuBP carboxylase oxygenase protein in leaves of several C3 and C4 plants. J. Exp. Bot. 30: 89-98.

Kuppers, U. and M. Weidner. 1980. Seasonal variation of enzyme activities in *Laminaria hyperborea*. Planta 148: 222-230.

Kursar, T.A., H. Swift and R.S. Alberte. 1981. Morphology of a novel cyanobacterium and characterization of light-harvesting complexes from it: implications for phycobiliprotein evolution. Proc. Natl. Acad. Sci. USA. 78 (11): 6888-6892.

Kwan, S.P., D.E. Yelton and M.D. Scharff. 1980. Production of monoclonal antibodies. IN: Genetic Engineering. J.K. Setlow and A. Hollaender (eds.). Plenum Publishing Corp. NY. 2: 31.

Lackie, A.M. 1977. Cellular recognition of "non-self" in insects. IN: Developmental Immunobiology. J.B. Soloman and J.D. Horton. (eds.). Elsevier/North-Holland Biomedical Press, Amsterdam. pp. 75-81.

Lackie, A.M. 1979. Cellular recognition of foreignness in two insect species, the American cockroach and the desert locust. Immunobiology. 36: 909-914.

Laemmli, U.K. 1970. Cleavage of structural proteins during the assembly of the head of bacteriophage T4. Nature. 227: 680-685.

Lande, R. 1976. Natural selection and random genetic drift in phenotypic evolution. Evolution. 30: 314-334.

Lande, R. 1977. Statistical tests for natural selection on quantitative traits. Evolution. 31: 442-444.

Lande, R. 1979. Quantitative genetic analysis of multivariate evolution, applied to brain-body size allometry. Evolution. 33: 402-416.

Lande, R. 1980. The genetic covariance between characters maintained by pleiotropic mutations. Genetics. 94: 203-215.

Lande, R. and S.J. Arnold. 1983. The measurement of selection on correlated characters. Evolution. 37: 1210-1226.

Lane, D. and H. Koprowski. 1982. Molecular recognition and the future of monoclonal antibodies. Nature. 296: 200-202.

Lane, D.J., D.A. Stahl, G.J. Olsen, D.J. Heller and N.R. Pace. 1985. Phylogenetic analysis of the genera *Thiobacillus* and *Thermomicrospira* by 5S rRNA Sequences. J. Bact. 163: 75-81.

Langlet, C. and J. Bierne. 1977. The immune response to xenografts in nemertines of the genus *Lineus*. IN: Developmental Immunobiology. J.B. Solomon and J.D. Horton. (eds.). Elsevier/North-Holland Biomedial Press, Amsterdam. pp. 17-26.

Lanier, L.L. and M.R. Loken. 1984. Human lymphocyte subpopulations identified by using three color immunofluorescence and flow cytometry analysis. J. Immunol. 132: 151.

Lanier, L.L., E.G. Engleman and P. Gatenby. 1983. Correlation of functional properties of human lymphoid cell subsets and surface marker phenotypes using multiparameter analysis and flow cytometry. Imnol. Rev. 74: 143.

Lannan, C.N., J.R. Winton and J.L. Fryer. 1984. Fish cell lines: establishment and characterization of nine cell lines from salmonids. In Vitro. 20: 671-676.

Lapointe, B.E. and C.S. Duke. 1984. Biochemical strategies for growth of *Gracilaria tikvahiae* (Rhodophyta) in relation to light intensity and nitrogen availability. J. Phycol. 20: 488-495.

Lasker, R. 1975. Field criteria for survival of anchovy larvae: the relation between inshore chlorophyll maximum layers and successful first feeding. Fish. Bull. U.S. 73: 453-462.

Laurell, C.-B. 1966. Quantitative estimation of proteins by electrophoresis in agarose gel containing antibodies. Anal. Biochem. 15: 45-52.

Laurell, C.-B. and E.J. McKay. 1981. Electroimmunoassay. Methods in Enzymology 73: 339-369.

Laurie-Ahlberg, C.C. 1985. Genetic variation affecting the expression of enzyme-coding genes in *Drosophila*: an evolutionary perspective. IN: Isozymes. Current Topics In Biol. and Med. Res. 12: 33-88.

Laurie-Ahlberg, C.C., G. Maroni, G.C. Bewley, J.C. Lucchesi and B.S. Weir. 1980. Quantitative genetic variation of enzyme activities in natural populations of *Drosophila melanogaster*.

Laws, E.A. and T.T. Bannister. 1980. Nutrient and light-limited growth of *Thalassiosira fluviatilis* in continuous culture, with implications for phytoplankton growth in the ocean. Limnol. Oceanogr. 25: 457-473.

Ledbetter, J.A., A.E. Frankel, L.A. Herzenberg and L.A. Herzenberg. 1981. Human Leu T Cell Differention Antigens: Quantitative Expression on Normal Lymphoid Cells and Cell Lines. IN: Monoclonal Antibodies and T Cell Hybridomas, Perspectives and Technical Notes. G. Haemmerling, U. Haemmerling and J. Kearney. (eds.). Elsevier North Holland, New York, NY.

Lee, J.J. 1980. Nutrition and physiology of the foraminifera 43-46. IN: Biochemistry and Physiology of Protozoa. 2nd Edition. M. Levandowsky and S.H. Hutner (eds.). New York Academic Press.

Lee, J.J. 1983. Perspective on algal endosymbionts in larger foraminifera. Int. Rev. Cytol. Suppl. 14: 49-77.

Lee, J.J. and M.E. McEnery. 1983. Symbiosis in foraminifera 37-68. IN: Algal Symbiosis. L.J. Goff. (ed.). Cambridge University Press.

Lee, J.J., M.E. McEnery, R.F. Rockwell, J. Erez, B. ter Kuile, A. Lagziel and R. Röttger. Identification and distribution of endosymbiotic diatoms larger foraminifera. (in progress)

Lee, J.J., M.E. McEnery, R. Röttger and C.W. Reimer. 1980. The culture, isolation and identification of endosymbiotic diatoms for *Heterostegina depressa* d'Orgbigny and *Amphistegina lessonii* d'Orbigny (larger foraminifera from Hawaii). Botanica Marina 23: 297-302.

Lee, J.J. and E.W. Reimer. 1983. Isolation and identification of endosymbiotic diatoms from larger foraminifera of the Great Barrier Reef, Australia, Makapuu Tide Pool, Oahu, Hawaii, and the Gulf of Elat, Israel, with the description of new species, *Amphora rottgeri, Navicula hanseniana* and *Nitzschia frustrulum* variety *symbiotica*. Proc. VII Inter. symp. Living and fossil diatoms. D.G. Mann. (ed.). Koeltz Scientific Books.

Lee, J.J., C. Reimer and M.E. McEnery. 1980. The taxonomy of diatoms isolated as symbionts from the larger foraminifera from the Red Sea. Botanica Marina. 23: 42-48.

Leutenegger. S. 1984. Symbiosis in benthic foraminifera: specificity and host adaptations. Jour. Foraminifera Res. 14: 16-35.

Lewin, J.C. 1955. Physiological races of the diatom, *Navicula pelliculosa*. Biol Bul. 109: 343.

Lewin, J.C. and R.A. Lewin. 1960. Autotrophy and heterotrophy in marine littoral diatoms. Can. J. Microbiol. 6: 128-134.

Lewis, M.R. and T. Platt. 1982. Scales of variability in estuarine ecosystems. IN: Estuarine Comparisons. Proc. 6th Bien. Int. Est. Res. Conf. V. Kennedy. (ed.). New York. Academic Press. pp. 3-20.

Lewis, Marion, W.G. Harrison, N.S. Oakey, D. Herbert and T. Platt. 1986. Vertical nitrate fluxes in the oligotrophic ocean. Science. 234: 870-873.

Lewontin, R.C. 1974. The Genetic Basis of Evolutionary Change. New York. Columbia University Press.

Lewontin, R.C. 1985. Population genetics. Ann. Rev. Genetics. 19: 81-102.

Lobb, C.J. and L.W. Clem. 1984. Fish lymphocytes differ in the expression of surface immunoglobulin. Dev. Comp. Immunol. 6: 473.

Lobb, C.J., M.D. Olson and L.W. Clem. 1984. Immunoglobulin light chain classes in a teleost fish. J. Immunol. 132: 1917.

Loken, M.R., L.L. Lanier, D. Recktenwald and L. Warner. 1982. Flow cytometry monitoring leukocyte sets and subsets. Flow Cytometry and Monoclonal Antibodies for Therapy Monitoring. Montpellier, France. Oct 25-26.

Lopez-Ruiz, A., J.P. Verbelen, J.M. Roldan and J. Diez. 1986. Nitrate reductase of green algae is located in the pyrenoid. Plant Physiol. 79: 1006-1010.

Lorenzen, C. J. 1966. A method for the continuous measurement of *in vivo* chlorophyll concentration. Deep Sea Res. 13: 223-227.

Lukavsky, J. 1979. Algal cultures from the point of view of population genetics. IN: Algal Assays and Monitoring Eutrophication. P. Marvan, S. Pribil and O. Lhotsky. (eds.) Stuttgart. E. Schweizerbart'sche Verlagsbuchhandlung. pp. 137-139.

Lynch, M. 1984. The genetic structure of a cyclical parthenogen. Evolution. 38: 186-203.

Lynch, M. and W. Gabriel. 1983. Phenotypic evolution and parthenogenesis. Am. Nat. 122: 745-764.

Macdonald, R.D. and T. Yamamoto. 1977. The structure of infectious pancreatic necrosis virus RNA. Journal of General Virology. 34: 235-247.

Madhavan, S. and B.N. Smith. 1982. Localization of ribulose bisphosphate carboxylase in the guard cells by an immunofluorescence technique. Plant Physiol. 69: 273-277.

Makino, A., T. Mae and K. Ohira. 1983. Photosynthesis and ribulose-1,5-bisphosphate carboxylase in rice leaves. Plant Physiol. 73: 1002-1007.

Makino, A., T. Mae and K. Ohira. 1983. Effect of nitrogen, phosphorus or potassium on the photosynthetic rate and ribulose-1,5-bisphosphate carboxylase content in rice leaves during expansion. Soil Sci. Plant Nutr. 30: 63-70.

Manly, B.F.J. 1985. The Statistics of Natural Selection on Animal Populations. London/New York. Chapman and Hall.

Manderino, G.L., G.T. Gooch and A.B. Stavitsky. 1978. Preparation, characterization and functions of rabbit lymph node cell populations. I. Preparation of KLH primed T and B memory cells with anti-Fab' affinity columns. Cell. Immunol., 41: 264-275.

Marchalonis, J.J. 1977. Immunity in Evolution. Arnold, London, 238 pp.

Marcus, N. 1985. Population dynamics of marine copepods: the importance of genetic variation. Bul. Mar. Sci. 37: 684-690.

Marty, F., D. Branton and R.A. Leigh. 1980. Plant Vacuoles. IN: The Biochemistry of Plants. P.K. Stumpf and E.E. Conn. (eds.). Academic Press. 1: 625-658.

May, R.C. 1974. Larval mortality in marine fishes and the critical period concept. IN: The early life history of fish. J.H.S. Blaxter (ed.). Springer-Verlag, NY. pp 3-19.

Maynard-Smith, J. 1978. The Evolution of Sex. Cambridge. Cambridge Univ. Press.

Maynard-Smith, J. and R. Hoekstra. 1980. Polymorphism in a variable environment: how robust are the models? Genet. Res. Cambr. 35: 45-57.

McAllister, P.E. and W.B. Schill. 1986. Immunoblot assay: a rapid and sensitive method for identification of salmonid fish viruses. Journal of Wildlife Diseases. 22: 468-474.

McFadden, B.A. 1980. A perspective of ribulose bisphosphate carboxylase/oxygenase, the key catalyst in photosynthesis and photorespiration. Acc. Chem. Res. 13: 394-399.

McFarren, E.F., E.J. Schantz, J.E. Campbell and K.H. Lewis. 1958. Chemical Determination of Paralytic Shellfish Poison in Clams. J. Assoc. Off. Agric. Chem. 41: 168.

McIntyre, A.D. 1969. Ecology of marine meiobenthos. Biological Reviews 44: 245-290.

Melamed, M.R., P.F. Mullaney and M.L. Mendelsohn (eds.). 1979. Flow Cytometry and Sorting, John Wiley & Sons, NY, 716 pp.

Merril, C.R., D. Goldman, and M.L. Van Keuren. 1982. Simplified silver protein detection and image enhancement methods in polyacrylamide gels. Electrophoresis 3: 17-23.

Miller, T.J. and H.O Stone. 1978. The rapid isolation of ribonuclease-free immunoglobulin G by protein A-sepharose affinity chromatography. J. Immunol. Methods. 24: 111-125.

Miziorko, H.M. and G.H. Lorimer. 1983. Ribulose-1,5-bisphosphate carboxylase-oxygenase. Ann. Rev. Biochem. 52: 507-535.

Moller, H. 1984. Reduction of a larval herring population by jellyfish predator. Science 224: 621-622.

Morris, I. 1980. Paths of carbon assimilation in marine phytoplankton. IN: Primary productivity in the sea. P.G. Falkowski (ed.). New York, Plenum Press. pp. 139-159.

Muirhead, K.A., P.K. Horan and G. Poste. 1985. Flow cytometry: present and future. Bio/technology. 3: 337-356.

Muller, W.E.G. 1982. Int. Rev. Cytol. 77: 129-181.

Murphy, L. S. 1978. Biochemical taxonomy of marine phytoplankton by electrophoresis of enzymes. II. Loss of heterozygosity in clonal cultures of the centric diatoms *Skeletonema costatum* and *Thalassiosira pseudonana*. J. Phycol. 14: 247-250.

Murphy, L. S. and R.A. Belastock. 1980. The effect of environmental origin on the response of marine diatoms to chemical stress. Limnol. and Oceanogr. 25: 160-165.

Murphy, L.S. and R.R.L. Guillard. 1976. Biochemical taxonomy of marine phytoplankton by electrophoresis of enzymes. I. The centric diatoms *Thalassiosira pseudonana* and *T. fluviatilis*. J. Phycol. 12: 9-13.

Murphy, L.S., R.R.L. Guillard and J.F. Brown. 1984. The effects of iron and manganese on copper sensitivity in diatoms: differences in the responses of closely related neritic and oceanic species. Biol. Oceanogr. 3: 187-201.

Murphy, L.S., R.R.L. Guillard and J. Gavis. 1982. Evolution of resistant phytoplankton strains through exposure to marine pollutants. IN: Ecological Stress and the New York Bight: Science and Management. G.F. Mayer. (ed.). Columbia. Estuarine Research Federation. pp. 401-412.

Murphy, L.S. and E. Haugen. 1985. The distribution and abundance of phototrophic ultraplankton in the North Atlantic. Limnol. Oceanogr. 30: 47-58.

Nakane, P. and A. Kawaoi. 1974. Peroxidase-labelled Antibody, a New Method of Conjugation. J. Histo. Cyto. 22: 1084-1091.

Nature. 1975. 265: 495-497.

Navaluna, N., and M.J. Perry (submitted) The effect of chemical fixation on some optical properties of phytoplankton.

Nawata, T and T. Siboaka. 1979. Coupling between action potential and bioluminescence in *Noctiluca*: Effects of inorganic ions and pH in vacuolar sap. J. Comp. Physiol. 134: 137-149.

Necas, J. 1979. Genetic variability and the resulting nonhomogeneity in algal populations. IN: Algal Assays and Monitoring Eutrophication. P. Marvan, S. Pribil and O. Lhotsky. (eds.). Stuttgart. E. Schweivzerbart'sche Verlagsbuchhandlung. pp. 141-152.

Nelson, D.M. and L.E. Brand. 1979. Cell division periodicity in 13 species of marine phytoplankton on a light:dark cycle. J. Phycol. 15: 67-75.

Nelson, D.M., J.J. Goering, S.S. Kilham and R.R.L. Guillard. 1976. Kinetics of silicic acid uptake and rates of silica dissolution in the marine diatom *Thalassiosira pseudonana*. J. Phycol. 12: 246-252.

Newman, S.M. and R.A. Cattolico (in press) Is RuBP Case in *Olisthodiscus luteus* an evolutionary divergent enzyme? Plant. Physiol.

Newman, S.M., M.V. Orellana, R.A. Cattolico and M.J.Perry (in preparation) Structure of ribulose-1,5-bisphosphate carboxylase-oxygenase among plants.

Nicol, S. 1984. Cod end feeding by the euphausiid *Meganyctiptianes norvegica*. Mar. Biol. 80: 29-33.

Nicolas, M.-T., C.H. Johnson, J.-M. Bassot and J.W. Hastings. 1985. Immunogold labeling of organelles in the bioluminescent dinoflagellate *Gonyaulax polyedra* with anti-luciferase antibody. Cell Biology International Reports. 9: 797-802.

Nicolas, M.-T., B.M. Sweeney and J.W. Hastings. 1986. Ultrastructural localization of luciferase in *Phyrocystis* (Pyrrophyta) by immuno-gold labelling. J. Cell Sci. in press.

Nicholson, B.L. and E. Henchal. 1978. Rapid Identification of Infectious Pancreatic Necrosis virus in Infected Cell Cultures by Immunoperoxidase Techniques. J. Wild. Dis. 14: 465-469.

Nicholson, B.L., G.W. Thorne, C. Janicki and A. Hanson. 1979. Studies on a host range variant from different isolates of infectious pancreatic necrosis virus (IPNV). Journal of Fish Diseases. 2: 367-379.

Nisonoff, A. 1982. Introduction to Molecular Immunology. Sinauer Associates Inc.: Sunderland, MA. 204 pp.

Nukina, M., L.M. Koyangi and P.J. Scheuer. 1984. Two interchangeable forms of ciguatoxin. Toxicon 22: 169-176.

Ochman, H., R.A. Wilson, T.S. Wittam and R.K. Selander. 1984. Genetic diversity within serotypes of *Escherichia coli*. Proc. 4th Int. Symp. on Neonatal Diarrhea. Saskatchewan. VIDO Publications. University of Saskatchewan. pp. 202-224.

O'Connell, C.P. 1980. Percent of starving northern anchovy larvae (*Engraulis mordax*) in the sea as estimated by histological methods. Fish. Bull. U.S. 78: 475-489.

Oi, V.T., A.N. Glazer and L. Stryer. 1982. Fluorescent phycobiliprotein conjugates for analyses of cells and molecules. J. Cell. Biol. 93(3): 981-986.

Oi, V.T. and L.A. Herzenberg. 1980. Immunoglobulin producing hybrid cell lines. IN: Selected Methods in Cellular Immunology. B.B. Mishell and S.M. Shiigi (eds.). W.H. Freeman and Co., San Franscisco, CA. 351 pp.

Olafson, R.W., S. Loya and R.G. Sim. 1980. Physiological parameters of prokaryotic metallothionein induction. Biochem. Biophys. Res. Comm. 95: 1495-1503.

Olsen, G.J., D.J. Lane, S.J. Giovannoni and N.R. Pace. 1986. Microbial ecology and evolution: a ribosomal RNA approach. Ann. Rev. Microbiol. 40: 337-365.

Olson, R.J., S.W. Chisholm and E.R. Zettler. 1986. Dual-beam flow cytometry for distinguishing between phytoplankton pigment types. EOS. 67: 973.

Olson, R.J., D. Vaulot and S.W. Chisholm. 1985. Marine phytoplankton distributions measured using flow cytometry. Deep-Sea Res. 32: 1273-1280.

Ong, L.J., A.N. Glazer and J.B. Waterbury. 1984. An unusual phycoerythrin from a marine cyanobacterium. Science. 224: 80-82.

Orellana, M.V. and M.J. Perry. 1985. Development of an immunofluorescent assay for primary production. EOS 66: 1305 (abstract)

Oudin, J. 1980. Immunochemical analysis by antigen-antibody precipitation in gels. Methods in Enzymology. 70: 166-198.

Ouchterlony, O. 1968. Diffusion-in-gel methods for immunological analysis. Ann Arbor Science Publishers, Ann Arbor, Michigan.

Ouchterlony, O. and L.A. Nilsson. 1978. Immunodiffusion and immunoelectrophoresis. IN: Handbook of Experimental Immunology. D.M. Weir (ed.). Oxford, Blackwell Scientific Publications. 3: 38.

Paau, A.S. and J.R. Cowles. 1978. Applications of flow cytometry to the study of algal cells and isolated chloroplasts. J. Exp. Bot. 29: 1011-1020.

Packard, T.T. 1973. The light dependence of nitrate reductase in marine phytoplankton. Limnol. Oceanogr. 18: 466-469.

Packard, T.T. and D. Blasco. 1974. Nitrate reductase activity in upwelling regions 2. Ammonia and light dependence. Tethys. 6: 269-280.

Packard, T.T., D. Blasco, J.J. MacIsaac and R.C. Dugdale. 1971. Variations of nitrate reductase activity in marine phytoplankton. Inv. Pesq. 35: 209-219.

Packard, T.T., R.C. Dugdale, J.J. Goering and R.T. Barber. 1978. Nitrate reductase activity in the subsurface waters of the Peru Current. J. Mar. Res. 36: 59-76;.

Packard, T.T., P.C. Garfield and L.A. Codispoti. 1983. Oxygen consumption and denitrification below the Peruvian upwelling. IN: Coastal Upwelling, pt.A. E. Suess and J. Thiede (eds.). Plenum Publishing Corporation. pp. 147-173.

Paine, R.T. 1980. Food webs: Linkage, interaction strength and community infrastructure. Journal of Animal Ecology. 49: 667-685.

Palenik, B., O.C. Zafiriou and F.M.M. Morel. 1987. Hydrogen peroxide production by a marine phytoplankton. Limnol. Oceanogr. (submitted)

Park, S.S., H. Miller, A.V. Klotz, P.J. Kloepper-Sams, J.J. Stegeman and H. V. Gelboin. Monoclonal antibodies against cytochrome P-450E from the marine teleost *Stenotomus chrysops* (scup). Arch. Biochem. Biophys. 249: 339-350.

Paul, J.H. 1982. Use of Hoechst dyes 33258 and 33342 for enumeration of attached and planktonic bacteria. Appl. Environ. Microbiol. 43: 939-944.

Peoples, M.B., V.C. Beilharz, S.P. Waters, R.J. Simpson and M.J. Dalling. 1980. Nitrogen redistribution during grain growth in wheat (*Triticum aestivum* L.) II. Chloroplast senescence and the degradation of ribulose-1,5-bisphosphate carboxylase. Planta. 149: 241- 251.

Perchorowicz J. T., D.A. Raynes and R.G. Jensen. 1981. Light limitation of photosynthesis and activation of ribulose bisphosphate carboxylase in wheat seedlings. Proc. Natl. Acad. Sci. USA. 78: 2985- 2989.

Perry, M.J., M.C. Talbot and R.S. Alberte. 1981. Photoadaptation in marine phytoplankton: Response of the photosynthetic unit. Mar. Biol. 62: 91-101.

Perry, M.J., J. Smith, M.C. Talbot and N. Welschmeyer. (submitted) Nitrogen limitation in marine phytoplankton: response of the photosynthetic apparatus. Mar. Ecol. Prog. Ser.

Perry, M.J., J. Smith, and N. Welschmeyer. (submitted) Resource exploitation in phytoplankton: the role of photosynthetic nitrogen. Amer. Nat.

Perkins F.O., L.W. Haas, D.E. Phillips and K.L. Webb. 1981. Ultrastructure of a marine *Synechococcus* possessing spinae. Can. J. Microbiol. 27: 318-329.

Peterson, C.H. and B.P. Bradley. 1978. Estimating the diet of a sluggish predator from field observations. Journal of the Fisheries Research Board Canada 35: 136-141.

Peterson, D.H., M.J. Perry, K.E. Bencala and M.C. Talbot (in press). Phytoplankton productivity in relation to light intensity: a simple equation. Est. Coastal Shelf Sci.

Pharmacia Fine Chemicals AB. 1980. Cell Affinity Chromatography: Principles and Methods. Pharmacia Fine Chemicals, Uppsala, Sweden.

Phinney, D.A., T.L. Cucci and C.M. Yentsch. (submitted). Perspectives on aquatic flow cytometry. I: Instrumentation and analysis.

Phinney, D.A., T.L. Cucci and C.M. Yentsch. (submitted). Perspectives on aquatic flow cytometry. II: Importance of standards and controls.

Pierce, R. 1986. Red tide (*Ptychodiscus brevis*) aerosol toxins: A review of current knowledge. Toxicon. 24: 955-966.

Pierce, R., R.C. Brown and J.R. Kucklick. 1985. Analysis of *Ptychodiscus brevis* toxins by reverse phase HPLC. IN: Toxic Dinoflagellates, D.M. Anderson, A.W. White and D.G. Baden. (eds.). Elsevier Science Publishers, NY. pp. 309-314.

Pimm, S.L. 1982. Food webs. Chapman and Hall, London.

Pimm, S.L. 1984. Food chains and return times. IN: Ecological communities: conceptual issues and the evidence, D.R. Strong, D. Simberloff, L.G. Abele, and A.B. Thistle, (eds.). Princeton University Press, Princeton, N.J. pp. 397-412.

Pinkas, L., M.S. Oliphant and I.L.K. Iverson. 1971. Food habits of albacore, blue fin tuna, and bonito in California waters. Fish. Bull. Calif. 152: 1-105.

Plumley, F.G., D.L. Kirchman, R.E. Hodson and G.W. Schmidt. 1986. Ribulose biphosphate carboxylase from three chlorophyll *c*-containing algae: physical and immunological characterizations. Plant Physiol. 80: 685-691.

Poli, M.A., T.J. Mende and D.G. Baden, 1985. Characterization of the *Ptychodiscus brevis* polyether binding component in excitable membranes. IN: Toxic Dinoflagellates, D.M. Anderson, A.W. White and D.G. Baden. (eds.). Elsevier Science Publishers, NY. pp. 357-362.

Poli, M.A., T.J. Mende and D.G.Baden. 1986. Brevetoxins, unique activators of voltage-sensitive sodium channels, bind to specific sites in rat brain synaptosomes. Mol. Pharmacol. 30: 129-135.

Porter, K.G. and Y.S. Feig. 1980. The use of DAPI for identifying and counting aquatic microflora. Limnol. Oceanogr. 25: 943-948.

Prezelin, B.B. 1982. Effects on light intensity on aging of the dinoflagellate *Gonyaulax polyedra*. Mar. Biol. 69: 129-135.

Prezelin, B.B. and F.T. Haxo. 1976. Purification and characterization of peridinin-chorophyll *a*-proteins from the marine dinoflagellates *Glenodinium* sp. and *Gonyalax polyedra*. Planta. 128: 133-141.

Prezelin, B.B. and H.A. Matlick. 1980. Time course photoadaptation in the photosynthesis-irradiance relationship of dinoflagellate exhibiting photosynthetic periodicity. Mar. Biol. 58: 85-96.

Prezelin, B.B. and H.A. Matlick. 1983. Nutrient dependent low-light adaptation in the dinoflagellate *Gonyaulax polyedra*. Mar. Biol. 74: 141-150.

Prezelin, B.B. and B.M. Sweeney. 1979. Photoadaptation of photosynthesis in *Gonyaulax polyedra*. Mar. Biol. 48: 27-35.

Provasoli, L. and I.J. Pintner. 1953. Ecological implications of in vitro nutritional requirements of algal flagellates. Ann. N.Y. Acad. Sci. 56: 839-851.

Purcell, J.E. 1981. Feeding ecology of *Rhizophysa eysenhardti*, a siphonophore predator of fish larvae. Limnol. Oceanogr. 26: 424-432.

Purcell, J.E. 1984. Predation on larval fish by the Portuguese man of war, *Physalia physalis*. Mar. Ecol. Prog. Ser. 19: 189-191.

Pyke, G.H., H.R. Pulliam and E.L. Charnov. 1977. Optimal foraging: a selective review of theory and tests. Quarterly Review of Biology 51: 137-154.

Rabinowitz, H., A. Reisfeld, D. Sagher and M. Edelman. 1975. Ribulose diphosphate carboxylase from autotrophic *Euglena gracilis*. Plant Physiol. 56: 345-350.

Rantz, L.A. and E. Randall. 1955. Use of autoclaved extract of hemolytic *Streptococci* for serological grouping. Stanford Med. Bull. 13: 290-291.

Raven, J.A. 1984. Energetics and transport in aquatic plants. MBL Lectures in Biology. Vol 4, 587 pp.

Recktenwald, D.J. 1985. Phycobiliproteins as multi-color probes for flow-cytometry. Federation of American Societies for Experimental Biology Annual Meeting, Anaheim, CA.

Recktenwald, D., H.C. Chen, J. Kimura and N. Warner. 1984. Phycobiliprotein IgG conjugates as tools for multi-color cytometric analysis. International Conference on Analytical Cytology X, Asilomar, CA. June 3-8.

Redinbaugh, M.G. and W.H. Campbell. 1983. Purification and squash NADH: nitrate reductase by zinc chelate affinity chromatography. Plant Physiol. 71: 205-207.

Reed, J.E. and R. Chollet. 1985. Immunofluorescent localization of phosphoenolpyruvate carboxylase and ribulose-1,5-bisphosphate carboxylase/oxygenase proteins in leaves of C3, C4 and C3-C4 intermediate *Flaveria* species. Planta. 165: 439-445.

Reinisch, C.L., A.M. Charles and J. Froutner. 1983. Dev. Comp. Immunol. 7: 33-39.

Reno, P.W. 1976. Qualitative and Quantitative Aspects of Infectious Pancreatic Necrosis Carrier State. Ph. D. Thesis, University of Guelph, Ontario.

Reyment, R.A. 1982. Quantitative-genetic analysis of evolution in two late Cretaceous species of ostracods. Proc. III N. Am. Paleontol. Conv. J. Paleontol. 56(2nd Suppl.): 21-22.

Reyment, R.A. 1982. Application of quantitative genetics to evolutionary series of microfossils. IN: Nordic Symp. of Appl. Statistics and Data Processing. Copenhagen. Forlag NEUCC. pp. 307-325.

Reyment, R.A. 1982. Phenotypic evolution in a Cretaceous foraminifer. Evolution. 36: 1182-1199.

Reyment, R.A. 1983. Phenotypic evolution in microfossils. Evolutionary Biol. 16: 209-254.

Reynoldson, T.B. and R.W. Davies. 1970. Food niche and coexistence in lake-dwelling triclads. J. Anim. Ecol. 39: 599-617.

Rhiel, E., E. Morschel and W. Wehrmeyer. 1985. Correlation of pigment deprivation and ultrastructural organization of thylakoid membranes in *Cryptomonas maculata* following nutrient deficiency. Protoplasma 129: 62-73.

Rippka, R., J. Deruelles, J.B. Waterbury, M. Herdman, R.Y. Stanier. 1979. Generic assignments, strain histories and properties of pure cultures of cyanobacteria. J. Gen. Microbiol. 111: 1-61.

Rivkin, R.B. and H.H. Seliger. 1981. Liquid scintillation counting for ^{14}C uptake of single algal cells isolated from natural samples. Limnol. Oceanogr. 26: 780-785.

Rivkin, R.B., D.A. Phinney and C.M. Yentsch. in press. Effects of flow cytometry on photosynthetic carbon uptake by phytoplankton in culture and from natural populations. Appl. Environm. Microbiol.

Robertson, A. 1959. Experimental design in the evaluation of genetic parameters. Biometrics. 15: 219-226.

Roch, M., J.A. McCarter, A.T. Matheson, M.J.R. Clark and R.W. Olafson. 1982. Hepatic metallothionein induction in rainbow trout (*Salmo gairdneri*) as an indicator of metal pollution in the Campbell River System. Can. J. Fish. Aquatic. Sci. 19: 1596-1601.

Rodricks, J.V. and H. Rapoport. 1971. Synthesis of Cyclic Guanidines. J. Org. Chem. 36: 46.

Roesijadi, G. 1981. The significance of low molecular weight, metallothionein-like proteins in marine invertebrates: current status. Mar. Environ. Res. 4: 167-179.

Roesijadi, G. 1982. Uptake and incorporation of mercury into mercury-binding proteins of gills of *Mytilus edulis* as a function of time. Mar. Biol. 66: 151-157.

Roesijadi, G. and A.S. Drum. 1982. Influence of meracaptoethanol on the isolation of mercury-binding proteins from the gills of *Mytilus edulis*. Comp. Biochem. Physiol. 71B: 455-459.

Roitt, I., J. Brostoff and D. Male. 1985. Immunology. E.V. Mosby. St. Louis.

Roldan, J.M., V.P. Verbelen, W.L. Butler and K. Tokuyasu. 1982. Intracellular localization of nitrate reductase in *Neurospora crassa*. Plant Physiol. 70: 872-874.

Rosen, B.H. and R.L. Lowe. 1984. Physiological and ultrastructural responses of *Cyclotella meneghiniana* (Bacillariophyta) to light intensity and nutrient limitation. J. Phycol. 20: 173-183.

Rosenberg, J.S., E. Weiss and P. Wilding. 1984. Immunogold staining: adaptation of a cell-labeling system for analysis of human leukocyte subsets. Clin. Chem. 30(9): 1462-1466.

Rothbarth, Ph.H., H.J. Tanke, N.A.J. Mul, J.S. Ploem, J.F.C. Uliegenthart and R.E. Ballieux. 1978. Immunofluorescence studies with 4-acetamido-4'-isothiocyanato stilbene-2,2' disulfonic acid (SITS). J. Immunol. Methods. 19: 101.

Rothschild, G.H.L. 1966. A study of a natural population of *Conomelus anceps* (Germar) (Homoptera: Delphacidae) including observations on predation using the precipitin test. J. Anim. Ecol. 35:413-434.

Rouse, H. 1938. Fluid mechanics for hydraulic engineers. Dover Publications, Inc. New York.

Rowe, G.T., N. Merrett, J. Shepherd, G. Needler, B. Hargrave and M. Marietta. 1986. Estimates of direct biological transport of radioactive waste in the deep sea with special reference to organic carbon budgets. Oceanologica Acta 9: 199-208.

Rubinson, K.A. 1982. HPLC Separation and Comparative Toxicity of Saxitoxin and its Reaction Products. Biochem. Biophys. Acta. 687: 315.

Salzman, A. 1986. Analysis of natural selection on a quantitative physiological trait: salt tolerance in Western Ragweed (*Ambrosia pfilostachya*). Ph.D. Dissertation. Department of Biology. University of Chicago.

Samuel, D., P.L. Amlot and R.A. Abuknesha. 1985. A new method of iodinating ovalbumin, a protein which lacks accessible tyrosine groups, by conjugation to a highly fluorescent coumarin active ester, CASE. J.Immunol. Methods. 81(1): 123-130.

Sandstrom, G. and H.W. Watz. 1984. The duct ELISA: A new technique to identify low numbers of *Fransiscella tularensis*. FOA Report C 40202-133. National Defense Research Institute, S90182 Umea, Sweden.

Sano, T., N. Okamoto and T. Nishimura. 1981. A new viral epizootic of *Anguilla japonica* Temminck and Schlegel. Journal of Fish Diseases. 4: 127-139.

Sarokin, D.J. and E.J. Carpenter. 1981. Cyanobacterial spinae. Bot. Mar. 24: 389-392.

Scheffė, H. 1959. The Analysis of Variance. New York. Wiley.

Scheuer, P.J., W. Takahashi, J. Tsutsumi, and T. Yoshida. 1967. Ciguatoxin: Isolation and chemical nature. Science 155: 1267-1268.

Schmidt, R.J. and A.R. Loeblich III. 1979. Distribution of paralytic shellfish poison among Pyrrhophyta. J. Mar. Biol. Assoc. U.K. 59: 479-487.

Schmidt, R.J., V.D. Gooch, A.R. Loeblich III and J.W. Hastings. 1978. Comparative study of luminescent and nonluminescent strains of *Gonyaulax excavata*. J. Phycol. 14: 5-9.

Schreier, M., G. Kohler, H. Hengartner, C. Berek, M. Trucco and L. Forni. 1980. Hybridoma Techniques. Cold Harbor Laboratory, Cold Spring, NY.

Schwinghamer, P., B. Hargrave, D. Peer and C.M. Hawkins. 1986. Partitioning of production and respiration among size groups of organisms in an intertidal benthic community. Mar. Ecol. Prog. Ser. 31:131-142.

Secombes, C.J., J.J. Van Gronigen and E. Egberts. 1983. Separation of lymphocyte subpopulations in carp *Cyprinus carpio* L. by monoclonal antibodies: immunohistochemical studies. Immunol. 48: 165.

Selander, R.K. 1985. Protein polymorphism and the genetic structure of natural populations of bacteria. IN: Population Genetics and Molecular Evolution. T. Ohta and K. Aoki. (eds.). Tokyo/Berlin. Japan Sci. Soc. Press/Springer Verlag. pp. 85-106.

Selander, R.K., D.A. Caugant and T.S. Whittam. 1987. Genetic structure and variation in natural populations of *Escherichia coli*. IN: *Escherichia coli* and *Salmonella typhimurium*, Cellular and Molecular Biology. J.L. Ingraham *et al*., (eds.). ASM Publications, Washington, D.C. (in press)

Selander, R.K. and B.R. Levin. 1980. Genetic diversity and structure in *Escherichia coli* populations. Science. 210: 545-547.

Selander, R.K. and T.S. Whittam. 1983. Protein polymorphism and the genetic structure of populations. IN: Evolution of Genes and Proteins. M. Nei and R.K. Koehn. (eds.). Sunderland. Sinauer.

Shapiro, H.M. 1985. Practical Flow Cytometry, Alan R. Liss, Inc., NY, 295 pp.

Shapiro, L.P. and R.R.L. Guillard. 1987. Physiology and ecology of the marine eukaryotic ultraplankton. IN: T. Platt and W. Li. (eds.). Physiological Ecology of Picoplankton. Can. J. Fish. Aq. Sci., Bul. 214. in press.

Sharkey, R.G., E. Jover, E., F. Courand, D.G. Baden and W.A. Catterall. 1986. Allosteric modulation of neurotoxin binding to voltage-sensitive sodium channels by *Ptychodiscus brevis* toxin 2. Mol. Pharmacol. in press.

Sheldon, R.W. and T.R. Parsons. 1967. A continuous size spectrum for particulate matter in the sea. J. Fish Res. Bd. Can. 24: 9-9-915.

Shimizu, Y., H.N. Chou, H. Bando, G. VanDuyne and J.C. Cardy. 1986. Structure of brevetoxin-a (GB-1), the most potent toxin in the Florida red tide organism *Ptychodiscus brevis*. J. Am. Chem. Soc. 108: pp. 514-515.

Shimizu, Y. and E. Ragelis. 1979. Alternatives to the Mouse Assay. IN: Toxic Dinoflagellate Blooms. D.L. Taylor and H.H. Seliger (eds.). Elsevier North Holland, Inc., New York, NY. pp. 453-455.

Shumway, S.E., T.L. Cucci, R.C. Newell and C.M. Yentsch. 1985. Particle selection, ingestion, and absorption in filter-feeding bivalves. J. Exp. Mar. Biol. Ecol. 91: 77-92.

Signorella, A.P. and W.C. Hymer. 1984. An enzyme-linked immunosorbent assay for rat prolactin. Anal. Biochem. 136: 372-381.

Simmonds, N.W. 1979. Principles of Crop Improvement. New York. Longman.

Smarrelli Jr., J. and W.H. Campbell. 1981. Immunological approach to structural comparisons of assimilatory nitrate reductases. Plant Physiol. 68: 1226-1230.

Smith, A.D. 1982. Immunofluorescence of sulphate-reducing bacteria. Arch. Microbiol. 133: 118-121.

Smith, J.C., T. Platt and W.G. Harrison. 1983. Photoadaptation of carboxylating enzymes and photosynthesis during a spring bloom. Prog. Oceanogr. 12: 425-459.

Snyder, H.R. and H.C. Geller. 1952. Cleavage of Sulfonamides with Aqueous Hydrobromic Acid and Phenol. II. J. Am. Chem. Soc. 74: 4864.

Snyder, H. R. and R.E. Heckert. 1952. A Method for the Rapid Cleavage of Sulfonamides. J. Am. Chem. Soc. 74: 2006.

Sober, E. 1984. The Nature of Selection: A Philosophical Inquiry. Cambridge, MA. MIT Press.

Solomonson, L.P. 1975. Purification of NADH-NR by affinity chromatography. Plant Physiol. 56: 853-855.

Soudek, D. and G.G.C. Robinson. 1983. Electrophoretic analysis of the species and population structure of the diatom *Asterionella formosa*. Can. J. Bot. 61: 418-433.

Springer, T.A. Editor. 1985. Hybridoma Technology in the Biosciences and Medicine. Plenum Press. NY.

Stahl, D.A., D.J. Lane, G.J. Olsen and N.R. Pace. 1984. Analysis of hydrothermal vent-associated symbionts by ribosomal RNA sequences. Science. 224: 409-411.

Stahl, D.A., D.J. Lane, G.J. Olsen and N.R. Pace. 1985. Characterization of a Yellowstone hot spring microbial community by 5S rRNA sequences. Appl. and Env. Microbiol. 49: 1379-1384.

Steeman-Nielsen, E. 1952. The use of radioactive carbon (^{14}C) for measuring organic production in the sea. J. Cons. Int. Explor. Mer. 18: 117-140.

Stegeman, J.J. 1981. Polynuclear aromatic hydrocarbons and their metabolism in the marine environment. IN: Polycyclic Hydrocarbons and Cancer. H.V. Gelboin and P.O.P. Ts'o (eds.). Academic Press. 3(Chapter 1): 1-60.

Stegeman, J.J., P.J. Kloepper-Sams and J.W. Farrington. 1986. Monooxygenase induction and chlorobiphenyls in the deep sea fish *Coryphaenoides armatus*. Science 231: 1287-1289.

Steidinger, K.A. and E.A. Joyce. 1973. Florida red tides. Florida Department of Natural Resources Laboratory Education Series, 17: pp. 1-26.

Stevens, S.E., D.L. Balkwill and D.A.M. Paone. 1981. The effects of nitrogen limitation on the ultrastructure of the cyanobacterium *Agmenellum quadruplicatum*. Arch. Microbiol. 130: 204-212.

Stoecker, D.K., T.L. Cucci, E.M. Hulburt and C.M. Yentsch. 1986. Selective feeding by *Balanion* sp. (Citiata:Balonionidae) on phytoplankton that best support its growth. J. exp. Mar. Biol. Ecol. 95: 113-130.

Strottmann, J.M., J.B. Robinson, Jr. and E. Stellwagen. 1983. Advantages of preelectrophoretic conjugation of polypeptides with fluorescent dyes. Anal.Biochem. 132(2): 334-337.

Strickland, J.D.H. and T. Parsons. 1972. A Practical Handbook of Seawater Analysis, Second edition. Fish. Res. Board Can., Ottawa, Bull. 167: 310 pp.

Sugiyama, T., M. Mizuno and M. Hayashi. 1984. Partitioning of nitrogen among ribulose-1,5-bisphosphate carboxylase/ oxygenase, phosphoenolpyruvate carboxylase, and pyruvate orthophosphate dikinase as related to biomass productivity in maize seedlings. Plant. Physiol. 75: 665-669.

Sullivan, J.J. and W.T. Iwaoka. 1982. High Pressure Liquid Chromatographic Determination of Toxins Associated with Paralytic Shellfish Poisoning. J. Assoc. Off. Anal. Chem. 66: 297.

Sullivan, J.J., M.M. Wekell and L.L. Kentala. 1985. Application of HPLC for the Determination of PSP Toxins in Shellfish. J. Food Sci. 50: 26.

Sunderland, K.D. and S.L. Sutton. 1980. A serological study of arthropod predation on wood lice in a dune grassland ecosystem. J. Anim. Ecol. 49: 987-1004.

Suzuki, D.T., R.C. Lewontin and J. Miller. 1985. An Introduction to Genetic Analysis. New York. Freeman.

Symington, J. 1984. Electrophoretic transfer of proteins from two-dimensional gels to sheets and their detection. IN: Celis, Bravo. (ed.). Two dimensional gel electrophoresis of proteins: Academic Press. pp 127-168.

Syrett, P. 1981. Nitrogen metabolism of microalgae. IN: Physiological Bases of Phytoplankton Ecology. T. Platt (ed.). Can. Bull. Fish. Aquat. Sci. 210: 182-210.

Tachibana, K., M. Nukina, Y.G. Joh and P.J. Scheuer. 1986. Recent developments in the molecular structure of ciguatoxin. Biol. Bull., in press.

Tachibana, K., P.J. Scheuer, Y. Tsukitani, H. Kikuchi, D. Van Engen, J. Cardy, Y. Gopichand and F. Schmitz. 1981. Okadaic acid, A cytotoxic polyether from two marine sponges of the genus *Halichondria*. J. Am. Chem. Soc. 103: 2469-2471.

Talbot, M.C., D.S. Thoreson and M.J. Perry. 1985. Photosynthesis vs. light intensity measurements: A miniaturized incubator. ONR Technical Report 406: 1-13.

Taubman, M. A. and D.J. Smith. 1974. Effects of local immunization with *Streptococcus mutans* on induction of salivary IgA antibody and experimental dental caries in rats. Infect. Immunity 9: 1079-1091.

Tenore, K.R. 1981. Organic nitrogen and caloric content of detritus. I. Utilization by the deposit-feeding polychaete *Capitella capitata*. Estuarine Coastal Shelf Sci. 12: 39-47.

Terry, K.L., J. Hirata and E.A. Laws. 1983. Light-limited growth of two strains of the marine diatom *Phaeodactylum tricornutum* Bohlin: chemical composition, carbon partitioning and the duel periodicity of physiological processes. J. Exp. Mar. Biol. Ecol. 68: 209-227.

Theilacker, G.H. 1986. Starvation-induced mortality of young sea-caught jack mackerel, *Tranchurus symmetricus*, determined with histological and morphological methods. Fish. Bull. U.S. 84: 1-17.

Theilacker, G.H., A.S. Kimball and J.S. Trimmer. 1986. Use of an ELISPOT immunoassay to detect euphausiid predation on larval anchovy. Mar. Ecol. Prog. Ser. 30: 127-131.

Theilacker, G.H. and R. Lasker. 1974. Laboratory studies of predation by euphausiid shrimps on fish larvae. IN: The early life history of fish. J.H.S. Blaxter (ed.). Springer-Verlag, Berlin. pp. 287-299.

Thirkill, C.E. and G.E. Kenny. 1974. Serological comparison of five arginine-utilizing *Mycoplasma* species by two-dimensional immunoelectrophoresis. Infection and Immunity. 10: 624-632.

Thirkill, C.E. and G.E. Kenny. 1975. Antigenic analysis of three strains of *Mycoplasma arginini* by two-dimensional immunoelectrophoresis. The Journal of Immunology. 114: 1107-1111.

Thornber, J.P and R.S. Alberte. 1976. Chorophyll-proteins: membrane-bound photoreceptor complexes in plants. IN: The Enzymes of Biological Membranes. M. Martonosi (ed.). 3: 163-190.

Tilman, D. 1977. Resource competition between planktonic algae: an experimental and theoretical approach. Ecology. 58: 338-348.

Titman, D. 1976. Ecological competition between algae: an experimental confirmation of resource-based competition theory. Science. 192: 463-465.

Titus, A.J., R.P. Haugland, S.O. Sharrow and M.D. Segal. 1982. Texas Red, a hydrophillic, red-emitting fluorophore for use with fluorescein in dual parameter: Flow microfluorometric and fluorescence microscopic studies. J. Immunol. Methods. 50: 193.

Towbin, H.,T. Staehelin and J. Gordon. 1979. Electrophoretic transfer of proteins from polyacrylamide gels to nitrocellulose sheets: procedure and some applications. Proc. Natl. Sci. USA 76: 4350-4354.

Trask, B.J., G.J. van den Engh and J.H.B.W. Elgershuizen. 1982. Analysis of phytoplankton by flow cytometry. Cytometry 2: 258-264.

Trask, B.J., G.J. van den Engh and J.H. Elbershuizen. 1982. Analysis of phytoplankton by flow cytometry. Cytometry. 2: 258-264.

Trask, B.J., G.J. van den Engh, J. Landegent, N. Jansen in del Wal and M. van der Ploeg. 1985. Detection of DNA sequences in nuclei in suspension by *in situ* hybridization and dual beam flow cytometry. Science. 230: 1401-1403.

Trumps, G.N. 1970. Goldfish immunoglobulins and antibodies to bovine serum albumin. J. Immunol. 104: 1267.

Truneh, A., P. Machy and P.K. Horan. (in press) Antibody-bearing liposomes as multicolor immunofluoro markers for flow cytometry and imaging. J. Immnol. Methods.

Twarog, B.M. 1974. "Immunity" to paralytic shellfish toxin in bivalve molluscs. Proc. Second Intnl. Coral Reef Symposium. Creat Barrier Reef Committee, Brisbane, Australia. pp. 505-512.

Twarog, B.M., T. Hidaka and H. Yamaguchi. 1972. Resistance to tetrodotoxin and saxitoxin in nerves of bivalve molluscs. Toxicon. 10: 273-278.

Twarog, B.M. and H. Yamaguchi. 1975. Resistance to paralytic shellfish toxins in bivalve molluscs. Proc. of the First Intnl. Conf. on Toxic Dinoflagellate Blooms. Mass. Sci. Tech. Fdn. Wakefield, Ma. pp. 381-393.

Underhill, P.A. 1977. Nitrate uptake kinetics and clonal variability in the neritic diatom *Biddulphia aurita*. J. Phycol. 13: 170-176.

Vaitukaitis, J.L. 1981. Production of antisera with small doses of immunogen: multiple intradermal injections. Methods in Enzymology. 73: 46-52.

Vaitukaitis, J.L. 1981. Production of antisera with small doses of immunogen: Multiple intradermal injections. Methods in Enzymology. 73: 46-75.

Vaitukaitis, J., J.B. Robbins, E. Nieschlag and G.T. Ross. 1971. A method for producing specific antisera with small doses of immunogen. Journal of Clinical Endocrinology. 33: 988-991.

Valembois, P. 1973. Quelques aspects phylogénéteiques de la réaction d'incompatibilité aux greffes chez les Métazoaires. Ann. Biol. 12: 1-26

van Baalen, C. 1962. Studies on marine blue-green algae. Bot. Mar. 3: 129-139.

vandenPol, A.N. 1984. Colloidal gold and biotin-avidin conjugates as ultrastructural markers for neural antigens. Q. J. Exp. Physiol. 69(1): 1-33. (REVIEW).

Varanasi, U.S., T.K. Collier, D.E. Williams and D.R. Buhler. 1986. Hepatic cytochrome P-450 isozymes and aryl hydrocarbon hydroxylase in English sole (*Parophrys vetulus*). Biochem. Pharmacol. 35: 2967-2971.

Vaughn, K.C., S.O. Duke and E.A. Funkhouser. 1984. Immunochemical characterization and localization of nitrate reductase in norflurazon-treated soybean cotyledons. Physiol. Plant. 62: 481-484.

Verbruggen, R. 1975. Quantitative immunoelectrophoretic methods: a literature survey. Clin. Chem. NY. 21: 5-43.

Via, S. and R. Lande. 1985. Genotype-environment interaction and the evolution of phenotypic plasticity. Evolution. 39:505-522.

Viarengo, A., M. Pertica, G. Mancinelli, S. Palermero, G. Zanicchi and M. Orunesu. 1981. Synthesis of Cu-binding proteins in different tissues of mussels exposed to the metal. Mar. Bull. Pollut. 12: 347-350.

Vierling, E. and R.S. Alberte. 1983. P_{700} chlorophyll *a*-protein, purification, characterization, and antibody preparation. Plant Physiol. 72: 625-633.

Visser, J.W.M. and G.R. Van den Engh. 1982. IN: Immunofluorescence Technology Selected Theoretical and Clinical Aspects. Wick. (ed.). Elsevier, Amsterdam. pp. 95-128.

Voller, A., A. Bartlett and D.E. Bidwell. 1978. Enzyme immunoassay with special references to ELISA techniques. J. Clin. Path. 31: 507-520.

Voller, A., D.E. Bidwell and A. Bartlett. 1979. The Enzyme Linked Immunosorbent Assay. Dynatech Laboratories, Inc., Alexandria. 125 pp.

von Caemmerer, S. and G.D. Farquhar. 1981. Some relationships between the biochemistry of photosynthesis and the gas exchange of leaves. Planta 153: 376-387.

Waaland, S.D. and B.A. Watson. 1980. Isolation of a cell-fusion hormone from *Griffithsia pacifica* Kylin, a red alga. Planta 149: 493-497.

Waalkes, M.P., J.S. Garvey and C.D. Klaasen. 1985. Comparison of methods of metallothionein quantification: cadmium radioassay, mercury radioassay, and radioimmunoassay. Toxicol. Appli. Pharmacol. 79: 524-527.

Wade, M.J. 1978. A critical review of the models of group selection. Quart. Rev. Biol. 53: 101-113.

Wallace, D.C., K. Garrison and W.C. Knowler. 1985. Dramatic founder effects in Amerindian mitochondrial DNAs. Am. J. Phys. Anthro. 68: 149-155.

Wang, A.-C. 1982. Methods of immune diffusion, immunoelectrophoresis, precipitation, and agglutination. IN: Antibody as a Tool, the Applications of Immunochemistry. J.J. Marchalonis and G.W. Warr. (eds.). John Wiley & Sons, Chichester. pp. 139-161.

Ward, B.B. 1982. Oceanic distribution of ammonium-oxidizing bacteria determined by immunofluorescent assay. Jour. Mar. Res. 40: 1155-1172.

Ward, B.B. 1984. Combined autoradiography and immunofluorescence for estimation of single cell activity by ammonium-oxidizing bacteria. Limnol. Oceanogr. 29: 402-410.

Ward, B.B. 1984. Autotrophic activity of ammonium-oxidizing bacteria: combined autoradiography and immunofluorescence for estimation of single cell activity in the primary nitrate maximum off the coast of Washington. Limnol. Oceanogr. 29: 402-410.

Ward, B.B. and A.F. Carlucci. 1985. Marine ammonium- and nitrite-oxidizing bacteria: serological diversity determined by immunofluorescence in culture and in the environment. Appl. Environ. Microbiol. 50: 194-201.

Ward, B.B. and M.J. Perry. 1980. Immunofluorescent assay for the marine ammonium-oxidizing bacterium *Nitrosococcus oceanus*. Appl. Environ. Microbiol. 39: 913-918.

Warner, N.L., J.Y. Kimura and D.J. Recktenwald. 1983. Multiparameter flow cytometry analysis of normal and neoplastic human monocytes with leu monoclonal antibodies. IN: Monoclonal Antibodies. F. Dammaco, G.Doria and A. Pinchera (eds.). Elsevier Biomedical Press. NY.

Warr, G.W., D. DeLuca and B.R. Griffin. 1979. Membrane immunoglobulin is present on thymic and splenic lymphocytes of the trout, *Salmo gairdneri*. J. Immunol. 123: 910.

Waterbury, J.B., S.W. Watson, R.R.L. Guillard and L.E. Brand. 1979. Wide-spread occurrence of a unicellular, marine, planktonic cyanobacterium. Nature 277: 293-294.

Waterbury, J.B., S.W. Watson, F.W. Valois and D.G. Franks. 1986. Biological and ecological characterization of the marine unicellular cyanobacterium *Synechococcus*. IN: Physiological Ecology of Picoplankton. T. Platt and W. Li (eds.). C.J. Fish. Aqua. Sci. in press.

Waterbury, J.B., J.M. Willey, D.G. Franks, F.W. Valois and S.W. Watson. 1985. A cyanobacterium capable of swimming motility. Science 230: 74-76.

Watson, D.A. and A.R. Loeblich III. 1983. The application of electrophoresis to the systematics of the marine dinoflagellate genus *Heterocapsa*. Biochem. Syst. and Ecol. 11: 67-71.

Weaver, G. 1984. PCB contamination in and around New Bedford Mass. Environ. Sci. Technol. 18: 22A-27A.

Webb, M. 1979. The metallothioneins. IN: The Chemistry, Biochemistry, and Biology of Cadmium. M. Webb (ed.). Elsevier North-Holland Biomedical Press, NY. pp. 195-266.

Weir, E.E., T. Pretlow, A. Pitts, E. Williams. 1974. A More Sensitive and Specific Histochemical Peroxidase Stain for the Localization of Cellular Antigen by the Enzyme-antibody Conjugate Method. J. Histo. Cyto. 22: 51.

Weir, M.D. (editor) 1978. Handbook of Experimental Immunology. 1:Chapter 7.

Westernhagen, H. von and H. Rosenthal. 1976. Predator-prey relationship between pacific herring, *Clupea harengus* pallasi, larvae and a predatory hyperiid amphipod, *Hyperoche medusarum*. Fish. Bull. 74: 669-674.

Westley, J. W. 1975. The polyether antibiotics: Monocarboxcylic acid ionophores. Ann Rep. Med. Chem. 10: 246-256.

Wetherell, J.R., Jr. and A.S. Bleiweiss. 1975. Antigens of *Streptococcus mutans*: Characterization of a polysaccharide antigen from walls of strain G5-5. Infect. Immun. 12: 1341-1348.

Wheeler, P.A. 1983. Phytoplankton nitrogen metabolism. IN: Nitrogen in the Marine Environment, E.J. Carpenter and D.G. Capone (eds.). Academic Press, NY. pp. 309-346.

Whittam, T.S., H. Ochman and R.K. Selander. 1983. Multilocus genetic structure in natural populations of *Escherichia coli*. Proc. Nat. Acad. Sci. 80: 1751-1755.

Widder, E.A., and J.F. Case. 1982. Luminescent microsource activity in bioluminescence of the dinoflagellate *Pyrocystis fusiformis*. J. Comp. Physiol. 145: 517-527.

Wigzell, H. 1976. Specific affinity fractionation of lymphocytes using glass or plastic bead columns. Scand. Jour. Immunol. 5(suppl. 5): 23-30.

Williams, D. and D. Buhler. 1983. Comparitive properties of purified cytochrome P-448 from -naphthoflavone treated rats and rainbow trout. Comp. Biochem. Physiol. 75C: 25-32.

Wittenbach, V.A., R.C. Ackerson, R.T. Giaquinta and R.R. Hebert. 1980. Changes in photosynthesis, ribulose bisphosphate carboxylase, proteolytic activity and ultrastucture of soybean leaves during senescence. Crop Sci. 20: 225-231.

Wolf, K. and M.C. Quimby. 1962. Established Eurythermic Line of Fish Cells In Vitro. Sci. 135: 1065-1066.

Wood, A.M. 1982. Occurrence and ecological significance of different pigment types of marine *Synechococcus*. EOS 63: 960.

Wood, A.M. 1985. Adaptation of photosynthetic apparatus of marine ultraphytoplankton to natural light fields. Nature. 316: 253-255.

Wood, A.M., P.K. Horan, K. Muirhead, D.A. Phinney, C.M. Yentsch and J.B. Waterbury. 1985. Discrimination between types of pigments in marine *Synechococcus* spp. by scanning spectroscopy, epifluorescence microscopy, and flow cytometry. Limnol. Oceanogr. 30: 1303-1315.

Wood, A.M., R. Lande and G.A. Fryxell. 1987. Quantitative genetic analysis of morphological variation in an antarctic diatom grown at two light intensities. J. Phycol. (in press).

Wright, S. 1921. Systems of mating. Genetics. 6: 111-178.

Wright, S. 1978. Evolution and the Genetics of Populations. Variability Within and Among Natural Populations. Chicago. University of Chicago Press. Chicago. Vol. 4.

Wybran, J., J. Rosenberg and F. Romasco. 1985. Immunogold staining: an alternative method for lymphocyte subset enumeration. Comparison with immunofluorescence microscopy and flowcytometry. J. Immunol. Methods. 76(2): 229-238.

Wysocki, L.J. and V.L Stao. 1978. Panning for lymphocytes: a method for cell selection. Proc. Natl. Acad. Sci. 75: 284.

Yamaga, K.M., R.T. Kubo and H.M. Etlinger. 1978. Studies on the question of conventional immunoglobulin on thymocytes from primitive vertebrates I. Presence of anti-carbohydrate antibodies in rabbit anti-trout Ig sera. J. Immunol. 120: 2068.

Yasumoto, T. 1985. Recent progress in the chemistry of dinoflagellate toxins. IN: Toxic Dinoflagellates. D.M. Anderson, A.W. White and D.G. Baden. (eds.). Elsevier Science Publishers, NY. pp.259-270.

Yentsch, C.M., T.L. Cucci and D. A. Phinney. 1984. Flow cytometry and cell sorting: problems and promises for biological ocean science research. IN: Marine Phytoplankton and Productivity. Holm-Hansen, O., L. Bolis and R. Gilles, (eds.). Springer-Verlag, Berlin.

Yentsch, C.M., T.L. Cucci, D.A. Phinney, R. Selvin and H.E. Glover. 1985. Adaptation to low photon flux densities in *Protogonyaulax tamarensis* var. *excavata*, with reference to chloroplast photomorphogensis. Mar. Biol. 89: 9-20.

Yentsch, C.M., T.L. Cucci, D.A. Phinney and J.A. Topinka. 1986. Real-time characterization of individual marine particles at sea: flow cytometry. IN: Tidal Mixing and Plankton Dynamics, Bowman, M., Yentsch, C.M. and Peterson, W.T. (eds.), Lecture Notes on Coastal and Estuarine Studies. Springer-Verlag, NY, 17: 414-448.

Yentsch, C.M., P.K. Horan, K. Muirhead, Q. Dortch, E. Haugen, L. Legendre, L.S. Murphy, M.J. Perry, D.A. Phinney, S.A. Pomponi, R.W. Spinrad, M. Wood, C.S. Yentsch and B.J. Zahuranec. 1983. Flow cytometry and cell sorting: a technique for analysis and sorting of aquatic particles. Limnol. Oceanog. 28: 1275-1280.

Yentsch. C.M. and S.A. Pomponi. 1986. Automated individual cell analysis in aquatic research. Int. Rev. Cytol. 105: 183-243.

Yentsch, C.S. and C.M. Yentsch. 1984. Emergence of optical instrumentation for measuring biological properties. Oceanogr. Mar. Biol. Ann. Rev. 22: 55-98.

Young, J.O. 1980. A serological investigation of the diet of *Helobdella stagnalis* (Hirudinea:Glossiphonudae) in British lakes. J. Zool. 192: 467-488.

Zagursky, G. and R.J. Feller. 1985. Macrophyte detritus in the winter diet of the estuarine mysid, *Neomysis americana*. Estuaries. 8: 355-362.

Zemel, E. and S. Gepstein. 1985. Immunological evidence for the presence of ribulose bisphosphate carboxylase in guard cells chloroplasts. Plant. Physiol. 78: 586-590

SUBJECT INDEX*

*In most cases, the page numbers listed in the subject index indicate only the earliest references of that topic within each chapter. Reading further in the chapter(s) will generally yield more information on the subject desired.